# What Species Mean

## A User's Guide to the Units of Biodiversity

W0113102

# SPECIES AND SYSTEMATICS

The *Species and Systematics* series will investigate the theory and practice of systematics, phylogenetics, and taxonomy and explore their importance to biology in a series of comprehensive volumes aimed at students and researchers in biology and in the history and philosophy of biology. The book series will examine the role of biological diversity studies at all levels of organization and focus on the philosophical and theoretical underpinnings of research in biodiversity dynamics. The philosophical consequences of classification, integrative taxonomy, and future implications of rapidly expanding data and technologies will be among the themes explored by this series. Approaches to topics in *Species and Systematics* may include detailed studies of systematic methods, empirical studies of exemplar taxonomic groups, and historical treatises on central concepts in systematics.

For more information visit:
www.crcpress.com/Species-and-Systematics/book-series/CRCSPEANDSYS

# What Species Mean
## A User's Guide to the Units of Biodiversity

Julia D. Sigwart

CRC Press
Taylor & Francis Group
Boca Raton London New York

CRC Press is an imprint of the
Taylor & Francis Group, an **informa** business

CRC Press
Taylor & Francis Group
6000 Broken Sound Parkway NW, Suite 300
Boca Raton, FL 33487-2742

First issued in paperback 2022

ISBN-13: 978-1-498-79937-9 (hbk)
ISBN-13: 978-1-03-233884-2 (pbk)
DOI: 10.1201/9780429458972

**Library of Congress Cataloging-in-Publication Data**

Names: Sigwart, Julia, author.
Title: What species mean : a user's guide to the units of biodiversity / Julia Sigwart.
Other titles: Species and systematics.
Description: Boca Raton : Taylor & Francis, 2018. | Series: Species and systematics series
Identifiers: LCCN 2018013228 | ISBN 9781498799379 (hardback : alk. paper)
Subjects: | MESH: Biological Evolution | Phylogeny | Biodiversity | Species Specificity
Classification: LCC QH380 | NLM QH 380 | DDC 576.8/6--dc23
LC record available at https://lccn.loc.gov/2018013228

**Visit the Taylor & Francis Web site at
http://www.taylorandfrancis.com**

**and the CRC Press Web site at
http://www.crcpress.com**

*For Rory, Xochi, Áine, Aza, Cormac, Cozy, Kate, Ezra, and Molly*

# Contents

# Series Preface

The *Species and Systematics* book series is a broad-ranging venue where authors can provide the scientific community with comprehensive treatments of the history and philosophy of fundamental concepts in systematic biology, phylogenetics, and the science of taxonomy. The series also intends to connect historical perspectives to new ideas and emerging technologies that have implications for the field. The series is committed to stimulating discussion among students and researchers in biology on controversial and clarifying ideas related to the future course we are charting in biodiversity research.

There are many approaches to the study of biological diversity and to embrace this, future volumes in *Species and Systematics* may include detailed development and comparisons of existing and novel methods in systematics and biogeography, empirical studies that provide new insight into old questions and raise new questions for biologists and philosophers of science, and historical treatises on central and reoccurring concepts that benefit from both a retrospective and a new perspective. Some volumes will address a single important concept in great depth, giving authors the freedom to present ideas with their own slant, while others will be edited collections of shorter papers intended to place alternative views in sharp contrast.

**Kipling Will**
*Berkeley, CA*

# Acknowledgements

This book was written during a sabbatical supported by funding from the European Union's Horizon 2020 research and innovation programme under grant agreement H2020-MSCA-IF-2014-655661. That award allowed me to spend an extended period working in the University of California, Berkeley, Museum of Paleontology. I am eternally grateful to my host and guru, David Lindberg, for his dauntless encouragement and positivity. I remain unconvinced of the merits of book writing. My colleagues in UC Berkeley were supportive and encouraging, as well as my collaborators in UC Davis, Bodega Marine Lab, and Lawrence Berkeley National Laboratory who all welcomed me into their scientific communities and made me feel at home. I must also thank the staff of the Queen's University Marine Laboratory and my research group in Portaferry, N. Ireland, for their patience with my long absence. An earlier version of this book was used as the basis for a graduate seminar in UC Berkeley, Department of Integrative Biology, and the members of that group deserve special thanks: Carl Rothfels who co-organised the seminar, and Allan Cabrero, Gabriel Damasco, Kyle DeMarr, Jesus Martinez-Gomez, Michael Song, Elisa Visher, Daniel Wait, Adiël Klompmaker, and Lucía Lohmann (University of São Paulo) all contributed to improving the text. Interest around this project also resulted in a symposium, 'Measuring Biodiversity and Extinction—Present and Past', at the Society for Integrative and Comparative Biology annual meeting in San Francisco (2018). I am indebted to the symposium participants for their contributions to that gathering, and inspiration from all of them filters into this book: Keith Bennett (University of St. Andrews), Stewart Edie (University of Chicago), Luke Mander (Open University), Beth Okamura (Natural History Museum, London), Kevin Padian (UC Berkeley), Dan Rabosky (University of Michigan), Quentin Wheeler (SUNY ESF), Judith Winston (Smithsonian Marine Station), and Norine Yeung (Bishop Museum, Hawai'i). Those people and innumerable other friends and colleagues gave me everything in the range of profound wisdom, advice, sympathy, alcohol, moral support, data, different points of view, many of the images reproduced here, and conversations that motivated and enriched this text even if they did not know it at the time, including: Noor Adelyna Mohammed Akib (Universiti Sains Malaysia), Caroline Ang (University of Bath), Wallace Arthur (NUI Galway), Daryl Birkett (Queen's University Belfast), Shannon Cassidy, Chong Chen (JAMSTEC), Robert Cowie (University of Hawai'i), Jo Denyer (Denyer Ecology), Jaimie Dick (Queen's University Belfast), Ivo Duijnstee (UC Berkeley), Greg Edgecombe (Natural History Museum, London), Amy Garbett (Queen's University Belfast), Madeleine Geiger (University of Cambridge), Michael Ghiselin (California Academy of Sciences), Terry Gosliner (California Academy of Sciences), Bea Hartman, Maria Herranz (University of British Columbia), Conor Houghton (University of Bristol), Izwandy Idris (Universiti Malaysia Terengganu), David Jablonski (University of Chicago), Patrick Krug (California State University, LA), Andrés Lira-Noriega (Instituto de Ecología INECOL), Cindy Looy (UC Berkeley), Leigh Marsh (University of Southampton), Charles Marshall (UC Berkeley), Marjolaine Matabos (IFREMER), Ralph and

Deborah McCaskey, Hugh McIsaac (University of Windsor), Christine Morrow (Queen's University Belfast), Diana Niemoeller, Patrick Orr (University College Dublin), Dula Parkinson (Advanced Light Source, Lawrence Berkeley National Laboratory), Brian Penney (St Anselm College), Bernard Picton (Ulster Museum), Mary Lou Pojeta, Jennifer Pollard, Anne Pringle (University of Wisconsin, Madison), Denis Riek, Jonathan Rolland (University of Lausanne), Barry Roth, Catherine Rushworth (UC Davis), Michael Schrödl (Bavarian State Collection of Zoology), Enrico Schwabe (Bavarian State Collection of Zoology), Brad Seibel (University of South Florida), Prashant Sharma (University of Wisconsin, Madison), Charles and Gretchen Sigwart, Boris Sirenko (Zoological Institute, St Petersburg), Lauren Sumner-Rooney (Oxford University), Mark Sutton (Imperial College, London), Burke Trieschmann, Jim Valentine (UC Berkeley), Geerat Vermeij (UC Davis), Paolo Viscardi (National Museum of Ireland), Janet Voight (The Field Museum), Bob Vrijenhoek (MBARI), David Wake (UC Berkeley), Erin Watson, Anne Weigert (Max Planck Institute for Evolutionary Anthropology), Caroline Williams (UC Berkeley), Susan Williams (UC Davis), Vera Wright, Leena Wong (Universiti Putra Malaysia), Jon Yearsley (University College Dublin), and Frank Zachos (Natural History Museum, Vienna). This does not pretend to be a complete list and I am grateful to everyone who helped me see this book completed.

# Author

**Dr. Julia D. Sigwart** is the director of the Queen's University Marine Laboratory, an interdisciplinary institute for marine science in Portaferry, N. Ireland. She completed her undergraduate education at the University of Victoria, and then took up a position on the scientific staff at the American Museum of Natural History. She later moved to Ireland to manage a research programme linking the National Museum of Ireland (Natural History) and University College Dublin. In 2009, she moved to her faculty position in Queen's University, Belfast. She has published over 70 papers on diverse topics and organisms. Her research examines the evolution of diversity in molluscs and other animal groups, spanning both living and fossil species.

# 1 Introduction

This book is mainly concerned with bridging the gap between those who name species, a specialist field of science, and those who use species, which includes absolutely everyone. The intended audience for this work is those of you who work on biodiversity and may have occasionally had some niggling doubts about what, exactly, is going on with the whole species thing.

Like technology, species are all around us, an intrinsic part of how we interact with the world. But there is a gap in communication between species as most people experience them, walking outdoors or examining an organism in a laboratory, compared to the conception of species by the experts who identify and name new taxa, and perhaps different again to people who consider species evolution or the basis of species in the philosophy of science. Discovering and describing species has been an active pursuit of scholars through the whole history of science. The Darwinian revolution shifted toward a general awareness that species are not fixed essences, but rather they change through evolutionary time. This creates a conundrum, where species and species groups are defined by observations that are mostly fixed in geological time, but we expect them to apply to the long-term evolutionary trajectories of an ever-changing thing. This is an issue of evolutionary theory that has substantial consequences in the experience of other branches of science.

The processes we use to identify species have become increasingly formalised and better resolved over more than 300 years of development. The study of biodiversity and species discovery is a genuinely integrative science, implicitly or explicitly using a total evidence approach that draws on a broad suite of tools to test and describe the similarities and differences among organisms. The judgement of those similarities requires detailed, technical knowledge of the species under consideration, and a level of specialist knowledge that is sometimes difficult to articulate or to transfer to a nonspecialist audience. But among all those detailed meticulous observations of variation among organisms, how do we separate signal from noise? And more broadly still, what are the emergent patterns that could help make sense of chaotic global biodiversity?

There are intrinsic mathematical patterns in nature, which serve to describe the order of natural processes. A Fibonacci sequence (1,1,2,3,5,8 ....) describes the arrangements of seeds in a sunflower and the spiral of a *Nautilus* shell. Species are natural units that populate the world around us, and they are formed through evolutionary processes. The branching processes of evolution also have a kind of mathematical nature. So it follows that we should be able to develop a set of general principles that describe global patterns of species groupings. This is hampered by two significant dilemmas: first, the questions of what species mean, or the technical details of what anyone means when they refer to a species or group of species, and second, the ability of available data to resolve these questions satisfactorily.

There is substantial variation among different disciplines in the life sciences about the ontology of species and species groupings. That is, when we use the word 'species', it is not at all clear that we are all talking about the same type of thing. The so-called 'species problem' is the worrying thought that a given group of organisms could be classified as one, or two, or many species, depending on the standard of evidence used or the personal inclination of the classifier. In fact, the scientific practice of identifying, describing, and naming species and their inter-relationships has a huge and well-developed theoretical foundation. A decision about how to conscribe the identification of a particular taxon depends on a total evidence assessment, incorporating many lines of evidence. Species are dynamic entities that are subject to evolutionary change over very long spans of time, so a universal equivalency or litmus test is not plausible. The nature of species has been considered in great detail by philosophers of science, but everyone in the life sciences uses species in subtly different ways, and at different scales in space and time. I contend that the enduring frustration about the 'species problem' is in part rooted in miscommunication, or a lack of understanding, about the way species concepts are applied in different areas of science.

Second dilemma: living biological diversity is significantly under-sampled. New species are discovered constantly, but the gap in understanding includes both not knowing how many species there are, and the very limited knowledge we have of many known species; it is not at all clear how big the data gap really is. This inspires reasonable scientific doubt that we do not have enough data coverage to make generalised conclusions about global biodiversity. As a corollary to this, scholars who have the most knowledge about species—taxonomists—tend to have expertise limited to one particular group of organisms. As they know their limits, those specialists would not tend to extend the observations of their own favourite clade to generalities.

There are undoubtedly clear patterns in global biodiversity. For example, there are generally more species in tropical environments than temperate or polar areas. The fossil record shows past crises that reduced global biodiversity, and many of those past events were correlated with changes in climate. However, natural systems are complex and it is easy to dig up counter examples or weaknesses that cast doubt on almost any grand conclusions. Causative arguments that explain these large-scale patterns are sometimes rejected by the community because of doubts about the completeness or relevancy of the underlying datasets that come from a necessarily limited sampling of species. As such, both these dilemmas are rooted in real issues about scientific data, as well as issues about communication among disparate disciplines and user groups.

While everyone uses species names, those who study macroevolutionary processes, and describe and name lineages of organisms, are the 'makers of names'. In this book, we will explore the different perspectives of name 'makers' and 'users'. Scientists who name species can be compared to engineers or inventors of technology, developing products that are adopted by a wider community of users. In this era of technology, we are increasingly separated from the workings of the objects we depend on every day, phones, computers, cars, and corporations, which seem like so many 'black boxes'. But it is useful to have some insights into how things work. Where is the 'on' button? What is the scope and limitation of this technology? I might only use a tablet to play a video game, but it would probably benefit me to know what else it is capable of in case extra features turn out to be interesting or useful. It is equally important for developers to understand the real motivation and interests of their users, so they can promote new content appropriately. Developers are themselves also users. But what developers create, if it is successful, works for a much wider audience than just their own use.

I should pause this metaphor to emphasise that I am comparing species *names* to technology, not species *per se*. As we will discuss later, some researchers have argued that our recognition of species is a human artefact or invention. Many disagree, and a comparison to human-fabricated technology is not intended to provide ammunition to so-called species nominalists, who view names as real but species as artificially classified assemblages. In my view, species are the primary units of evolutionary processes. A species is a real unit of biological organisation, a group of populations that are interconnected through ancestor-descendent relationships. Species groups (such as a genus, family, etc.) reflect and help us to communicate about the relationships among those larger units. The names we give to species and species groups, however, are part of language, which is our own human invention. The metaphor linking taxonomy to technology has another flaw: in the development of a commercial product, there is an assumption of competition, and that some products will survive and others will wane away to obsolescence, and thus go extinct. This may sound like an evolutionary process, and corporate managers are fond of co-opting and abusing evolutionary terms. All valid species names are valuable. But among ideas about species, like software, whether they get widely discussed and studied depends in part on interactions between makers and users.

Unlike some disciplines, working with species has arguably not been revolutionised by a single obvious technological advancement. The universality of species, a long continuous history of study, and a catholic approach to relevant data means that it is almost impossible to pinpoint a single technique that prompted any dramatic transformation of how species are described. Linnaeaus developed the foundation for the modern formulation of names; he introduced standardisation, but the actual process of identifying separate species and justifying their description as new taxa has developed continuously over three centuries into a robust discipline. Later, the Darwinian revolution—and Darwin was not the first to promote the mutability of species, just the first to make it stick—still did not really change the nuts and bolts of how species are described. Molecular biology is another significant advancement, but again this has primarily added another important character set that can be used in delimiting the boundaries between species. Electron microscopy similarly illuminated never-before-seen differences among similar species, yet SEM has become such a commonplace technology that it is taken for granted as part of morphological descriptions.

Researchers working in systematics and taxonomy—two not-quite-equivalent terms for the science that covers the practice of discovering, describing, and assigning classification to species—are acutely aware of the faults, weaknesses, and bottlenecks of the species discovery process. In my opinion, this has led to a culture of borderline self-flagellation that is holding back the progress of taxonomy as a discipline. In fact, systematics has made huge strides in increasing rigour, and practising taxonomists and systematists are advancing courageously into a future full of threats to biodiversity.

Modern biodiversity loss is a major concern for all biologists and all humans. Anthropogenic changes to our planet are decreasing species diversity through the negative impacts of pollution, habitat destruction, direct extirpation of species, and climate change. Mass extinction events (and subsequent recoveries) have happened before in Earth's history, and these provide important context to understand ecological responses to modern environmental change. The work of assessing biodiversity is woven into ecology, environmental science, conservation, phylogenetics, and many other disciplines; yet, ultimately the way we measure diversity depends on species.

Species are incredibly complicated. The so-called 'species problem' has confounded great thinkers for hundreds of years. This is not because our intellectual ancestors (in the 1990s, or the 1790s, or earlier) lacked the technology to solve the problem, or because they were working under outmoded world views, it is because the problem is *hard*. The modern solution and a holistic view of species with fuzzy boundaries, population-lineages linked by ancestor-descendant relationships, remains frustratingly difficult to explain. Serious thinking about species requires confronting multi-dimensional lineages that move chaotically through space and time. This is not what people expect about the ordinary creatures we see around us every day.

For name makers, the review here might serve as part reassurance—I emphasise that an environment of limited data and high uncertainty is a natural part of working in complex systems—and part reminder that there is a global audience that needs the products of this science. The rest of life sciences, and in a very real way the global economy, depends on identifying and understanding species.

For all users of species names, this book is for you. I hope this book can provide a translation, or a peek inside the black box, of what exactly is going on with species. There is a rich and fascinating history and robust scientific theories that underpin our understanding of species and their evolution. Understanding a bit of the mechanics of how taxonomy works might even make it a bit less maddening when species names keep changing (Figure 1.1).

## A BRIEF GUIDE

Much of the literature on biodiversity and the nature of species focusses on terrestrial vertebrate animals; the 'charismatic megafauna' are the organisms that are best studied and best understood of all the species we share a planet with. Observations of these species cannot necessarily be extended to generalisations that apply to the two other broad groups of large-bodied organisms, plants and fungi, or even to other animals. Other, important work has grappled with how the idea of species identity can be translated to less intuitive systems, in microbes, or any organisms that reproduce asexually, or those that form colonies. My own research interests sit in a place midway in this spectrum of accessibility, as most of my experience has been in the study of marine, non-vertebrate (but non-colonial) animals that mostly reproduce by external fertilisation. The scope of this book is explicitly limited to the biodiversity of macroscopic, eukaryotic organisms. I apologise if plants and fungi may feel their presence is more like tokenism at some points. Our subjects—animals, plants, and

"I don't know much about taxonomy,
but I know what I like"

**FIGURE 1.1**   Cartoon by Barry Roth (previously published in 1998 in *The Veliger*, 41: 294).

fungi—are organisms that primarily reproduce by sexual recombination, and most of them have an identifiable, individual identity.

This book will not, unfortunately, be able to solve any particular issue over how to manage the identification or classification of species. My intention is to provide guidance and background to understand the scope of these problems, and some tools and advice about how to think about them. This short book can only offer a rapid and therefore rather superficial overview of many complex ideas, multiple areas of research that each have extensive scientific literature. A number of key principles are laid out in their own section (Chapter 2) as a kind of conceptual glossary, which is intended to frame the point of view taken here.

The remainder of the book is divided into ten chapters (Chapters 3–12). The chapters were written such that each can stand on its own; it is not necessary to read them in the order presented. The current sequence seems to be the best version for using this volume as a textbook, but casual readers may find another order more appealing. (I like some of the later chapters much better than the middle ones.) The general idea is that all of the varied subjects covered here are inter-connected, and founded on how we think about species. Chapter 3 highlights the broad application of species and the relevance of understanding species in so many fields outside of biodiversity. The idea of a coherent acceptance of species is taken for granted in almost all fields outside of taxonomy and systematics. Chapter 4 addresses the perennial frustration over taxonomic revision, when species users sometimes find themselves baffled by an apparently spontaneous change in taxonomic names. In order to understand how and why names change, we have to know a bit about the mechanisms of taxonomy and 'type' specimen material, and the sources of uncertainty around species identification. These ideas are unpacked further in following chapters. Chapter 5 discusses the distinction between micro- and macroevolutionary processes, separating the variation within species from the differences that divide independent evolutionary lineages. Chapter 6 returns to the issue of classification of species into a ranked taxonomic hierarchy, and the interpretation and application of ranked groups in an evolutionary phylogenetic framework. Ranks (e.g., genus, family, order, class, phylum) are also units that are taken for granted by many people, for example when an organism cannot confidently be identified to species level and the family or genus name is substituted as a surrogate. It is important to consider what impact taxonomic resolution has on the way that we measure biodiversity. Those chapters provide some background for a question that otherwise remains an elephant in the room, the philosophical problem of whether species are 'real', discussed in Chapter 7.

The latter part of the book is more focussed on species as units of global biodiversity. Chapter 8 provides a brief primer on the process of species description, and a guide to reading and finding taxonomic literature for the non-taxonomist. Chapter 9 focusses on the contribution of the fossil record to interpreting living biodiversity and explores history on both human and geological timescales as guidance for understanding the current extinction crisis. The following chapter, Chapter 10, considers the problem of estimating the number of living species on Earth. Chapter 11 considers the global distribution patterns of species (such as the latitudinal diversity gradient) and what issues, on short and long temporal scales, influence our perception of what is 'normal' in ecosystems. The final chapter, Chapter 12, expands on these human issues of how we study biodiversity, the dangerous fallacy of environmental economics, the global disparity between the distributions of species diversity and scientific resources, and the ongoing global endeavour to discover and protect Earth's species.

# 2 General Concepts

This is a brief conceptual glossary to summarise and clarify the foundational ideas adopted in this book and themes that occur repeatedly across the subjects covered in different chapters. Many of the concepts below are inter-related and interdependent.

*Sufficiency in the moment*—there is no evolutionary quest for perfection. Evolution does not have an end point; natural selection favours those that are 'good enough' by selecting against those that are deficient in present circumstances. What qualifies as good enough is likely to change in unexpected and unpredictable ways over time and space. Species are participants in, not products of, evolution.

*Time-dynamic species*—species are evolutionary units that are variable in time and space. This is conceptually important, that species are constantly changing, moving through time; they are not simply fixed endpoints that were arrived at by the preceding evolutionary history. The rate of change over time is different among various organisms, geographical areas, and eras.

*Horizontal species and vertical species*—species have a geographical range, but they also have a range through time; both aspects add natural variation. If we think of evolution recorded in geological series, adding more to the top of the rock record as time goes on, it is easy to visualise time as a 'vertical' dimension cutting through multiple geological layers. In one horizontal slice or snapshot of time, such as the present day, each species occupies a range in geographic area and environmental conditions, and has certain physiological limitations and morphological and genetic traits. Time-horizontal species are observed in the living biota, and can be observed in former time slices through palaeontological evidence. Species lineages also exist in a time-vertical dimension, linking past and present populations through ancestor-descendant relationships.

*Data-limited systems*—the assessment of species *always* has some constraint or some data that are inaccessible. Fossil species are usually more data-limited that living species, because various kinds of biological data are not preserved in fossils. But living species are also intrinsically data-limited, to a varying extent, because many biological activities and features cannot be observed. We cannot fully know the experience of another organism.

*Tropical biodiversity and understudied systems*—the 'discovery gap', between complete knowledge and available data can cripple understanding of a natural system. It is intrinsically difficult to estimate what is unknown. This is particularly acute in understudied regions or groups of organisms, where biological diversity overwhelms the available scientific labour force. A lack of knowledge about species (globally, not by individual scientists) can hamstring the development of ambitious research programmes or conservation strategies.

*Tree thinking*—evolution is best depicted and understood through branching patterns, beginning with the stem of the common ancestor, growing vertically through time and branching outward through expanding diversity. Some branches terminate (an extinction), or split into descendent species (speciation), or persist unchanged. The distal tips of these branches are the diversity of life on Earth today, or in whatever horizontal time slice a phylogeny is sampled. Although evolutionary rates vary widely, everything on Earth has been evolving for the same length of time.

*Natural phenomena follow mathematical patterns*—Many natural systems are characterised by skew distributions, also called long-tailed or hollow-curve distributions. These patterns are universal, and they are concordant with evolutionary processes. Few species or groups are explosively diverse or abundant. Most clades are small. Most species are rare.

*Relativity, not universality*—our understanding of evolutionary relationships, and the way they are translated into taxonomic classification, is based in a comparative framework, not on an absolute single metric that can be applied across all forms of Life. Since the evolutionary distance separating lineages is wildly variable, there cannot be a satisfactory litmus test to determine the limits of a species or higher level grouping. The separation among species and species groups can only be recognised by context, in their relative similarity or dissimilarity to other closely related species.

*Taxonomic ranks are important for communication*—The relative inter-relatedness of organisms is expressed by a nested, relativistic classification based on the ranks proposed by Linnaeus and adapted over 300 years of refinement: phylum, class, order, family, genus, and species. The use of these ranks began before evolutionary thinking, yet the distribution of species in higher taxonomic ranks (genera, families, etc.) follows a pattern that strongly correlates with predictions from phylogenetic theory. Ranks provide a clear, universally understood surrogate to rapidly explain relationships and relative diversity to a broad audience including nonscientists.

*Name-makers and name-users*—Everyone uses species, including specialist scientists who engage in identifying and describing new species or their evolutionary relationships, and many audiences including those outside of science. The users of species names, in other scientific disciplines, and outside of science, far outnumber the makers.

*Variable variability*—species are mutable, and a lineage can vary substantially in response to environmental change or other influences. Such *microevolutionary* changes often are reversible in subsequent generations and do not accumulate to a sufficient extent to generate *macroevolutionary* or lineage-splitting change. The transition between these levels causes species lineages to have 'fuzzy boundaries'.

The 'fuzziness' is the natural, expected variability within a cohesive but dynamic species lineage. Variation in rates of genetic and morphological change and plasticity mean it is impossible to define a universal cut-off for species.

***Species names are hypotheses***—assigning a species name to an individual is a statement of a hypothesis about that individual's relatedness to other individuals (and in particular to the name-bearing type specimens). The description of a species as a unit is a hypothesis about what features (genetic, morphological, and other biologically relevant data) identify a group of organisms, distributed in space and time, that represents a lineage with an independent evolutionary trajectory. This does not mean that species themselves are subjective, only that our judgement is necessarily based on limited evidence and subject to future refinement.

***Species are real***—just as evolution is real. Their reality is demonstrable by the existence of the products of evolutionary radiations: identifiable genetic and morphological types among living organisms. In practical terms, it is impossible to achieve a fully resolved genealogy of all individuals of all species that have ever lived; therefore, multiple lines of evidence are usually required to resolve a robust understanding of what conscribes a lineage. Species lineages have their own, independent evolutionary trajectory, and may be evolving at rates faster or slower than other species. Lineages have 'fuzzy' boundaries, natural variation among individuals, and this does not undermine the reality of macroevolution nor the reality of species lineages themselves. Adjacent lineages include those with clear-cut separation, but also incipient species, hybrids, and intermediate types; this is an expected state of ongoing dynamic evolution.

# 3 Everyone Uses Species

## THE ADVANTAGES OF UNDERSTANDING SPECIES

If you look out a window almost anywhere, you will see several species. From my desk in urban California, I can see a lemon tree, a magnolia, several shrubs whose names I do not know, and in the distance a hill covered in eucalyptus and other trees. Populations of the more dominant species fade together into 'landscape', but up close they have their own individual lives and identity. Species are components of our aesthetic world, and these species are also the components of ecosystems, even within a disrupted human-dominated world. Closer to the window, there are predator-prey interactions unfolding between yet more species, my cat (*Felis catus* Linnaeus, 1758) is watching a hummingbird (*Calypte anna* (Lesson, 1829)) in the tree (*Citrus × meyeri* Tanaka, 1946), and under the landscaping bark between the shrubs there are salamanders (*Aneides lugubris* (Hallowell, 1849)), pill bugs (*Littorophiloscia richardsonae* (Holmes & Gay, 1909)), and ants (*Linepithema humile* (Mayr, 1868)).

It is nice to be able to put names on these things. There is a warm satisfaction in knowing the identity of objects around us, a behaviour that begins when we first learn language. We take it for granted that species have names, in the same way that objects, like a book or a chair, have identity as nouns.

Species are recognised in many, or perhaps all, languages. Anthropologists and linguists have shown strong parallels between scientific taxonomy and the 'folk' taxonomy of indigenous peoples isolated from Western scientific traditions (Berlin, 1973). Recently, many scientists have recognised that indigenous knowledge could help biodiversity inventories in inaccessible areas such as the Amazon basin (Begossi et al., 2008). The concordance between folk taxonomy and scientific recognition of species does tend to fall down when comparing smaller, inconspicuous animals and plants, the same organisms that modern students need training and practice to recognise accurately (Briggs & Walters, 1997). Knowing the species identities of organisms we encounter carries substantial scientific meaning, and links to further information and knowledge; it is the aim of this book to help you access as much of that information as possible (Chapter 2).

Species are components of landscapes, but they are also the fundamental units of evolutionary biology and biodiversity. There is a significant interaction between the way we recognise and name species and groups of species, and the scientific process of understanding their evolutionary origins. Being able to identify species, to tell them apart, and know what they are is important for a number of reasons with wide implications. To begin with, it is worth reflecting on some more or less familiar examples of species names and the ways in which both scientists and nonscientists encounter them.

## PUTTING NAMES TO FACES

There are many purely practical reasons why it is important to identify species and to be able to do so reliably and repeatably. The quest to understand all species on Earth is a grand intellectual challenge, but a microscopic species that only lives in a place you will never visit does not rouse the same level of enthusiasm in everyone. Even to an unabashedly anthropocentric perspective, the identity of species is critically important: we need to know the species we share our homes with, that we consume for food, others that cause or cure diseases, or that can inspire new technology.

Identifying something to species becomes especially important when there may be similar looking things that have very different attributes. The domestic dog represents a very wide range of morphological types, which have been developed through selective breeding, a form of genetic modification (Figure 3.1; Larson & Burger, 2013). How then can we confidently distinguish between a Grey Wolf (*Canis lupus* Linnaeus, 1758) and the domesticated subspecies, the dog, *Canis lupus familiaris* (Linnaeus, 1758)? There are a number of features we could use to separate them. In mammals, we often use features of the skull and dentition to separate species, and the incisors of wolves are longer than dogs, and they also differ in other aspects of morphology, as well as DNA, physiology, habitat, and behaviour. Despite some differences, dogs and wolves share a common ancestor as recently as 15,000 years ago, which is a short blink in geological time (Wayne & Ostrander, 2007; Ersmark

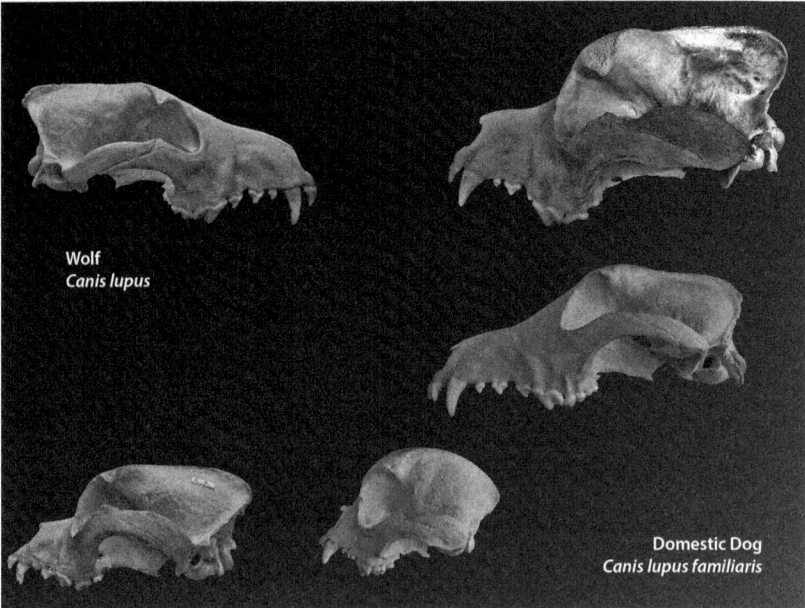

**FIGURE 3.1**   A side-by-side comparison of the wolf (*Canis lupus*, top left) and domestic dog breeds (*Canis lupus familiaris*), showing the cranial morphological plasticity in (clockwise from top right) the St. Bernard, Greyhound, French Bulldog, and Boxer breeds. (Photos by Madeleine Geiger.)

et al., 2016). Strong genetic similarities and mounting evidence for a very recent divergence between their lineages, has led to the current consensus that dogs represent a domesticated subspecies, rather than a species fully separate from wolves.

The scientific names of species are actually a two-part name, the genus epithet and species epithet. The first part, the genus (e.g., *Canis*) names a group of species that are more closely related to each other than to anything outside that group; the second, species name, refers to just one lineage (*Canis lupus*, or one of about eight other living species). So the two-part scientific name contains a lot of embedded comparative information.

Although both dogs and wolves are very cute as puppies, only one is legally allowed inside your house, and the other one is an endangered species. It is clearly important to distinguish between these two forms, and they make a good illustration that it is useful and important to identify organisms with precision. Most reasonable people do not want to share their living space with a wolf. Wolves are apex predators that shape ecosystems by removing the weakest individuals from local populations of their prey. The reintroduction of wild Grey Wolf populations into Yellowstone National Park in the USA in the 1990s, after their extermination 50 years earlier, was politically controversial, and many people feared that these predators could exterminate native ungulates or attack livestock outside the park (Bath, 1989). Fifteen years after the reintroduction of wolves found significant ecosystem recovery; the wolves limited browsing elk, allowing natural restoration of dense willow and aspen-dominated forest ecosystems. The reduction in elk activity allowed trees to regrow, and new forests in turn supported additional tree-dependent biodiversity, including beavers. The introduction of a predator species resulted in an overall increase in total diversity and biomass (Ripple & Beschta, 2012). In an alternative world where we only considered the strong genetic similarities between *Canis lupus* and *Canis lupus familiaris*, and ignored their differences, the large population of dogs in the US could be abused as a disincentive to conserve wolves, and lose their benefits to wild places.

Species that are genetically similar can nonetheless have very distinct ecological and cultural roles. Conversely, widely separated species could be seen as functionally equivalent. But all species have their own separate histories and niches, so there is never a true equivalency.

## IDENTITY POLITICS: WHEN SPECIES MATTER TO US

There are many kinds of fish in the sea, around 30,000 species of fish in freshwater and marine habitats worldwide, and many yet to be discovered. In Western cuisine, fish as food are mostly used interchangeably, with the exception of some large carnivorous species such as tuna (multiple species in the genus *Thunnus*) and salmon (species in the family Salmonidae). Less than 400 finfish species are the targets of large-scale commercial fishing (FAO, 2017), but many more are collected as catch by industrial vessels or through smaller-scale traditional fisheries. Advancements in maritime technology in the 20th century produced vessels that indiscriminately scoop up whole schools of fish in one haul and store them in onboard freezers, while the boats continue accumulating catch at sea for weeks at a time. This is in stark contrast to historical fishing practices, by small vessels without extensive refrigeration with limited range and capacity. The increase in mechanisation and technological reach starting in the 1960s led to a dramatic increase in fish landings, for example, in the North Atlantic cod (*Gadus morhua* Linnaeus, 1758) fishery. Landings in that case dramatically exceeded the reproductive capacity of the species and the stocks collapsed catastrophically in 1992 (Bavington, 2010). This was a painful experience for communities in maritime Canada, and for the cod, and shocked many consumers of fish around the world. *Gadus morhua* is not extinct, and commercial landings have continued since then at comparatively very low levels; over 25 years later, there is still a slow drip of only cautiously optimistic reports that the stocks may recover in the future (e.g., Hutchinson et al., 2003; Svedäng & Hornborg, 2015).

With increasing awareness about conservation of marine ecosystems, and especially the impacts of fishing, many people have become aware of which species appear in fish sticks or fish and chips. Advocates for ethical consumption of seafood have raised awareness about fish that are under serious threat of over-exploitation (e.g., swordfish, cod, tuna). Non-ichthyologists can be forgiven for not understanding that many of the most delicious species of fish are themselves apex predators: tuna and salmon depend on the underlying ecosystem to provide sufficient prey to maintain their populations. There are far more elk than wolves, so there are far more sardines than tuna. It follows logically that there are not enough tuna to survive the rapacious appetites of humans without technological intervention such as aquaculture. Over-exploitation is dangerous for the species but at least as much for the economy that supports their harvest.

Prepared and filleted 'whitefish' of many species can be indistinguishable, except by a true gourmet palate or a DNA test. In the US, the name 'Pacific red snapper' was invented to increase the value of market fish by alluding to the high value Red Snapper (*Lutjanus campechanus* (Poey, 1860)) from the Gulf of Mexico. Because of the popularity and historical ubiquity of cod, processed fish products continued to claim 'cod' was their primary ingredient, long after it became less available (Miller & Mariani, 2010). Cod is a large and distinctive fish when alive, with a relatively distinct flavour and texture when cooked. The common name 'cod' has been applied to various other unrelated species, which may be confusing, misleading or fraudulent. Genetic analyses of fish sold in only 27 businesses in California and Washington identified nine separate fish species, all sold under the name 'Pacific red snapper' (Logan et al., 2008). *Species* names are intrinsically linked with other biological data about evolutionary relationships and natural history. Pacific Red Snapper is not a species, it is a made-up name (Cronin & Johnson, 1985). The mislabelling of fish products continues to be a widespread problem even as public opinion demands sustainable seafood and clear product information.

An equivalent problem in livestock supply chains introduced horse meat (*Equus ferus caballus* Linnaeus, 1758) to prepared food products made of ground beef (*Bos taurus* Linnaeus, 1758). This was first uncovered in January 2013 by DNA tests conducted by the Food Safety Authority of Ireland, and rapidly expanded to a Europe-wide scandal (O'Mahony, 2013). Both horsemeat and fish mislabelling represent fraudulently introducing a cheaper food product, from a different species with a lower market value, on the assumption that consumers will not be able to make species-level identifications by taste. The horsemeat scandal generated more media attention, perhaps because of the much greater rates of consumption of beef than fish in the UK and Ireland, and that horses and cows have different cultural values there. The worst contamination levels found traces of horsemeat mixed in combination with beef in around 30% of supermarket burgers. Fish fraud uses a wholesale replacement of one species for another, and several studies have found up to 60% of commercial products to be mislabelled (Jacquet & Pauly, 2008; Armani et al., 2012). Mammal species are not interchangeable. Fish species are also not interchangeable, nor are any other group (Dentinger & Suz, 2014) and consumers have a right to know the contents of their food.

Identifying food organisms may hold the earliest history of taxonomy, as early humans developed signifiers and names to communicate information about the relative quality of food targets or to avoid toxic or venomous species. The family Solanaceae contains many economically important food plants—potatoes (*Solanum tuberosum* Linnaeus), tomatoes (*Solanum lycopersicum* Linnaeus), and peppers (species in the genus *Capsicum*), among others—but also closely related, yet toxic species. Some wild Solanaceae are potentially lethal, such as the deadly nightshades (including *Atropa belladonna* Linnaeus). The Carolina Horsenettle (*Solanum carolinense* Linnaeus) from the southeastern USA produces a fruit that looks quite convincingly like a wild tomato but it is poisonous.

Natural toxins can also be harnessed for beneficial therapies; toxic compounds in deadly nightshade were used as an early anaesthetic for surgery in the 2nd century BCE (Askitopoulou et al., 2000). Now, the search for novel chemical compounds with beneficial properties has expanded to species all over the Earth and in the oceans (Sipkema et al., 2005). Some of the simplest animals, the sponges (phylum Porifera), lack well defined tissues or organs, and are sessile, living attached to the seafloor where they cannot escape predators and are prone to smothering by biofilms and surrounding benthic animals competing for limited space (Müller et al., 2003). As part of a complex of defences against predation and fouling, living sponge species have evolved a 'shock and awe' approach with chemical weapons.

There are many species of sponges, perhaps 10,000 are known and many more remain undiscovered. Sponges may seem similarly spongey to most of us, but they occupy what seem (to sponges) to be very different ecological niches, filtering different size particles of food from the water, living on slightly different substrates, being composed of different sorts of cells, and creating very different chemical compounds. Sponge species are traditionally differentiated by microscopic characters, including the spicules that form their skeleton (Figure 3.2). More recent work incorporating molecular evidence has shown that the basic types and forms of spicules have evolved many times within the living diversity of sponges (Morrow et al., 2013). These primitive animals have a complex history and are still constantly evolving. This underpins some of the complexity and variety and the power of secondary metabolites that sponges use to protect themselves from marine viruses, bacteria, fouling organisms, and predators. The chemical innovations that sponges have developed over 500 million years of evolution to ward off microbes can provide new treatments that are effective against antibiotic-resistant infectious strains in humans.

Where a particular chemical compound has been identified from a living organism, and that compound turns out to be potentially important to humans, it is useful to be able to go back to the source. This requires being able to identify the source species. Arguably, one does not necessarily require knowing the *name* of the source

**FIGURE 3.2** The skeletons of sponges are composed of delicate silicious or calcitic spicules, with microscopic features that are used in taxonomic identification. Fabricated glass model of hexactinellid sponge tissue, *Pararete semperi* (Schulze, 1886) by Leopold and Rudolf Blaschka, in the collection of the National Museum of Ireland, Natural History Division, NMINH 1888.326.1. (Photo by Paolo Viscardi.)

species to be able to find it again. It may be enough just to reliably distinguish it from other similar things, but names make this much easier to communicate reliably. Ethnobotanical studies have carefully demonstrated the multitude of reasons to attach species names to the products derived from plants or animals, ranging from being able to search a database for a name that is useful and spelled correctly, to not confusing the species of interest with another one that could kill you (Bennett & Balick, 2014).

Assays for bioactivity require relatively large volumes of tissue, and most sponges are small. Many small organisms turn to chemical defences as a way to punch above their weight in warding off predators. Because there are so many similar sponge species, so many remain undiscovered, and sponges are animals that do not necessarily have well defined boundaries, it would be extremely easy to make a mistake and combine multiple species in a single sample. It is clearly imperative to be able to identify the specific source of a potential new miracle cure, and that depends on systematics.

Sometimes, identifying a species could save your life. Another illustration of highly variable natural toxic compounds are the venoms of snakes; many species of snakes look similar, but closely related species may have very different venom composition (Wüster & McCarthy, 1996). The most effective antivenoms are keyed against a specific poison, so species identification of the snake becomes important to medical treatment.

Species are the basis for novel inventions, through extraction or replication of natural chemical properties or inspiring biology-mimicking or 'biomimetic' designs. Most examples of bio-inspired design are not necessarily restricted to a single species, but are features that distinguish groups of species. The strength and agility of snakes have been used as the basis for several robotic designs, to develop machines that can traverse complex landscapes or swim without the need for articulating limbs (Hopkins et al., 2009). A vermiform robot successfully entered an unexcavated part of the Great Pyramid of Giza, Egypt, and photographed writing that had been sealed for millennia without damaging the surrounding structure (Richardson et al., 2013). Species' morphological adaptations have been transferred to large and small inventions, including hypodermic syringes, Velcro, flight, beautiful architecture, and the 'lotus effect' that characterises self-cleaning material surfaces.

Design translations inspired by biology often mimic a 'key innovation' that is common to many species, such as leglessness in snakes, an adaptation that may have been pivotal in cladogenesis or success of that species group. In these cases, the original source of design inspiration may be observations of a single species, but a feature that is common to many. To be brutal about the natural world's servitude to humans, if we use the really successful ideas for designing our buildings and robots, why do we need to name individual species, or study multiple examples? Our human-centric interpretation could be misleading. Tiny, perfect crystal structures in brittlestars were first thought to be lenses, but actually have nothing to do with their ability to see using sensors within their skin (Sumner-Rooney et al., 2018). Yet that misconception inspired interesting work on the optical properties of brittlestar skeletons. More importantly, many species remain undiscovered and unnamed, and among the 1.8 million that have been described so far a large portion are obscure to people outside the group of specialist experts that study them. There are whole phyla that remain unfamiliar to most people, but still can inspire, like the tiny kinorhynch that can retract its head through an articulating hatch that looks like a space-port (Figure 3.3; Herranz &

50 µm                                                                100 µm

**FIGURE 3.3**   Two species of kinorhynchs. Left, head detail—*Echinoderes ohtsukai* Yamasaki and Kajihara, 2012 (collected in Boundary Bay, BC, Canada). Right, overview with complex head retracted, showing only a little tuft at top left—*Echinoderes kozloffi* Higgins, 1977 (collected from Victoria, BC, Canada). (Scanning electron microscope images by Maria Herranz.)

Leander, 2016). Even totally unfamiliar species have adaptations for their particular niches, and there were extinct fossil species that had forms unlike anything on Earth today, any of which could provide the next solution for a human problem.

Our relationships with other species are, admittedly, not wholly positive. When humans move species into different environments, or provide new potential habitats through urban development, the consequences can be damaging for both the ecosystem and the economy. A number of human problems are caused by other species when they act as pests or vectors for diseases. Termites are devastating pests that cause damage costing billions of dollars per year in the US alone (Gross, 2017). Efficient management and local eradication of pests depends on being able to identify the enemy. A broader offensive may cause more collateral damage than the original problem, in contrast to the targeted eradication of a well-identified enemy. Termites are not all the same; the group Isoptera includes over 3000 termite species that live in drywood, leaf litter or soil. Attacking ground nests with insecticide is not effective for termite species that live only within wooden structures, and vice versa. In tropical Southeast Asia, standard references listed three different pest termite species across the region, but in fact only one is a pest and two are harmless forest species (Kirton, 2005). The Asian Subterranean Termite *Coptotermes gestroi* (Wasmann, 1896) is responsible for attacks on artificial structures. The vast majority of termite species live in natural forest settings, making important contributions to carbon cycling, some engineering gigantic high-rise nests, living fascinating lives with complex eusocial behaviours, and showing no interest in eating houses (Abe et al., 2000).

Invasive species, animals or plants have been transplanted outside their natural range and can disrupt or take over a new ecosystem; this is another realm where species identification comes into play (Chapter 11). Some local native species are simply naturally abundant and invasive species may be superficially similar to a non-threatening or naturally rare native.

## HOW TO IDENTIFY DIFFERENTNESS

Thinking about lists of species names might invoke memories of a biology classroom or heading off on a field outing armed with dichotomous identification keys. Some of us who enjoyed that first experience of keying out organisms go on to do it for a living. Many more people who have no professional interest in science actually do this for fun.

Birdwatchers are the most visible such constituency, a global community of passionately interested people, and a force to be reckoned with. One attraction of this hobby is learning to appreciate that even backyard birds rapidly reveal a surprising number of species, animals that share our living space. With a little practice, most birds are surprisingly easy to tell apart, and this is a skill that empowers you to recognise not only birds in your own neighbourhood, but anywhere that birds can be found. The key features that separate birds—beak shape, colour patterns, body shape—can be applied to distinguish bird species anywhere in the world that one might see while travelling or in videos of faraway places. Even if you do not know their names immediately, knowing which features to look for gives you the power to tell species apart and to identify the species later from reference materials that stay still.

Birders refer to lists of species names on their 'life lists' or check lists of the species found in particular places. Among birders, as distinct from professional ornithologists, there is lively ongoing debate about the relative merits and appeal of scientific names and standard English common names for recording observations. Different species can have the same vernacular name in different regions, or a common name could apply to a group of many similar species. The scientific name applies unambiguously to a species, and to the technical information that underpins its description. Given the vast information available via the Internet, scientific names are a better starting point for an unambiguous search. Some people are emotionally opposed to scientific 'Latin' names because they feel constrained or formal, but the justification usually offered by intelligent birders for their bewildering preference for vernacular names is that they are more 'stable'. Scientific names are occasionally subject to taxonomic revision, which means the genus or species epithet may be revised to reflect new understanding of the evolutionary relationships in a group (Chapter 4). Hobbyists have the freedom to choose what aspects of science they engage with. Although they may use the standard vernacular names rather than the scientific names, both systems are still significant in identifying and enumerating the specific kinds of birds observed, and this creates that link to other knowledge about each species.

The process of honing identification skills in birds has an exact equivalence in plants, or mushrooms, or insects, or molluscs, and these and other groups of organisms have their followers, too. The same approach, transferable skills, and generalities about observations and identification can be said about any of these very different types of living things. And even better, learning to identify birds will make it easier to identify mushrooms, through the practice of understanding that there are key features important within the group and practising good noticing. In the current decade, this art of noticing the world around us has become fashionably known as 'mindfulness'.

Birds are accessible to a wide constituency of admirers, we generally see them every day (even if we do not notice them), they have a compelling aesthetic value, and when they retreat to the skies it only makes them more alluring. In terms of species identification, these organisms are relatively easy to understand because in fact they are all broadly similar, variations on a theme. Even an ostrich and a hummingbird can be compared using the basic features that are familiar to all birders: beak shape, wing shape, feet, plumage, tail feathers, an eye ring.

Different features are used for classification at different levels of resolution. Broad taxonomic categories can seem trivially different—that Apodiformes (hummingbirds) and Sphenisciformes (penguins) are different types of beast apparently requires no further explanation. It may seem more difficult to explain their similarities than their differences. They have feathers and beaks and scaly feet, but the details of those morphological aspects also shows the adaptations that define these clades. We can easily discriminate a hummingbird from a penguin, but features with more subtle variation would be needed to identify particular species of penguins.

Most people's experience of identification is restricted within a particular local fauna, in the places they live or work. Hummingbirds and penguins are enchantingly exotic to people who did not grow up watching them (and really, everyone), but it is possible to see them as a normal part of a landscape. Differences are relative, and perceptual, and depend a lot on what you are used to. When the whole category is strange, all individual objects within a group can seem very similar. Visiting a market in a new country, at first, everything is remarkable and mostly looks simply foreign. After living in that country for a while, variations in region, style, and quality start to reveal themselves. Likewise, in California, a visiting scientist from overseas might be delighted to see tiny birds stationary in mid air between the blur of their wings, occupying the same ecological niche as large beetles in other regions. Some local biologists with a practised eye for colour, size, shape, and behaviour are more focussed on whether it is one of 14 different regional species. Little differences stand out more among things that are familiar.

As a marine biologist, I usually maintain a satirical disdain for terrestrial vertebrates (though secretly I go armed with binoculars and tips from birding friends whenever I travel to a new place). In the more interesting and diverse marine realm, a look in a tide pool reveals a much broader variety of life forms: anemones, snails, hermit crabs, red and green seaweeds—these are all from different *phyla*, the broadest divisions of animal life (Chapter 6). While it is easy to tell the broad categories apart, it takes a little education to understand what features are important and why. But, once learned, again the understanding of key features can be transferred to understand some basic forms in rocky shores worldwide.

The broader case to be made here is that there is no universal measure of differentness, and our perception of what features are relevant for separating categories is dependent on familiarity, practical constraints, and the knowledge base. Learning to discriminate between similar things is a lifelong process of improvement, and mindfulness as every observation gives another point for comparison. This is true for skills of individual people, but it is also true at a global scale for the scientific community and the total body of data assembled by science to date. There are a very few groups or organisms, such as birds, where most species have been described and many have been studied intensively. In most clades, the organisms are far less accessible or less familiar, and this fundamental understanding of what does or does not constitute a critical difference has often not yet been well established.

## DIFFERENCES OF OPINION WITH CONSEQUENCES

The potential for uncertainty in species identification or delineation is one reason why some conservation biologists prefer to focus on overall ecosystem health and the preservation of wild places, to avoid becoming mired in seemingly irresolvable debates about individual lineages (Rapport et al., 1999). However, designating a species as officially 'vulnerable' or 'endangered' is recognised worldwide, and the International Union for Conservation of Nature (IUCN) Red List is a powerful tool to gain attention and support for specific conservation efforts.

Rockhopper Penguins in the *Eudyptes chrysocome* species complex comprise multiple lineages, distributed in sub-Antarctic islands. In animals with small ranges, such as these seabirds, species-specific knowledge is crucial for conservation (Figure 3.4). For example, three species were named historically based on the birds' morphology: *Eudyptes chrysocome* (Forster, 1781), *Eudyptes filholi* Hutton, 1879, and *Eudyptes moseleyi* Mathews and Iredale, 1921. These were originally named as species (e.g., Hutton, 1879), but the differences seemed inconsequential and some

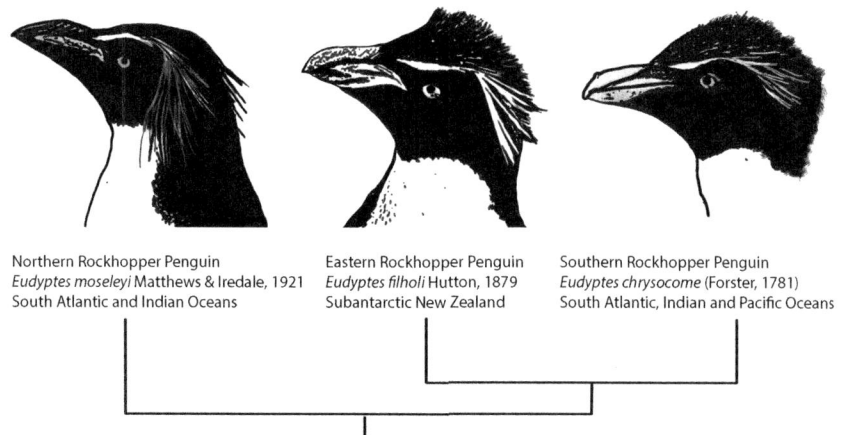

Northern Rockhopper Penguin
*Eudyptes moseleyi* Matthews & Iredale, 1921
South Atlantic and Indian Oceans

Eastern Rockhopper Penguin
*Eudyptes filholi* Hutton, 1879
Subantarctic New Zealand

Southern Rockhopper Penguin
*Eudyptes chrysocome* (Forster, 1781)
South Atlantic, Indian and Pacific Oceans

**FIGURE 3.4** The *Eudyptes chryosocome* species complex of sub-Antarctic Rockhopper Penguins. All three species were originally recognised by morphology, later considered to be different forms and populations of one species, and more recently recognised as at least two, probably three separate lineages. (Phylogenetic relationships as inferred by de Dinechin et al., 2009.)

considered them to be subspecies (subdivisions of *Eudyptes chrysocome*), and later the names fell into obscurity. More recently, the names were brought back into use to recognise separate types of penguins based on genetic evidence. The publication that showed the Northern Rockhopper Penguin, *Eudyptes moseleyi*, was a valid distinct lineage focussed on volcalisations, head-tuft morphology, and nesting seasonality just for the penguins that live in northern sub-Antarctic islands (Jouventin et al., 2006). That is, it was a scientific argument based in natural history and biogeography, as well as genetics. Later, another paper by a different team further split the Eastern Rockhopper Penguin *Eudyptes filholi* from the main lineage *E. chrysocome*, in a study that primarily emphasised genetic distances (Banks et al., 2006). The genetic method used in that study has been criticised by others working on DNA barcoding in ornithology (Fregin et al., 2012), but to date it has not been repeated, and different experts are equivocal about whether it is a subspecies (Del Hoyo et al., 2014) or a species (Thiebot et al., 2013). The rank in this case is perhaps less important than the data that indicate it is a distinct lineage.

The IUCN Red List committee for birds uses a standardised assessment to determine which species are accepted as valid, and that method rather downgrades the importance of genetic bar coding (Tobias et al., 2010). This could be viewed as old-fashioned, but it is also arguably pragmatic. Published genetic studies may include DNA from a combination of field-collected and historical specimens held in museum collections. To confirm identification in the field, obtaining a tissue sample for DNA is often impractical and could require harassing the bird. Additional abundant data on morphology, behaviour, and other types of information are more available for birds than most other groups of organisms. On this basis, the Eastern Rockhopper Penguin is not currently recognised or assessed by the IUCN.

Species-specific knowledge is crucial for conservation. In this case, New Zealand has recognised *E. filholi* as a distinct species, and it is considered endangered with 'nationally critical' status (Banks et al., 2016). Local populations of *E. filholi* have declined dramatically, losing 90% of historical population numbers (Cuthbert et al., 2009). If this is a local population of a penguin species that ranges over many sub-Antartic islands, then it is still a problem for New Zealand; if *E. filholi* is endemic to New Zealand territorial waters and these are the only populations, then an entire species is facing extinction.

## HOW TO SEPARATE SPECIES

Since species are so ubiquitous and familiar, it seems like a conundrum that there is so much uncertainty about species identification. The differences that separate species are real, they are the products of evolution. Our ability to understand these differences is limited by the available comparative framework. This 'framework' applies to our own personal knowledge—I am much better at identifying mollusc species (especially polyplacophoran molluscs, my favourite organisms; Figure 3.5) than insects or mammals. At another level, this applies to the global scientific knowledge base. We might encounter uncertainty about a species identification because the recorder simply had not learned the details of that group, or it could be that no one can identify it because we are missing information, or because that specimen turns out to be a new species. In some cases, different species really are difficult to tell apart.

**FIGURE 3.5**   A chiton, *Tonicella lineata* (Wood, 1815). Chitons, class Polyplacophora, are marine molluscs that have a soft foot protected by eight articulating shell plates.

Most species are rare (e.g., Stork, 1997). This is a simple, though perhaps unexpected, fact of natural systems. Within any ecosystem, there are a few abundant species, and relatively many more rare species. This pattern expands to global biodiversity. In terms of global diversity, it is relatively easier to discover a species that is common than one that is rare, so the project of finding Earth's species is increasingly restricted to a race to find all of the many rare species before they go extinct (Costello et al., 2013). Issues of limited data, limited observations, and limited specimens for potential new species are not the exception, but the rule.

Taxonomists have the inside scoop about where this global knowledge base is a bit thin—they know a few things about what we do not know. There are staggering numbers of species on Earth that have not been discovered or described (Chapter 10). Specialists on most groups of organisms routinely discover new species, to the point where the moment of discovery is less 'Eureka!' and more 'Here we go again!' The rate of discovery depends on that knowledge base for a particular group or geographical region. New frontiers of diversity have been opened with new technologies, for example, in DNA sequencing, microscopic imaging, and access to sampling deep-sea diversity. These are examples not just of minor revolutions, but pathways that continue to accelerate discovery as the technology improves. Among complex but inconspicuous organisms, like the sponges mentioned above or fungi and many others, DNA sequencing was confounded for many years by the close relationships between bacteria and the host tissue, where the microbes were much easier to sequence. More knowledge about diversity, an improved global knowledge base, provides context to assess and discover more diversity. The rate of discovery is a problem in regions where the balance between diversity and scientific infrastructure is tilted toward diversity (Chapter 12).

The more pressing problem is that the separation of species is difficult to demonstrate; even where such distinctions are obvious to a specialist, the process of distinguishing a new species still involves a lot of data and a lot of technical expertise. Can technology solve all of our problems? When fish are mislabelled in food products, DNA sequences can reveal their correct species identity. Logically, we could hope to extend this to a universal tool that could identify anything with just a quick lab test. It is tempting to imagine a DNA barcode as an objective assay that sidesteps the difficulties of training human scientists to recognise and assess complex morphological variation. Molecular genetic sequences have had a huge influence on the identification of new species, uncovering 'cryptic' species complexes that look sufficiently alike that they could not be distinguished without this independent source of data to separate them (Knowlton, 1993; Bickford et al., 2007). So what stops this technology from offering a complete panacea to the species identification problem?

There are three substantial impediments which permanently limit a 'quick fix' identification solution (Sigwart & Garbett, 2018). First, using a genetic sequence to identify a species depends entirely on someone else having sequenced the same species before. If I discover a new species of chiton (Figure 3.5), sequence its DNA, and then compare that sequence to a reference library to get my identification, I would either find that there is no good match, or, worse, a computer program may misleadingly suggest the most similar available data, which would be an incorrect identification. The vast majority of species on Earth have never been sequenced.

There are close to 250,000 eukaryotic species in the GenBank taxonomy database (Federhen, 2011), and while it is growing constantly, that is only about 14% of the 1.8 million species described, and a much more tiny fraction of the total diversity of life on Earth (Chapter 10). Second, there is a problem of data availability; it should be noted that you cannot get a genetic sequence out of everything. DNA is a pretty unstable molecule that breaks down quickly as tissue decays (Allentoft et al., 2012), and it falls apart with exposure to certain chemicals used as fixatives, such as formalin (formaldehyde in aquaeous solution). But those fixatives are important for studying microscopic cell structures. The same issue of degradation limits the potential to extract DNA from many older specimens, and in deeper time no DNA remains when tissues have been replaced by minerals in fossils. Third, there is the overarching issue of what level of differentness qualifies the separation of species.

## SPECIES CRITERIA

Species are not equally different. In part, this is an issue of perception and our relative familiarity with diagnostic characteristics of wolves, or birds, compared to sponges, or fungi. But the fossil record and genetic data both show us that various species are evolving at different rates and were separated from their nearest species relatives shorter or longer ago in evolutionary history (Chapter 5). Dogs and wolves share a common ancestor perhaps 15,000 years ago, and there are strong genetic similarities (the popular press has widely reported a figure of over 99% similarity, based on pairwise differences of mitochondrial partial genomes), though the intervention of genetic modification through selective breeding has rapidly expanded the morphological variety, or 'disparity' shown in domestic dogs. These lines of evidence support the view that dogs are a subspecies of wolves, distinct but not a fully separate lineage on their own evolutionary trajectory. But morphologically and ecologically, they are compellingly different. Which is correct? As we will discuss in later chapters, species are not fixed entities waiting to be classified, but dynamic patterns that we catch in a particular moment of evolutionary time.

One simplistic metric offered in school textbooks is that species are groups of interbreeding organisms that can produce viable offspring (Mayr, 1942). Reproductive isolation is a useful shorthand, but it is an incomplete and misleading explanation. This idea is presented in school textbooks to explain, for example, that horses and donkeys are different species because their crossbred offspring are infertile, though there are also quite obvious differences in appearance. Mules (and hinnies) are infertile because of parental chromosomal incompatibility (a horse has 64 chromosomes; a donkey has 62); having 63 chromosomes with incomplete pairing does not impede normal mitotic division of somatic cells, but they cannot produce gametes (Trujillo et al., 1962). This type of infertile hybrid is actually a relatively unusual case; other cross-breeding experiments among closely related species, such as many ornamental plant hybrids, or interbreeding tigers and lions (*Panthera tigris* (Linnaeus, 1758) *Panthera leo* (Linnaeus, 1758)), often result in fertile offspring. Clearly tigers and lions are still separate species, that look different, behave differently, live in different habitats, and are native to different continents. The commonness of hybrids is familiar to botany, but its frequency in an animal context often surprises people (Gardner, 1997). Importantly, over-simplistic ideas about biology can impact policy decisions (Fairbanks, 2006). In more rigorous scientific debates, the concept of 'interbreeding' refers to a natural long-term tendency for genetic exchange; geographic and behavioural isolation contribute to maintaining separate species. Reproductive isolation is one line of evidence within our modern understanding of species, and offspring in violation of this 'rule' do not mean the parents are not separate species (Chapter 7).

Species can only be distinguished using the best available data. We might accurately identify a bird species from its song, or a photograph, without access to its DNA. Fossils are always intrinsically data limited, and must be identified based on the parts that are preserved. Many deep sea animal species have only been seen a handful of times. There is a species of palm tree, Tahina spectabilis Dransf. & Rakotoarinivo, discovered in 2008 in Madagascar, that had been confidently but incorrectly identified for 20 years until it produced a distinctive flower that betrayed its unique identity (Dransfield et al., 2008). Some plants only flower once in their lifetime, which may be more than 80 years in the case of *Puya raimondii* Harms, Queen of the Andes, the world's largest bromeliad species (Sgorbati et al., 2004). Really, the idea of interbreeding populations can never be tested with most species, fossil or living. The approach to understanding species requires all available evidence be taken into consideration in an informed way.

Species are complicated, but in fact a lot of problems are removed if the 'rule' about interbreeding is lifted or accepted to be inadequate. We will return to this discussion (Chapter 7), but it is important to confront this because interbreeding is what most people have in mind, consciously or not, when thinking about species. In short, the idea of interbreeding is one source of data about species, but such data are not available for the vast majority of species, and it is not always particularly informative from experimental evidence. More broadly, the exchange of genetic information, through ancestor-descendant lines, is an essential aspect of the evolution of species. Species are independent functioning evolutionary lineages, with a clear evolutionary trajectory that is independent of any individual constituent population. Individuals or populations within a species may go extinct, or hybridise, or indeed adapt and split away from the primary species trajectory. There are many lines of evidence that can be applied to identify such population-lineages as species: shared characteristics, including appearance, life history traits, genetic similarity, and more, are all relevant data that help distinguish species.

## EVERYONE USES SPECIES

We see species every day. Identifying species unlocks important information and metadata about diversity and ecosystem functioning and enables the repeatability of key findings in the properties of individual species. The process of reporting reliable identification guidelines to recognise these units, species, is taxonomy.

The species name is a practical shorthand for a technical description of important distinguishing features, understood by everyone. The name also provides a direct link to additional information about the species in the scientific literature, and other sources, removing doubt about whether we are comparing like with like. But while some species are obviously distinguishable, there are sometimes disagreements about what does and does not qualify as a separate species. The name, and the associated description and evidence, allow other scientists to independently replicate observations. Expert opinions about what is or is not a species can have significant practical consequences for others who are the users of species names.

Most biologists are probably already convinced that species are interesting for their own sake, but there are some doubters and other people who assume they have no opinion about the usefulness or not of taxonomy. The specialist discipline of discovering, naming, and classifying species—taxonomy— has a reputation as chronically under-funded and under threat. Its reach, however, is mighty. These examples touch on some of the many synergies between taxonomy and other disciplines. Various scholars have suggested, or assumed, that taxonomy is a subsidiary part of other kinds of biology. In the mid-20th century, life sciences moved away from descriptive work and toward more population-based or experimental approaches. More recently, others have tried to co-opt taxonomy toward refocussing on phylogenetics, or on conservation biology (Golding & Timberlake, 2003). The variety of research applications mentioned in this chapter just scrapes the surface. Taxonomy is a rich field in and of itself that cannot be wholly subsumed into another area of biology. The process of identifying species units is relevant to all areas that use species. The users that depend on species come from all branches of life sciences and beyond.

## REFERENCES

Abe T, Bignell DE, Higashi M (editors). 2000. *Termites: Evolution, Sociality, Symbioses, Ecology.* Springer.

Allentoft ME, Collins M, Harker D, Haile J, Oskam CL, Hale ML, Campos PF, Samaniego JA, Gilbert MT, Willerslev E, Zhang G. 2012. The half-life of DNA in bone: Measuring decay kinetics in 158 dated fossils. *Proceedings of the Royal Society B.* 279: 4724–33.

Armani A, Castigliego L, Guidi A. 2012. Fish fraud: The DNA challenge. *CAB Reviews.* 7(71): 1–12.

Askitopoulou H, Ramoutsaki IA, Konsolaki E. 2000. Analgesia and anesthesia: Etymology and literary history of related Greek words. *Anesthesia & Analgesia.* 91: 486–91.

Banks J, Van Buren A, Cherel Y, Whitfield JB. 2006. Genetic evidence for three species of rockhopper penguins. *Eudyptes chrysocome. Polar Biology.* 30: 61–7.

Banks JC, Clark JA, Nield P, Stanton JA, Wagner E. 2016. Haplotyping cryptic Adélie penguin taxa using low-cost, high-resolution melt curves. *New Zealand Journal of Zoology.* 43: 163–70.

Bath AJ. 1989. The public and wolf reintroduction in Yellowstone National Park. *Society & Natural Resources.* 2: 297–306.

Bavington D. 2010. *Managed Annihilation: An Unnatural History of the Newfoundland Cod Collapse.* UBC Press.

Begossi A, Clauzet M, Figueiredo JL, Garuana L, Lima RV, Lopes PF, Ramires M, Silva AL, Silvano RA. 2008. Are biological species and higher-ranking categories real? Fish folk taxonomy on Brazil's Atlantic forest coast and in the Amazon. *Current Anthropology.* 49: 291–306.

Bennett BC, Balick MJ. 2014. Does the name really matter? The importance of botanical nomenclature and plant taxonomy in biomedical research. *Journal of Ethnopharmacology.* 152: 387–92.

Berlin B. 1973. Folk systematics in relation to biological classification and nomenclature. *Annual Review of Ecology & Systematics*. 4: 259–71.

Bickford D, Lohman DJ, Sodhi NS, Ng PK, Meier R, Winker K, Ingram KK, Das I. 2007. Cryptic species as a window on diversity and conservation. *Trends in Ecology & Evolution*. 22: 148–55.

Briggs D, Walters SM. 1997. *Plant Variation and Evolution*. Cambridge University Press.

Costello MJ, May RM, Stork NE. 2013. Can we name Earth's species before they go extinct? *Science*. 339: 413–6.

Cronin I, Johnson P. 1985. 'Red Snapper Is Not a Local Rockfish'. *LA Times*. May 9, 1985.

Cuthbert R, Cooper J, Burle MH, Glass CJ, Glass JP, Glass S, Glass T, Hilton GM, Sommer ES, Wanless RM, Ryan PG. 2009. Population trends and conservation status of the Northern Rockhopper Penguin *Eudyptes moseleyi* at Tristan da Cunha and Gough Island. *Bird Conservation International*. 19: 109–20.

De Dinechin M, Ottvall R, Quillfeldt P, Jouventin P. 2009. Speciation chronology of rockhopper penguins inferred from molecular, geological and palaeoceanographic data. *Journal of Biogeography*. 36: 693–702.

Del Hoyo J, Collar NJ, Christie DA, Elliott A, Fishpool LDC. 2014. *Illustrated Checklist of the Birds of the World. Volume 1: Non-passerines*. Lynx Edicions/BirdLife International.

Dentinger BT, Suz LM. 2014. What's for dinner? Undescribed species of porcini in a commercial packet. *PeerJ*. 2: e570.

Dransfield J, Rakotoarinivo M, Baker WJ, Bayton RP, Fisher JB, Horn JW, Leroy B, Metz X. 2008. A new Coryphoid palm genus from Madagascar. *Botanical Journal of the Linnean Society*. 156: 79–91.

Ersmark E, Klütsch C, Chan YL, Sinding MH, Fain SR, Illarionova NA, Oskarsson M, Uhlén M, Zhang YP, Dalén L, Savolainen P. 2016. From the past to the present: Wolf phylogeography and demographic history based on the mitochondrial control region. *Frontiers in Ecology & Evolution*. 4: 00134.

Fairbanks P. 2006. Blending in: Addressing hybridization in the endangered species act. *Journal of Natural Resources & Environmental Law*. 21: 51–9.

Federhen S. 2011. The NCBI taxonomy database. *Nucleic Acids Research*. 40: D136–43.

[FAO] Food & Agriculture Organisation of the United Nations. 2017. FAO Fish Finder. www.fao.org accessed November 2017.

Fregin S, Haase M, Olsson U, Alström P. 2012. Pitfalls in comparisons of genetic distances: A case study of the avian family Acrocephalidae. *Molecular Phylogenetics & Evolution*. 62: 319–28.

Gardner JP. 1997. Hybridization in the sea. *Advances in Marine Biology*. 31: 1–78.

Golding JS, Timberlake J. 2003. How taxonomists can bridge the gap between taxonomy and conservation science. *Conservation Biology*. 17: 1177–8.

Gross M. 2017. How insects shape our world. *Current Biology*. 27: R283–5.

Herranz M, Leander BS. 2016. Redescription of *Echinoderes ohtsukai* Yamasaki and Kajihara, 2012 and *E. kozloffi* Higgins, 1977 from the northeastern Pacific coast, including the first report of a potential invasive species of kinorhynch. *Zoologischer Anzeiger*. 265: 108–26.

Higgins RP. 1977. Redescription of *Echinoderes dujardinii* (Kinorhyncha) with descriptions of closely related species. *Smithsonian Contributions to Zoology*. 248: 1–26.

Hopkins JK, Spranklin BW, Gupta SK. 2009. A survey of snake-inspired robot designs. *Bioinspiration & Biomimetics*. 4: 021001.

Hutchinson WF, van Oosterhout C, Rogers SI, Carvalho GR. 2003. Temporal analysis of archived samples indicates marked genetic changes in declining North Sea cod (*Gadus morhua*). *Proceedings of the Royal Society B*. 270: 2125–32.

Hutton FW. 1879. On an apparently new species of Penguin, from Campbell Island. *Proceedings of the Linnean Society of New South Wales*. 3: 334–5.

Jacquet JL, Pauly D. 2008. Trade secrets: Renaming and mislabeling of seafood. *Marine Policy*. 32: 309–18.

Jouventin P, Cuthbert RJ, Ottvall R. 2006. Genetic isolation and divergence in sexual traits: Evidence for the northern rockhopper penguin *Eudyptes moseleyi* being a sibling species. *Molecular Ecology*. 15: 3413–23.

Kirton LG. 2005. The importance of accurate termite taxonomy in the broader perspective of termite management. In: *Proceedings of the Fifth International Conference on Urban Pests*, pp. 1–7. Penang, Malaysia: P&Y Design Network.

Knowlton N. 1993. Sibling species in the sea. *Annual Review of Ecology & Systematics*. 24: 189–216.

Larson G, Burger J. 2013. A population genetics view of animal domestication. *Trends in Genetics*. 2: 197–205.

Logan CA, Alter SE, Haupt AJ, Tomalty K, Palumbi SR. 2008. An impediment to consumer choice: Overfished species are sold as Pacific red snapper. *Biological Conservation*. 141: 1591–9.

Mayr E. 1942. *Systematics and the Origin of Species from the Viewpoint of a Zoologist*. Columbia University Press.

Miller DD, Mariani S. 2010. Smoke, mirrors, and mislabeled cod: Poor transparency in the European seafood industry. *Frontiers in Ecology and the Environment*. 8: 517–21.

Morrow CC, Redmond NE, Picton BE, Thacker RW, Collins AG, Maggs CA, Sigwart JD, Allcock AL. 2013. Molecular phylogenies support homoplasy of multiple morphological characters used in the taxonomy of Heteroscleromorpha (Porifera: Demospongiae). *Integrative & Comparative Biology*. 53: 428–46.

Müller WE, Brümmer F, Batel R, Müller IM, Schröder HC. 2003. Molecular biodiversity. Case study: Porifera (sponges). *Naturwissenschaften*. 90: 103–20.

O'Mahony PJ. 2013. Finding horse meat in beef products—a global problem. *QJM: An International Journal of Medicine*. 106: 595–7.

Rapport DJ, Böhm G, Buckingham D, Cairns J, Costanza R, Karr JR, De Kruijf HA, Levins R, McMichael AJ, Nielsen NO, Whitford WG. 1999. Ecosystem health: The concept, the ISEH, and the important tasks ahead. *Ecosystem Health*. 5: 82–90.

Ripple WJ, Beschta RL. 2012. Trophic cascades in Yellowstone: The first 15 years after wolf reintroduction. *Biological Conservation*. 145: 205–13.

Richardson R, Whitehead S, Ng TC, Hawass Z, Pickering A, Rhodes S, Grieve R, Hildred A, Nagendran A, Liu J, Mayfield W. 2013. The 'Djedi' Robot Exploration of the Southern Shaft of the Queen's Chamber in the Great Pyramid of Giza, Egypt. *Journal of Field Robotics*. 30(3): 323–48.

Sigwart JD, Garbett A. 2018. Biodiversity Assessment, DNA Barcoding, and the Minority Majority. *Integrative and Comparative Biology*. doi:10.1093/icb/icy076.

Sipkema D, Osinga R, Schatton W, Mendola D, Tramper J, Wijffels RH. 2005. Large-scale production of pharmaceuticals by marine sponges: Sea, cell, or synthesis? *Biotechnology & Bioengineering*. 90: 201–22.

Sgorbati S, Labra M, Grugni E, Barcaccia G, Galasso G, Boni U, Mucciarelli M, Citterio S, Benavides Iramátegui A, Venero Gonzales L, Scannerini S. 2004. A survey of genetic diversity and reproductive biology of *Puya raimondii* (Bromeliaceae), the endangered queen of the Andes. *Plant Biology*. 6(2): 222–30.

Stork NE. 1997. Measuring biodiversity and its decline. In: Reaka-Kudla ML, Wilson DE, Wilson EO (editors). *Biodiversity II: Understanding and Protecting our Biological Resources*, pp. 41–68. Joseph Henry Press.

Sumner-Rooney L, Rahman IA, Sigwart JD, Ullrich-Lüter E. 2018. Whole-body photoreceptor networks are independent of 'lenses' in brittle stars. *Proceedings of the Royal Society B*. 285: 20172590.

Svedäng H, Hornborg S. 2015. Waiting for a flourishing Baltic cod (*Gadus morhua*) fishery that never comes: Old truths and new perspectives. *ICES Journal of Marine Science*. 72: 2197–208.

Thiebot JB, Cherel Y, Crawford RJ, Makhado AB, Trathan PN, Pinaud D, Bost CA. 2013. A space oddity: Geographic and specific modulation of migration in *Eudyptes* penguins. *PLoS ONE*. 8: e71429.

Tobias JA, Seddon N, Spottiswoode CN, Pilgrim JD, Fishpool LD, Collar NJ. 2010. Quantitative criteria for species delimitation. *Ibis*. 152: 724–46.

Trujillo JM, Stenius C, Christian LC, Ohno S. 1962. Cromosomes of the horse, the donkey, and the mule. *Chromosoma*. 13: 243–8.

Wayne RK, Ostrander EA. 2007. Lessons learned from the dog genome. *Trends in Genetics*. 23: 557–67.

Wüster W, McCarthy CM. 1996. Venomous snake systematics: implications for snakebite treatment and toxinology. In: Bon C, Goyffon M (editors). *Envenomings and their Treatments*, pp. 13–23. Fondation Mérieux.

Yamasaki H, Kajihara H. 2012. A new brackish-water species of *Echinoderes* (Kinorhyncha: Cyclorhagida) from the Seto Inland Sea, Japan. *Species Diversity*. 17: 109–18.

# 4 Why Do the Names Keep Changing?

## TO IMPROVE IS TO CHANGE*

All scientific disciplines feel a certain push and pull, balancing growth and stability, anticipating improvement from new discoveries, yet waiting for well-grounded practical results to warrant change. Species names are slightly different than other scientific products, in that species are used by everyone, and it is not always obvious how a change in name offers any improvement. The explicit goal of nomenclature is to provide clear, stable information about species identities (e.g., Godfray & Knapp, 2004). This seems somewhat at odds with the goals of systematics, as with any scientific discipline, to iteratively test hypotheses about species identity and inter-relationships (Chapter 2).

Species names include two parts: the genus epithet and species epithet. Intuitively, species that share the same genus name are more similar to each other than anything outside their group. When new information changes understanding of these relationships, names and classification change to reflect the revised hypothesis. Frustrations always arise when knowledge is created by a group of specialists and it does not become accessible to the broad community of users. This chapter offers some explanation of the hidden framework that controls ever-shifting species nomenclature and a forensic guide to reading the added meaning in taxonomic names.

## COLONIALISM, REVISION, AND THE TYPE CONCEPT

Imagine the thrill of first opening a new edition of your favourite field guide, full of crisp colour images of your favourite regional species. But then, the initial thrill is followed rapidly by grim disappointment as you realise once-familiar species are presented with new and unfamiliar tongue-twister names. What has happened to change our old friends? And more betrayal, some others you were taught to call by species names now just have a genus epithet and a dissatisfying 'sp.' (or worse still 'species complex'). We sigh. We gather our resolve, accept that our friends are still the same even with their changed names, and we set out to practise these new phrases until they become comfortable again.

If this does not sound familiar to you, you probably grew up in Western Europe. A basic principle of taxonomic practice combined with historical Euro-centrism means that, in general across most organisms, European species names stay the same, while the much larger diversity in the rest of the world experiences these pains that come with modernisation.

---

* 'To improve is to change, so to be perfect is to have changed often.'—Winston Churchill, 1925

Taxonomy in most organisms was established long before there was any formal idea of evolution, and early hypotheses about similarity have been overturned. Taxonomic revision has been happening for as long as there has been taxonomy; the advent of phylogenetic systematics was not the first impetus for changes in names or group names. Importantly, phylogenetic systematics as a concept was formulated before *molecular* phylogenctics. Taxonomic revisions are frequently dismissed as some sort of conflict between morphology ('old' taxonomy) and molecules ('new' taxonomy), but this is not accurate (Mallet & Willmott, 2003). Systematic revisions began immediately at the dawn of taxonomy—as did grumpy arguments about differing interpretations (Harrison, 2009). Molecular tools have helped to resolve many long-standing conflicts or suspicions about relationships among species (Padial et al., 2010). And though DNA evidence has come to the fore of taxonomy, new information about anatomy and physiology of living species and the fossil record of extinct species also still play large roles in the ongoing process of revising old hypotheses about species and species relationships.

The earliest forays into formal taxonomy were led by Carl Linnaeus, of the eponymous Linnean System, and a host of other contemporary active scientists documenting the natural world (Chapter 7). Linnaeus documented the names of many European and global species, standardising many names that were already in use and adding many more. The museums, botanic gardens, and scholarly societies in the major capitals of Europe became the centres for comparative information and the sources for publications documenting new observations (Frodin, 2001; Paknia et al., 2015). European governments were engaged in extensive exploration and the active occupation of land and resources all over the rest of the world. Specimens of animals and plants from faraway lands were delivered to the scholars of Europe to describe, analyse, and classify (Chapter 8).

Several important points of context, then, remain relevant to classification and to species names we still have today. First, many of these early taxonomists responsible for the first attempts at naming specimens from Africa, South America, and Southeast Asia, never actually left Europe. Some of them had never experienced a tropical ecosystem first hand. There is a real, visceral experience for any biologist being immersed in a new type of habitat. We forget this, in part because travel is now relatively easy and available, and faraway places become familiar through nature documentaries and the principles of global ecology and biogeography as taught in undergraduate curricula. Wild places are also increasingly difficult to find; post-industrial era ecosystems are so damaged in most of the world, so reduced in diversity, and so brutalised by invasive species. It is quite possible now for even scientifically educated tourists to visit tropical Southeast Asia and come away thinking, there are nice white sand beaches, but I'm not sure what all the fuss is about over this so-called diversity. In a way, a modern tourist's observations of a slightly random subsample from awesomely diverse but possibly misunderstood ecosystems is not dissimilar to the exposure of historical 18th century European scientists, receiving samples from afar representing a scattering of the most dominant, or economically interesting, or brightly coloured, or easy to collect animals and plants.

European scholars generally named new species and put them into the system of classification that was already in place for European taxa, and North American scholars would use the local North American fauna as a reference frame. The major consequence of this is that many taxa in tropical countries, especially former European colonies, have names that reflect this historical perspective rather than relevant evolutionary relationships (Walters, 1986).

The process of taxonomic revision, in some cases, brings a more global perspective, but it also means that the fauna and flora of former European colonies change while European names more often stay the same. For example, the California market squid, *Doryteuthis opalescens* (Berry, 1911), is an important fisheries species, which was formerly and still often called '*Loligo*' *opalescens*, because it was originally described as a member of the genus *Loligo*. Starting in the late 1990s, phylogenetic analyses found natural, evolutionary groups of squid species corresponded to different ocean regions (e.g., Anderson, 2000). In the genus *Loligo*, the 'type species' of the genus is the European market squid *Loligo vulgaris* Lamark, 1798. The type species is a designated taxonomic reference point, usually the first species that was historically named. When it became clear that the market squid of the world are not all directly related to each other, *L. vulgaris* had to be the reference point. *Loligo vulgaris* and its nearest relatives stay still within the genus *Loligo*, while other species from other oceans have to change the first part of their names to new genera (Vecchione et al., 2005). Is the biology of a European squid a useful proxy for a California squid? If they share a genus name, it would make them seem more similar than they really are; it is useful to know they are not as closely related as an overstretched global genus group might imply.

I grew up and studied in North America, learning the names of species around me. I moved to Europe in my 20s and found myself surrounded by type species, the very ones that laid the reference frame for names I learned at home. Not just a naticid snail, but *Natica* itself. Not just a pycnogonid, *Pycnogonum*. Not just any littorinid, *Littorina littorea* (Linnaeus, 1758). The frogs are called *Rana*, the mice are called *Mus*, the earliest taxonomic names that adopted direct translations from Latin words. These taxa were the first to be described by early taxonomists and so the names are fixed reference points. Compared to other regions, European taxonomic names change relatively less frequently, but this does not make them immune from phylogenetic progress and also does not imply any special evolutionary significance.

*Rana temporaria* Linnaeus, 1758 is the type species of the genus *Rana*, and was the first frog species named in taxonomic nomenclature. This early recognition implies neither that it is a primitive ur-frog, nor an advanced or ideal species. In this case, the *Rana* lineage is in fact relatively recently derived (Bossuyt et al., 2006). Systematics does not *de facto* suggest anything about the phylogenetic or evolutionary position of the group. (Although the common name 'true frogs' for the family, Ranidae, has a somewhat judgemental tenor.) In a way, this is progressive; one could imagine a taxonomic system that expected the type taxon should be the ultimo, most complex, most advanced species (and all species from the colonies could be primitive versions striving to match their European counterparts), but this is emphatically not the case. The rules laid out in the international codes of nomenclature, discussed later in this chapter, establish a system based as much on finding shared similarities, as identifying differences.

## TAXONOMIC SURROGACY

Species names are relevant for identifying a particular organism, and for studying larger scale biological phenomena. Natural communities are composed of species, and the constituent species in a region vary across environments and habitats, and might change over time. The level of precision required in those identifications, to detect a pattern of interest, is called 'taxonomic sufficiency'. In most cases, environmental perturbations can be detected when the local flora or fauna is identified to genus level, rather than species level (Terlizzi et al., 2003). In part, this has to do with the magnitude of biotic change, but it is also largely dependent on natural evolutionary patterns that shape taxonomy. Any local habitat contains only a small fraction of the world's species. There are, usually, very limited numbers of co-occurring co-generic species. Each genus is only represented by one or two species in a local ecosystem, regardless of the total global size of the genus. Is it sufficient to identify things to genus level, or even higher clade level, to capture relevant information about biodiversity? Arguments in favour of surrogate approaches have a practical logic. If I want to quantify environmental impact or change, does it really matter that I cannot identify thousands of individual organisms to species level? Does that resolution improve our understanding of impact, when the entire forest has burned to the ground? The counter-argument is that this justification is potentially circular; it is impossible to detect changes that are finer than your measuring instrument, but those small changes may be unexpectedly important. The use of taxonomic surrogacy—substituting species with a higher-ranked group—conflates locally-occurring species with the globally-distributed group of their relatives. If a species is new to you, is it new to the area (an invasive species? a range shift?), something that occurs in such low abundance that you never had an opportunity to see it before, or is really new to science? Rare species may be less abundant but equally ecologically important as more abundant ones. Accepting coarser taxonomic resolution can occlude important data about rare species (Maurer, 2000).

Taxonomic surrogacy is extremely common across fields, and biodiversity is often measured or compared without acknowledging the difference between species and species groups. Invertebrate animals, herbaceous plants, and fungi are commonly left unidentified in environmental surveys. These may be identified to a group or to an ecological guild (e.g., Thorp & Covich, 2009). Not everything can be perfectly identified all of the time, but it is an unappreciated error to conflate species-level and higher-level taxa as units in the same analysis. Usually, we can identify large dominant species relatively easily, while the physically smaller or more rare species require further expertise. If some things are identified to species level, and others only to genus or family level, the specific identifications may be over-counted, duplicated in analysis.

Higher-ranked groups are more appropriate substitutes in groups that have a clear evolutionary kinship rather than polyphyletic assemblages, so the effectiveness of taxonomic surrogacy is linked to phylogenetic constraint (Timms et al., 2013). Evolutionary relatedness provides predictive power about the similarity among groups of species (such as shared trophic guilds). The potential need for phylogenetic systematic revision is not obvious from studying a single local species, it depends on comparison with the global diversity of the group. That is information that is not immediately accessible to a broad audience.

The common intertidal marine gastropods *Steromphala umbilicalis* (da Costa, 1778) and *Steromphala cineraria* (Linnaeus, 1758) occur on European coasts; these two species were formerly in the genus *Gibbula* as part of a larger NE Atlantic species complex that has long been debated. Based on molecular evidence and morphometric analysis, the clade was recently extensively and quite decisively revised (Affenzeller et al., 2017). These species have been observed by every European undergraduate marine biology field trip and appear in every field guide to seashore life of Britain and Ireland (Hayward & Ryland, 2017). These two *Steromphala* species also co-occur with the type species of the genus *Gibbula*: *G. magus* (Linnaeus, 1758). Prior to 2016, these three species would have accounted for only one genus, but now they are in two genera. This creates a minor problem using earlier beach surveys that identified gastropods to genus level, but it would not be a problem in species-level data. They are more like each other than to additional local snail species. Panning out to a more global perspective, *G. magus* is more closely related to other Gibbula species in other regions, than it is to the *Steromphala* species it lives with.

Species we see have other, unseen, global relatives, and those other species may make for more informative comparison than similar co-occuring taxa; shared evolutionary history can predict life history, physiology, and other traits not explained by local ecology (Chapter 5). An individual scientist may have a diverse research portfolio and study taxa and habitats in many regions, but most branches of science have a species focus (model organisms, for instance) or a regional focus (ecology) with finite numbers of taxa. These subsamples, from a single model organism or even a community assemblage, represent relatively few species per genus or per higher taxon, and the nearest relatives of your study organism may be out of view.

The effects of the reclassification of a species are felt locally by people who only interact with one species at a time. The data that lead to this type of revision usually come from much larger-scale studies of the phylogeny of a broadly distributed group. This is hugely frustrating—it can feel like revisions are done by people who do not care about our precious local fauna. Some phylogeneticist ran around all over the world collecting samples to do a big analysis about all these foreign snails, and suddenly I have to change the name I call my friend when I see it every day. This can seem terrible. But, it reminds us that the relations of our local species live all over the world, and it has a part to play not only in our local ecosystem, but the evolution of global biodiversity.

## NAMES CARRY IMPORTANT INFORMATION

Some names are descriptive, and books have been written about translating the Latin and Greek roots of formal species names (e.g., Borror, 1960; Stearn, 2004; Lederer & Burr, 2014). Many scientific names are literal descriptions; *Rana* is both the Latin word for and a genus of frog, while other organisms have genus names that have been adopted into English (*Hippopotamus*, *Iguana*, *Octopus*). Good scientific names can be evocative and verge on poetic. Most people have heard anecdotes or news stories of species named after celebrities and important people (patronyms). This is a tradition that is as long as taxonomy, starting with Linnaeus naming species after both supporters and at least one critic, purportedly naming the genus *Siegesbeckia* (St Paul's wort) to immortally associate the name of his rival Johann Georg Siegesbeck with a noxious weed (Linnaeus, 1753). Siegesbeck had viciously criticised Linnaeus's classification of plants based on their reproductive traits, calling it 'immoral' (Rowell, 1980; Jardine et al., 1996). Taxonomic names have continued to capture important contemporary figures. A set of beetles was named for conservative US politicians *Agathidium bushi, A. cheneyi,* and *A. rumsfeldi* Miller & Wheeler, 2005. These are fungus-eating slime mould beetles, and their names were widely reported in the media as subversive political commentary, but every interview with the authors affirmed that they actually did intend it as an honour. A series of species of trilobites in the genus *Mackenziurus* (Edgecombe & Chatterton, 1990) was named *M. joeyi, M. johnnyi, M. deedeei,* and *M. ceejayi* **Edgecombe & Chatterton, 1990,** for the members of the punk band the Ramones. *Gaga monstraparva* Pryer et al. 2012 (in Li et al., 2012) is a fern in a genus named for the musician Lady Gaga, to honour her 'active support of equality and individual expression'. And a nudibranch genus and family *Mandelia* Valdés and Gosliner, 1999 (family Mandeliidae) from South Africa were named after Nelson Mandela.

These names capture moments in the history of science, and it may be frightening to think that the species names could be subject to revision or change later, but even after revision the legacy may remain intact. Other names, however, we might wish could be changed or updated. *Puffinus* species (Shearwaters) are not puffins, *Penguinus* (the great auk) is not a penguin, *Echidna* is an eel, not an echidna; *Nasturtium* (watercress) is not a nasturtium, *Gymnosperma* (a Mexican flower called gumhead) is not a gymnosperm. Usually, these apparent misnomers reflect the changing use of words—common names that were applied or changed long after the scientific name was established.

Strange or seemingly misapplied names may reflect changing scientific data or interpretations. The specific name for the gooseneck barnacle *Lepas anatifera* Linnaeus, 1767, '*anatifera*', means 'goose-carrying', following the Mediaeval idea that these barnacles found on floating logs were the missing larval stage in the life cycle of migratory geese (Buckeridge & Watts, 2012). Barnacles are arthropods and geese are not, but the true depth of their dissimilarity was not wholly obvious before the advent of modern comparative anatomy. The seasonal disappearance of migratory European geese was a longstanding mystery until the early 1800s; the Barnacle Goose, *Branta leucopsis* (Bechstein, 1803), nests in the high Arctic so its chicks had never been seen. An idea emerged that the geese started life attached to coastal trees in a barnacle-like form, then metamorphosed and dropped into the sea as adults (Figure 4.1). The possibly incongruent explanation of a biphasic barnacle-goose lifecycle supposedly served as part explanation and part justification that the Barnacle Goose was not technically meat and so could be eaten during the Christian fasting month of Lent. Pelagic barnacles, often seen on logs washed up on shore, have a pedicle that could be compared to a bird neck and the feeding appendages are compared to wings (Dobson, 1959). A century before Linnaeus, Moray (1677) for example, published a scientific paper in the *Philosophical Transactions of the Royal Society of London* documenting pelagic barnacles 'having within them little Birds perfectly shap'd'. This is so horrifically wrong as to be charming. And the scientific name, honouring this idea, remains unchanged 250 years later.

The reason for retaining these names is driven not by nostalgia, but a long-term goal of attaining 'stability' in nomenclature. The agreed goal of taxonomy and systematics is continuity and clear communication about the meaning of names,

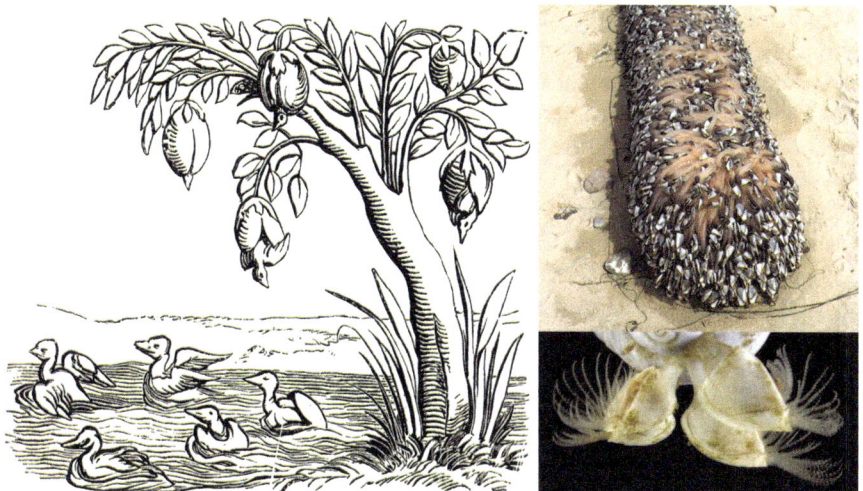

**FIGURE 4.1** Goose barnacles. Left, barnacle-like growths on trees, imagined as a larval stage of the Barnacle Goose, in a drawing dated 1552. Top right, a dense colony of pelagic *Lepas* washed ashore on a log on Malin Head, Ireland, showing their long pink fleshy stalks and dark shells. (Photo by K.D. Bennett.) Bottom right, close up of the feeding appendages of *Lepas pectinata* Spengler, 1793, reminiscent of wings. (Photo by Denis Reik.)

which underpins the rules set down in the international codes of nomenclature: the International Code for Zoological Nomenclature (ICZN, 1999) and the Botanical Code or the International Code of Nomenclature for algae, fungi, and plants (ICN; McNeill et al., 2012). The true meaning of names really refers to the association of a description and reference material to a species moniker. There are rules against names being 'offensive' but there are no rules that a name has to actually make literal sense. The choice of which names get retained for centuries, and which get changed, is not based on any judgement of the name itself, but the systematic position of the species with its nearest relatives and the rules agreed in the codes of nomenclature.

The fate of names to stay the same, or change, is controlled by several things, the why, who, and when of taxonomic revision. Why a name changes depends on shifts in knowledge about the inter-relationships of species. This arises from the fundamental nature of species, which is discussed in detail later in this book (Chapter 5, Chapter 6, Chapter 7). Knowledge of species is inevitably data limited, and progress always prompts reconsideration of the standing hypotheses about identity and inter-relationships. The 'who' question reflects the relationship of a species to the type specimen or type species, and 'when' is the influence of history, or the principle of priority.

## HOW DO NAMES CHANGE AND HOW CAN YOU TELL WHAT HAPPENED?

There are official rules for naming species and species groups, but they focus on technical implementation rather than guidelines for decision making. The rules are governed separately for animals and for plants, by the International Commission on Zoological Nomenclature (ICZN), and the International Association for Plant Taxonomy (IAPT), respectively. The way these two organisations manage their responsibilities are different in the details, but each of them publishes a rule book or 'code' for nomenclature. The ICZN publishes the International Code for Zoological Nomenclature (ICZN, 1999); the Botanical Code covers plants, seaweed, fungi, and some protists, for historical reasons (McNeill et al., 2012); and there is another separate governing body for the taxonomy of microbes (which falls outside the scope of this book). The codes are a regulatory framework, like legal guidelines, only concerned with the internal consistency of names as entities themselves. The judgment of what evidence is sufficient to describe or revise a species or species group is left to individual experts and the scientific publication process (also see Chapter 8).

The original name (genus and species) used in the first description of a species is called the 'basionym' (base name) or 'protonym' (first name) although in zoological nomenclature it is more usually just called the 'original combination'. The fact that there are several different ways to talk about the old name might be a hint that recombination or revision of names is very, very common (Table 4.1).

The main events that prompt nomenclatural changes are (1) species novelty, a new species requires a new name; (2) reclassification, including from phylogenetic analyses, where established species need a new name to maintain meaningful species-groups; and (3) nomenclature anomalies, where a name has been inappropriately used in the past and a correction is needed. This third category really upsets people. Change is not allowed just because a name later changes its meaning or interpretation, so

---

## TABLE 4.1
## Some Useful Definitions

Also see Chapter 8 for further information.

**Synonym**—A name that is not in current use for a species. Taxonomic synonyms are *not* interchangeable. They are the alternative names of species that are now known correctly by a different name; a 'junior' synonym refers to its younger age, when an older, original name has priority. Synonyms are not equivalent names and do not have the same status as the correct name.

**Synonymy**—A list of the history of the alternative names used for a particular species.

**Homonym**—The same word (or combination) used to name two different species; this is not allowed within animals or plants though it is possible to have the same name refer to an animal and to a plant species (for example, the genus *Leptochiton* refers to a mollusc and also an Amaryllis, *Morus* is a mulberry tree as well as a gannet).

**Tautonym**—The same word used for both genus and species name (e.g., *Mola mola*, the Ocean Sunfish); tautonyms are not allowed in botanical nomenclature.

**Authority**—The author(s) who described a species, and date of the original publication of that description.

**Revision**—Alterations to nomenclature, implemented to correct historical problems or to improve how names reflect evolutionary relationships.

**Combination**—A genus name and species name; if a species is revised into a different genus, the revised binomial is a 'new combination', and the binomial used in the original description is the 'original combination'.

---

nasturtiums are still in the genus *Tropaeolum* and *Lepas anatifera* remains nominally goose-bearing.

The main clue to understanding species names is in the authority and date, often overlooked and not included, but formally part of the species name. This is also the answer to the common question about whether you can name a species after yourself—although there is no explicit prohibition, it is taboo, and the name of the author is already permanently associated with the species name. The author citation is *not* a formal part of the taxon name but it is a useful tool and should always be cited (ICZN, 1999: Art. 51). Linnaeus himself advocated citing the author of a species name with every usage (Stearn, 1959). There is some confusion about whether the bibliographic details for taxonomic descriptions should be included among literature cited. This book uses an arbitrary threshold of desuetude; all taxonomic authorities are cited in full if they were published 50 years or less before the time of writing. Including taxonomic publications of course is the most effective way to ensure the authors are fully credited with the citations their contributions deserve (Werner, 2006).

Species epithets can be re-used repeatedly in different genera. Many species are '*C. elegans*', not only the rhabditid nematode model organism *Caenorhabditis elegans* (Maupas, 1900), but also a leaf-mining beetle *Callistola elegans* Gressitt, 1960, a copepod *Cyclopinodes elegans* (Scott T., 1894), and many others. The unifying feature of these species is only that the taxonomist thought the Latin word for 'elegant' was a good description. Even if the genus is abbreviated to the first initial 'C', the name of the author, the taxonomic authority, after the species shows they are different beasts.

The syntax of the authority is slightly different for animal and plant names, but key things to look for are the date and the use of parentheses (explained in Table 4.2). In non-taxonomic papers, this information is often omitted, or abbreviated, because authors from other fields either do not need or do not understand taxonomic details, but a quick Internet search of other uses of a species name in the scientific literature will usually reveal the authority and date relatively quickly. However, because the use of these extra data is a little obscure to most non-taxonomists, misuse of parentheses, in particular, is pretty widespread for many familiar species so it may demand some critical interpretation. Indeed, the production team that typeset this book initially attempted to 'correct' the use of parentheses, even in this chapter, which serves as a caution that it is always good advice to consult multiple sources.

The history of the multiple names used for a particular species is called a 'synonymy', and online taxonomic projects often have similar lists, which bear mentioning here. In primary taxonomic literature, this is frequently confused with a 'chresonymy', a list that includes all published mentions of a species, which serves to document and clarify any previous records for the newly-recognised taxon (Smith & Smith, 1972; see Chapter 8). Synonymies are not just a list of alternative name forms, they include misidentifications and published misspellings, which makes them very useful but also messy and difficult to understand (Froese & Pauly, 2000). Names in a chresonymy or synonymy may also include separate valid species that get listed because they were historically misapplied (Table 4.3).

## TABLE 4.2
## Syntax of Original and Revised Combinations

Some examples of original and revised combinations of genus and species names in animals and plants. The authority of a taxonomic name refers to the publication of the most specific epithet presented. If a subspecies is used, you should cite authority of the subspecies, not of the species name. The use of parentheses indicates a genus-level revision, showing the species is now classified in a different genus than when it was originally named.

Magnificent Frigatebird—***Fregata magnificens*** Mathews, 1914

    The species was named by Matthews in 1914 and is still used in the original combination (genus and species names).

Ascension Frigatebird—***Fregata aquila*** (Linnaeus, 1758)

    The species was named by Linnaeus in 1758 as ***Pelecanus aquilus*** and was subsequently recognised as part of a separate genus, *Fregata*, thus the authority is listed in parenthesis to indicate a revised combination. This could be abbreviated or shortened, as *F. aquila* (Linnaeus) or simply *F. aquila* (L.).

Northern White Cedar—***Thuja occidentalis*** Linnaeus, 1753

    For plant species, the earliest accepted names are from *Species Plantarum* by Linnaeus (1753), whereas zoological nomenclature begins with the 10th edition of his *Systema Naturae* in 1758 (Linnaeus, 1758). This too could be abbreviated or shortened, as *T. occidentalis* Linnaeus or simply *T. occidentalis* L.

Japanese Thuja—***Thuja standishii*** (Gordon, 1862) Carrière

    In botanical nomenclature, a revised combination is shown with the original author in parentheses, and the author of the current accepted name at the end. This tree was originally named ***Thujopsis standishii*** (Gordon, 1862), and was shortly thereafter revised to *Thuja standishii* by Carrière (1867). The syntax would typically be abbreviated as *T. standishii* (Gordon) Carr.

## SPECIES NOVELTY AND RECLASSIFICATION

The classification of species groups above the genus level can be revised without changes to any species names. Imagine yourself putting your collection of species in a basket and putting that basket on a shelf. That basket is a genus. Moving the basket to a different shelf, neither the basket nor the contents has changed, but its systematic position is very different. 'Pisces' was the former class rank name for fish, and it is still often referred to as a 'class' (Bone & Moore, 2008); however, the descendents of ancestral fish include many radiations of modern fish as well as tetrapods. In modern phylogenetic systematics, other subgroups of fish are considered to have the rank of 'class', such as the clade Actinopterygii (Nelson et al., 2010). This usage is more reflective of fish evolution and has steadily increased since the 1960s, but meanwhile, an ongoing back and forth about how the class rank is conscribed (all fish, or subgroups like the ray-finned fish) does not change the name of any fish species. Back to the basket of species, new species have to be put in an established basket, or we may need a new one; insights from the new species might prompt reorganisation of other species.

Revisions of genus-level groups are the most common reasons for changing species names, and because the genus name is the first part of a scientific name it may render the new name unrecognisable (Table 4.2). Changes to genus-level

## TABLE 4.3

## How to Read a Synonymy

A synonymy is not only a list of synonyms (see Table 4.1); below, this published chresonymy (list of cited uses of a taxon name) for the chiton *Leptochiton cascadiensis* includes other names that are valid separate species, historically confused with *L. cascadiensis* (Figure 4.2).

| Published Records for the Species, in Chronological Order of When a Combination First Appeared | Explanation of Records |
|---|---|
| ***Leptochiton cascadiensis*** sp. nov. Sigwart & Chen, 2017 | Name of the new species |
| *Leptochiton cancellatus*: Whiteaves 1887: 113, 125; Newcombe 1893: 56; Berry 1927: 160; Berry 1951: 215, 218.<br>*Lepidopleurus cancellatus*: Dall 1921: 186; Oldroyd 1927: 246. | The first records of what is now called *L. cascadiensis*, were identified as members of the species *L. cancellatus* (Sowerby, 1840). All records in the chresonymy include author, date, and pages from the publications. Two different genus names were used because of other taxonomic revision to the genera *Leptochiton* and *Lepidopleurus*.<br><br>*Leptochiton cancellatus* is still a valid name, but its range is restricted to the NE Atlantic species. |
| *Leptochiton rugatus* (partim): Ferreira 1979: 147; Kaas and Van Belle 1985: 85. | This 1979 study combined several species under the name '*L. rugatus*', and that description was used in a later study. The note 'partim' shows these records *partly* refer to *L. cascadiensis*. *Leptochiton rugatus* is still a valid name for a similar, but more southern species found in California. |
| *Leptochiton* sp.: Kelly and Eernisse 2007: appendix 1. | Molecular evidence indicated a northern species *L. cascadiensis* was separate from other more southern specimens in *L. rugatus*, but the authors did not name it and indicated its separation as *Leptochiton* sp. |
| *Leptochiton rugatus*: Sigwart et al. 2011: fig 1B; Lamb and Hanby 2005; Carey et al. 2012: fig 1D; Carey 2014: appendix B; Layton et al. 2014; Sigwart et al. 2014: figs 2, 3, 8, tables 2–3; Sumner-Rooney et al. 2014. | Without a name to distinguish the northern species, later authors continued to use '*L. rugatus*'. |
| ? *Leptochiton* sp.: Dell'Angelo et al. 2011. | The '?' here indicates uncertainty about the real identity of a fossil specimen described from Washington state, which the authors of that publication compared to *L. rugatus*, *L. cancellatus*, and other species; as it is from within the modern range of *L. cascadiensis*, that may be its identity. |
| non *Leptochiton alascensis*: Layton et al. 2014: table S2. | A 'non' is used to clearly exclude these records from further north, beyond the known range for *L. cascadiensis*. |

classification alter names with or without the discovery of new species. In botanical nomenclature, names may change more radically to accommodate an additional rule that does not allow tautonyms. Tautonyms are cases where the same word is used for both genus and species such as the animal species *Puffinus puffinus* (Brünnich, 1764) (the Manx Shearwater) or *Extra extra* (Jousseaume, 1894) (a tiny marginellid marine snail). The domestic tomato, *Solanum lycopersicum* Linnaeus, 1753 was historically revised to *Lycopersicon esculentum* Miller, 1768. Miller classified several tomato species in their own genus, *Lycopersicon*, based on morphological (and chemical) differences with other similar plant species in the genus *Solanum*, like the potato. The revised name for the common tomato, *Lycopersicon lycopersicum*, is too close to a botanically-forbidden tautonym, so it was given a new species epithet, *Lycopersicon esculentum* (esculentum meaning 'edible'). Later, after further study, the picture is a bit more complicated. Genetic evidence favours a very close relationship among all the *Solanum* and *Lycopersicon* species (Peralta et al., 2008). Indeed, if *Lycopersicon* spp. are separated in their own genus, it breaks up the remainder of the large *Solanum* group creating a problem of 'paraphyly' (Chapter 6). For now, these plants have been reunited into a morphologically diverse but monophyletic *Solanum* and the old original name for the tomato is back in use: *Solanum lycopersicum*.

The appearance of a new name for a local species may be a result of taxonomic 'splitting', where a widespread or diverse species is subsequently found to be multiple independent lineages. Along the Pacific coast of North America, there is a group of small, white-shelled chitons, small molluscs that live under stones in the low intertidal. Several different species had been named in the 19th century, but they all look extremely similar even in microscopic features, so in the 1970s, one researcher published a study concluding that all of these names represented variation within a single species called *Leptochiton rugatus* (Carpenter in Pilsbry, 1892) (Ferreira, 1979). Not all experts agreed with this conclusion, and some suspected that there really were probably multiple species. The presence of more than one species was supported by early molecular evidence (Kelly & Eernisse, 2007), and had been mentioned in faunal lists (e.g., Carlton, 2007). Later, more detailed observations of their ecology helped to prove that animals on the coasts from Oregon to southern Alaska really were different from more southerly populations. But did they need a new name, or would one of the older names fit this species? In fact, the name *L. cancellatus* had been used regularly up to the 1920s to recognise a northern coastal species and *L. rugatus* as the southern species. But *L. cancellatus* was also used for a quite different species in Europe, in the Atlantic Ocean. The Atlantic species called *L. cancellatus* had 'priority', meaning that was the original usage of the name (Table 4.3). Although this northern Pacific species had been recognised in the 19th century, it was not really recognised as its own species so it needed a new name. It is now called *L. cascadiensis* Sigwart & Chen, 2017, after the biogeographic province of 'Cascadia' and a peaceful separatist movement to politically unite the region (Figure 4.2).

Cases where a new species are discovered that have never been observed by science before do occur regularly. Far more frequently, a 'new' species had been observed but identified under the name of something that looks similar. Apart from establishing relevant lines of evidence that separate our new species, the first question

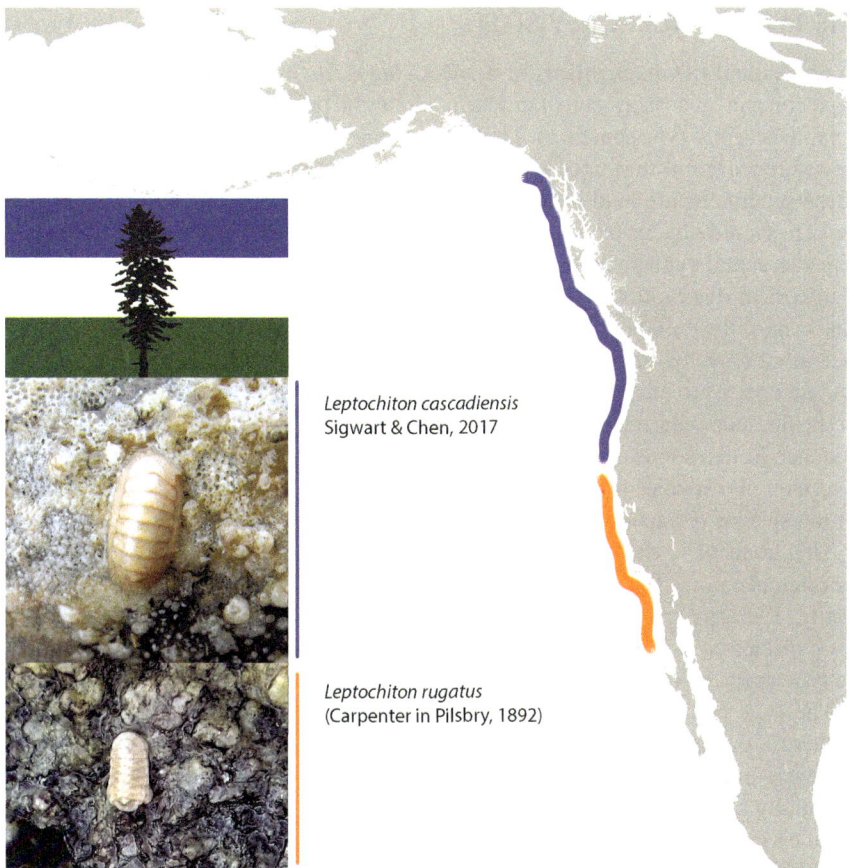

**FIGURE 4.2**   Distribution of two nearly indistinguishable species of chiton. *Leptochiton cascadiensis* is found on the coasts of 'Cascadia' (regional flag shown upper left), separated from *Leptochiton rugatus* in California and Mexico.

is, has anyone noticed this before? Very often, subtle differences in morphology were enshrined in taxonomic names that were later rejected as inadequate to distinguish separate species, but molecular evidence eventually shows that those subtle differences belie deeper genetic divergence. The controversial penguin species *Eudyptes filholi* Hutton, 1879 is recognised as a valid species by the government of New Zealand, but not (currently) by the IUCN (Chapter 3). The species was acknowledged as distinct from others in the genus by a recent molecular study (Banks et al., 2006), but that authority and date in 1879 after the scientific name immediately indicate that this species was originally discovered much earlier. The relevant point here is that those obsolete but now resurrected ideas are associated with established names, and if a name is already on the books, you are not allowed to make up a new one. As a species user, you can tell if this happened by the information in the authority and date that follow the species name.

## THE TYPE CONCEPT REVISITED

To understand the mechanisms of name changes, it is necessary to go back to the 'type concept', a cornerstone of modern taxonomic practice. First, every species has a 'name bearing' type specimen. That is the single preserved specimen that becomes the reference point that fixes the identification of a species in modern taxonomy. These specimens are held in the permanent care of natural history museums, and they are sacred objects of systematics and evolutionary biology. Direct comparison with the actual type specimen trumps any opinion of an expert or interpretation of historical descriptions. The 'type locality', the place where the type specimen was originally collected, also has a special significance. Most museums have special storage for their type specimens, held under extra security or in special climate control to ensure their long-term preservation. These specimens represent the bridge between the frustrating fuzziness of species definitions and the hard evidence from a single decisive data source (Farber, 1976). The designated holotype specimen of a species has a special taxonomic 'legal' status as the reference point for the species (Chapter 8).

This concept is extended in the taxonomic hierarchy by type *species* that are the reference point for each genus, and a type *genus* that is the reference point for a family. The type species is the pivot point, it fixes the name of a genus, and every other species has to be compared to that—if a candidate species is similar enough to the type species, it stays in the genus, if it is too different it has to go.

The type concept has a more recent history than other aspects of taxonomy, so there are some legacy cases where type specimens have not been designated. However, all species and all higher ranked groups (genera, families) 'has actually or potentially a name-bearing type' (ICZN, 1999: Art. 61). The purpose of types is to provide a clear reference point for future revisions; rather than a subjective reshuffling of all our metaphorical baskets, there is an established rule that the first species that got put into a given basket is the basis for deciding what else gets put there.

This system works well for making sense of a lot of undescribed biodiversity, building organically by comparison from whatever limited knowledge is currently available. It is not necessary to have discovered all the species before starting to classify them, you just use the first one you run into as a reference point and start expanding from there. The side effect of this is that the choice of reference point is only the best available example at the time. With more information, that type, the first reference point, might turn out to be entirely typical of a larger group, or it may equally be revealed as a strange aberration. With complete foreknowledge of diversity to classify, we might have chosen differently about some aspects of the organisation. Omniscience, however, is not usually an option.

The idea of type specimens, and type species, was developed and formalised long after Linnaeus. Designations of these specimens and species have been applied to a lot of historical literature *post hoc*. Early taxonomic works in the 18th and 19th centuries were usually associated with collections, which are still held by museums, and included illustrations of material that can often be clearly linked to an individual specimen. These objects become type specimens, by subsequent designation (Winston, 1999).

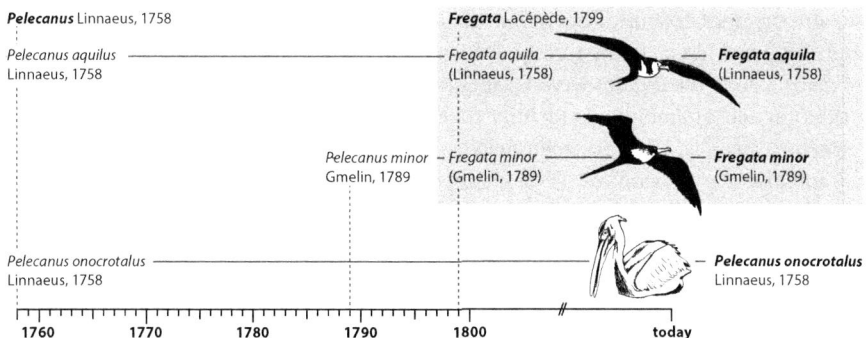

**FIGURE 4.3** Chronology of the names of three birds originally named in the genus *Pelecanus.*

Linnaeus named a bird genus *Pelecanus* that originally included a large heterogeneous group of seabirds, including pelicans, frigatebirds, cormorants, and boobies. Pelicans and frigatebirds are now in separate taxonomic families (Del Hoyo et al., 1992). Their nomenclature illustrates several key points about systematic revision, including the way names may naively seem inappropriate, the importance of type specimens, and how taxonomic issues may re-emerge centuries later (Figure 4.3).

Frigatebirds were moved to a separate genus *Fregata* (Lacépède, 1799), with early recognition of their distinctive features that clearly set them apart from pelicans; black feathers, unusually long wings, forked tails, and the bright red gular pouch on the throats of mature males. The first described or first-recognised species, chronologically, usually becomes the type species for a species-group. For *Fregata*, the type species is *Fregata aquila* (Linnaeus, 1758), the Ascension Island Frigatebird named originally by Linnaeus as *Pelecanus aquilus* in 1758. The root of the species name stays the same, but the suffix changes from *-us* to *-a* to match the Latin gender of the genus name (*Fregata* is feminine, *Pelecanus* is masculine). The establishment of this new genus *Fregata* also changed the name of the only other Frigatebird species known at that time in the 18th century (there are now 5 recognised species in the genus). The Greater Frigatebird was originally named *Pelecanus minor*, and its name changed to *Fregata minor*—it is larger than other Frigatebirds, but it is certainly smaller than a pelican, which was the consideration when it was named. A change in reference frame does not mean the name can change; a species epithet sticks with the species even if it is moved to a different genus.

For *Pelecanus*, the type specimens of several of these seabird species would be the specimens in Linnaeus's original collection; however, some of them have not survived, and/or the original description was based on other material Linnaeus never owned. In the case of *Fregata aquila* there was no specimen from Linnaeus's collections, but it became important to fix the identity of the species to the only known breeding population of the species associated with that name. Without a specimen that belonged to the original author, the next best thing would be to find a specimen of something that the author may have borrowed or examined, or specimens that fed into other contemporary or historical reports that provided reference information feeding into the original formal description of the species. Carlos and colleagues (2016) carefully considered the historical descriptions that could be attributed to *Fregata aquila* and concluded there were no specimens that satisfied the criteria of historical attribution and unambiguous identity that were definitely *F. aquila* and not another frigatebird species. Old specimen material is often damaged, and old accounts may be imprecise about localities. Without appropriate early historical material, the team selected and formally nominated a specimen in the collections of the Natural History Museum in London, UK as the 'neotype' for *Fregata aquila* (see also Chapter 8). Nomination of a neotype, a new type specimen of an already-described species, is not valid without proving the case that the original type material is lost, and that the new specimen is an appropriate choice in context of the history and the diagnosis of the species. Frigatebirds are distinctive animals and seem like they should be easily recognised from descriptions and modern photographs, which may explain why the lack of a type specimen was never solved before 2016—over 250 years after the species was first described. But other cases of the ongoing discovery of new cryptic species may prompt some concern; species identity is critical to establishing conservation priorities (Chapter 3) and the formal designation of a neotype fixes the identity to that specimen as a reference for the species.

## REVISION END GAME

The idea of the type taxon is a powerful practical tool for discovery-based science, but it is nonetheless at odds with phylogenetic or evolutionary systematics. The type is an arbitrary reference point, a legal definition rather than a biological hypothesis, and makes no assumptions about the position of the taxon within a phylogenetic tree. So, it may or may not be typical of its closest related group of species. This decouples taxonomy and phylogeny (or rather, it is a legacy of taxonomy pre-dating tree-thinking) and balancing both aspects is often awkward (Chapter 6). The advantage of this system is that new taxa can be described and established even when there is not sufficient data to generate phylogenetic hypotheses. This is far more relevant to understudied groups and may never be an issue to researchers who work on vertebrate animals, for example. To discuss and compare notes about a species or even particular specimens, it is useful to have names to call them by as a starting point. Later revisions, changing names, are usually motivated by improved classification to reflect global evolutionary relatedness among species. Use of the type taxon puts emphasis on the history of the name and our human discoveries rather than on the phylogenetic history of the clade.

The framework of the type concept facilitates discovery, but it also creates a major headache and a lot of forensic work for working taxonomists. Every new discovery has to be compared with the historical body of work and available names. The question 'which one is it?', matching new species to old names, can paralyse systematic study. The example of *Leptochiton cascadiensis* is a comparatively easy case, but the species could not confidently be given a new name until it was clear how it matched with historical records from the 1920s. Is taxonomic progress more straightforward in more severely neglected (or recently discovered) groups, with no historical baggage? The converse of a clean slate is a void of comparative material or established knowledge.

The scale of the 'discovery gap' is quite different among groups of organisms. New species are common in some groups, such as many invertebrate animals, fungi, and many plant groups; in others, discovery is complete and most taxonomic work is in revision. The discovery gap also varies geographically: it is smaller in Europe and much larger in tropical regions and remote places. Both activities—new species and revision—can change names. 'Understudied' groups are those where the species diversity is poorly described, and the volume of undescribed or poorly described species overwhelms the taxonomic expertise. As clades become progressively better known, and the diversity discovery gap closes, taxonomic effort shifts from describing species novelty toward revision and reclassification. Although both processes—discovery and revision—happen in tandem, the first step is often just to get names on species, and ongoing refinement of their systematic classification continues afterwards. The dates of publication of taxonomic names record this process (Figure 4.4). If a new genus was named later than the species it contains, that documents taxonomic revision. If a species publication date is later than the first publication of its genus name, this was a new species that was found and classified into an already established genus. The species diversity of birds is better described than any animal group, and the rate of discovery of new species of birds is low and dropping. In marine molluscs, the rate of discovery of new species is increasing and discovery still outweighs reclassification.

Revisionary systematics, names getting changed, is a familiar frustration to anyone interested in identifying species, but such revisions should lead us toward a common goal of improved classification that reflect real informative (phylogenetic) relationships. As systematists and taxonomists are human scientists, there is a range of quality and skill among practitioners (and even within the career of an individual). Some work has been labelled 'taxonomic vandalism', and it is clear that mischievous taxonomic changes can create a cascade of problems to other fields. But some caution is due over these accusations. The governing bodies for taxonomy deliberately allow for full academic freedom: one person's vandalism is another person's revolutionary hypothesis. Earlier, I mentioned a study in the 1970s (Ferreira, 1979) that 'lumped' many chiton species, a hypothesis now considered incorrect and decades later is only partially corrected, but no one questions the good intentions, skill, or honour of the original author. Change is in the nature of hypothesis-based science, but it is a profoundly uncomfortable part of taxonomy where we strive for nomenclatural stability. The charge of 'vandalism' has been levelled at a small number of people who revise the classification of well-studied groups and skirt the bare minimum of publishing standards, just enough to ensure the rest of the field takes their work seriously. This is not a widespread phenomenon; most of us who work on overwhelming undescribed diversity can see that sort of interference as a luxury problem.

Even if all species were named, which is far from finished, still taxonomy is not dying, it is not finished, and it does not really have an end point. Among fish, estimates from the comprehensive data projects within FishBase suggested that around 10% of the taxonomic names reported in any large work may be obsolete 10 years after publication (Froese & Pauly, 2000). This is mainly based the rate at which old names are reconsidered and found to be junior synonyms of other taxa. The large number of junior synonyms in some groups has been held up as evidence that the project of naming Earth's biodiversity is within reach (Costello et al., 2013a)—on

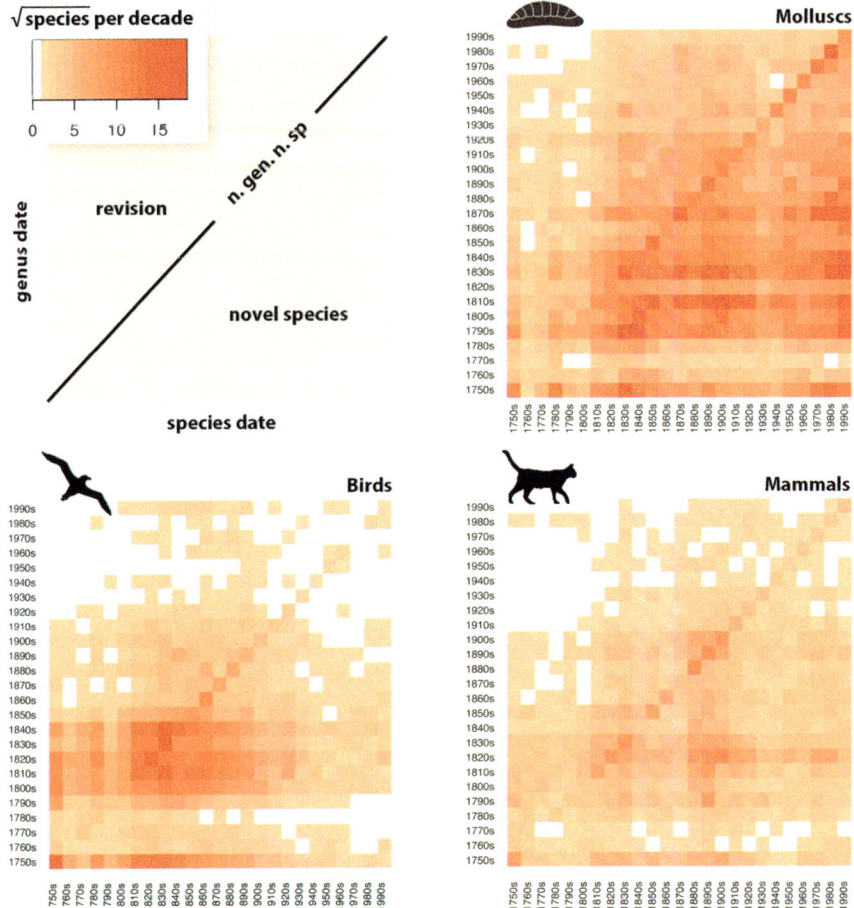

**FIGURE 4.4**   These heatmaps illustrate the chronology of species discovery and revision. Darker pixels to the right and toward the top of each diagram indicate recent, ongoing taxonomic work. Each cell represents a number of species per decade, where the horizontal axis is the date the species was named and the vertical axis represents the date the genus was established. So the main diagonal are squares when the genus and species were named in the same decade, usually the original combination. Squares below the diagonal are genera that were named before the species—new species put into existing genera; above the diagonal are newly-established genera that take in earlier-recognised species, or revisions. Each species in a dataset is represented only once, based on the genus date and species date of the current accepted combination. Counts were square-root transformed to facilitate comparisons across taxonomic groups. (Data reanalysed from Sigwart et al., 2018.)

the assumption that all those unused names must correlate to real species out there somewhere. But, like the example of *Leptochiton cascadiensis*, even if a species was recognised by some former worker, that may be more confusing than helpful. Synonyms are also not necessarily correlated to the regions or taxa with significant discovery gaps (Bouchet & Strong, 2010; Mora et al., 2013).

Among understudied groups of organisms, some large genera with hundreds of species are called 'bucket' or 'wastebasket' taxa—the genus you put a species in when there are no straightforward distinct diagnostic features, or when that feature does not reflect evolutionary relatedness, but there is no good alternative and we use it anyway. The number of species in these genera makes it almost impossible to study them comprehensively (Frodin, 2004). Some large genera, containing hundreds of species, represent large rapid radiations where others are a grab bag of species with multiple evolutionary origins. The largest non-insect animal genera (various groups of gastropods and ostracods) are all of this latter 'polyphyletic' sort, where taxonomists know there are problems that no one has fixed yet (Sigwart et al., 2018). Every year, more species are discovered and put into these bucket genera. Revising such groups presents a huge challenge with the promise of relatively little academic reward, and the use of these taxonomic group names often persists because there are no clear sub-groups that can be diagnosed by morphology. Integrating genetic data often breaks the deadlock, but it is difficult or even impossible to get genetic data for many rare species (and this problem applies equally to fossils). Classification of future new species still depends on morphological features, synapomorphies, that correlate with clades and allow for field identification.

There are a lot of popular misconceptions about taxonomy as a branch of science. (The main theme among them is that the discipline itself is either irrelevant and/ or being driven to extinction by external forces. Both ideas are pernicious and they feed into each other.) Wide dissemination of information increases popular support and decreases the continued use of obsolete ideas. Most people do not have a particular opinion about any taxonomic revisions, a rare point in modern life where people are generally willing to accept the pronouncements of experts. But people are resistant to change, and if they think they can get away with not changing, they will stick with the old way. It is a common joke among biological recorders that if they just hang on long enough, an old name will come back into fashion. That does sometimes happen, as in the tomato returning to *Solanum lycopersicum*. Taxonomists and systematists must keep in mind that responsibility to adopt the latest upgrade does not just rest with the user. It is the makers' responsibility to make sure the upgrade is put into the hands of every user. Most major software companies pursue this with campaigns of aggressive pestering, such that your life is not worth living until the update is finally installed. Modern digital resources provide comprehensive lists and classifications of global species (e.g., World Register of Marine Species; WoRMS) or regional identification guides and records based on public participation or citizen science (e.g., iNaturalist). While not authoritative in themselves, such projects can provide an accessible interface between large audiences of species-users and the smaller field of name-makers (Costello et al., 2013b). These projects are divisive, and their detractors quite reasonably view them as dangerously incomplete and detached from vetting by peer review. However, they exist, they are popular, they get updated faster than printed field guides, and they are available established tools to emphasise the acceptance of new taxonomic ideas.

Species are and will remain units subject to a very high degree of uncertainty. Later chapters in this book devoted space to discussion about the fuzzy boundaries of species and scope for uncertainty in identifying species and species groups (Chapter 6, Chapter 7). The names will keep changing, not because we keep getting it wrong, but because we keep learning new things.

## REFERENCES

Affenzeller S, Haar N, Steiner G. 2017. Revision of the genus complex *Gibbula*: An integrative approach to delineating the Eastern Mediterranean genera *Gibbula* Risso, 1826, *Steromphala* Gray, 1847, and *Phorcus* Risso, 1826 using DNA-barcoding and geometric morphometrics (Vetigastropoda, Trochoidea). *Organisms Diversity & Evolution.* 17: 789–812.

Anderson FE. 2000. Phylogenetic relationships among loliginid squids (Cephalopoda: Myopsida) based on analyses of multiple data sets. *Zoological Journal of the Linnean Society.* 130: 603–33.

Banks J, Van Buren A, Cherel Y, Whitfield JB. 2006. Genetic evidence for three species of rockhopper penguins, *Eudyptes chrysocome. Polar Biology.* 30: 61–7.

Bone Q, Moore R. 2008. *Biology of Fishes.* Taylor & Francis.

Borror DJ. 1960. *Dictionary of Word Roots and Combining Forms.* Mayfield Publishing Company.

Bossuyt F, Brown RM, Hillis DM, Cannatella DC, Milinkovitch MC. 2006. Phylogeny and biogeography of a cosmopolitan frog radiation: Late Cretaceous diversification resulted in continent-scale endemism in the family Ranidae. *Systematic Biology.* 55: 579–94.

Bouchet P, Strong EE. 2010. Historical name-bearing types in marine molluscs: An impediment to biodiversity studies? In: Polaszek A (editor). *Systema Naturae 250.* CRC Press.

Buckeridge J, Watts R. 2012. Illuminating our world: An essay on the unraveling of the species problem, with assistance from a barnacle and a goose. *Humanities.* 1: 145–65.

Carlos CJ, Voisin JF, Grouw HV, Moreno IB. 2016. A neotype designation for the Ascension Frigatebird *Fregata aquila* (Aves: Fregatidae). *Zoologia (Curitiba).* 33: e20160111.

Carlton JT (editor). 2007. *The Light and Smith Manual: Intertidal Invertebrates from Central California to Oregon.* University of California Press.

Costello MJ, May RM, Stork NE. 2013a. Can we name Earth's species before they go extinct? *Science.* 339: 413–6.

Costello MJ, Bouchet P, Boxshall G, Fauchald K, Gordon D, Hoeksema BW, Poore GC, van Soest RW, Stöhr S, Walter TC, Vanhoorne B. 2013b. Global coordination and standardisation in marine biodiversity through the World Register of Marine Species (WoRMS) and related databases. *PLoS ONE.* 8: e51629.

Del Hoyo J, Elliot A, Sargatal J. 1992. *Handbook of the Birds of the World.* Volume 1. Lynx Editions.

Dobson J. 1959. Facts and fallacies. *Annals of the Royal College of Surgeons of England.* 25: 331–5.

Edgecombe GD, Chatterton BD. 1990. *Mackenziurus,* a new genus of the Silurian *'Encrinurus' variolaris* Plexus (Trilobita). *American Museum Novitates.* 2968: 1–22.

Farber PL. 1976. The type-concept in zoology during the first half of the nineteenth century. *Journal of the History of Biology.* 9: 93–119.

Ferreira AJ. 1979. Family Lepidopleuridae (Mollusca, Polyplacophora) in the Eastern Pacific. *Veliger.* 22: 145–65.

Frodin DG. 2001. *Guide to Standard Floras of the World: An Annotated, Geographically Arranged Systematic Bibliography of the Principal Floras, Enumerations, Checklists and Chorological Atlases of Different Areas.* Cambridge University Press.

Frodin DG. 2004. History and concepts of big plant genera. *Taxon.* 53: 753–76.

Froese R, Pauly D (editors). 2000. *FishBase 2000: Concepts, Design and Data Sources.* ICLARM, Los Baños, Laguna, Philippines.

Godfray HCJ, Knapp S. 2004. Introduction: Taxonomy for the twenty-first century. *Philosophical Transactions of the Royal Society B.* 359: 559–69.

Harrison P. 2009. Linnaeus as a Second Adam? Taxonomy and the religious vocation. *Zygon.* 44: 879–93.

Hayward PJ, Ryland JS (editors). 2017. *Handbook of the Marine Fauna of North-West Europe,* Second Edition. Oxford University Press.

[ICZN] International Commission on Zoological Nomenclature. 1999. *International Code of Zoological Nomenclature,* Fourth Edition. International Trust for Zoological Nomenclature, London.

Jardine N, Second JA, Spary EC (editors). 1996. *Cultures of Natural History.* Cambridge University Press.

Kelly RP, Eernisse DJ. 2007. Southern hospitality: A latitudinal gradient in gene flow in the marine environment. *Evolution* 61: 700–7.

Lederer R, Burr C. 2014. *Latin for Bird Lovers: Over 3,000 Bird Names Explored and Explained.* Timber Press.

Li FW, Pryer KM, Windham MD. 2012. *Gaga,* a new fern genus segregated from *Cheilanthes* (Pteridaceae). *Systematic Botany.* 37: 845–60.

Linnaeus C. 1753. *Species Plantarum,* 2 volumes. Laurentii Salvii, Holmiae.

Linnaeus C. 1758. *Systema naturae per regna tria naturae, secundum classes, ordines, genera, species cum characteribus, differentiis, synonymis, locis.* Tenth edition. Laurentii Salvii, Holmiae.

Mallet J, Willmott K. 2003. Taxonomy: Renaissance or Tower of Babel? *Trends in Ecology & Evolution.* 18: 57–9.

Maurer D. 2000. The dark side of taxonomic sufficiency (TS). *Marine Pollution Bulletin.* 40: 98–101.

McNeill J, Barrie FR, Buck WR, Demoulin V, Greuter W, Hawksworth DL, Herendeen PS, Knapp S, Marhold K, Prado J, Prud'homme Van Reine WF (editors). 2012. *International Code of Nomenclature for algae, fungi and plants (Melbourne Code) adopted by the Eighteenth International Botanical Congress Melbourne, Australia, July 2011.* Regnum Vegetabile 154.

Miller KB, Wheeler QD. 2005. Slime-mold beetles of the genus *Agathidium* Panzer in North and Central America, part II. Coleoptera: Leiodidae. *Bulletin of the American Museum of Natural History.* 291: 1–67.

Mora C, Rollo A, Tittensor DP. 2013. Comment on 'Can we name Earth's species before they go extinct?' *Science* 340: 237.

Moray R. 1677. A relation concerning barnacles. *Philosophical Transactions of the Royal Society of London.* 12: 925–7.

Nelson JS, Schultze HP, Wilson MV. 2010. *Origin and Phylogenetic Interrelationships of Teleosts.* Verlag.

Padial JM, Miralles A, De la Riva I, Vences M. 2010. The integrative future of taxonomy. *Frontiers in Zoology.* 7: 16.

Paknia O, Sh HR, Koch A. 2015. Lack of well-maintained natural history collections and taxonomists in megadiverse developing countries hampers global biodiversity exploration. *Organisms Diversity & Evolution.* 15: 619–29.

Peralta IE, Spooner DM, Knapp S. 2008. Taxonomy of wild tomatoes and their relatives (*Solanum* sect. Lycopersicoides, sect. Juglandifolia, sect. Lycopersicon; Solanaceae). *Systematic Botany Monographs.* 84.

Rowell M. 1980. Linnaeus and botanists in eighteenth-century Russia. *Taxon.* 1: 15–26.

Sigwart JD, Chen C. 2017. Life history, patchy distribution, and patchy taxonomy in a shallow-water invertebrate (Mollusca: Polyplacophora: Lepidopleurida). *Marine Biodiversity.* doi.10.1007/s12526-017-0688-1

Sigwart JD, Sutton MD, Bennett KD. 2018. How big is a genus? Towards a nomothetic systematics. *Zoological Journal of the Linnean Society.* 183: 237–52.

Smith HM, Smith RB. 1972. Chresonymy ex synonymy. *Systematic Biology.* 21: 445–5.

Stearn WT. 1959. The background of Linnaeus's contributions to the nomenclature and methods of systematic biology. *Systematic Zoology.* 8: 4–22.

Stearn WT. 2004. *Botanical Latin*, Fourth Edition. Timber Press.

Terlizzi A, Bevilacqua S, Fraschetti S, Boero F. 2003. Taxonomic sufficiency and the increasing insufficiency of taxonomic expertise. *Marine Pollution Bulletin.* 46: 556–61.

Thorp JH, Covich AP (editors). 2009. *Ecology and Classification of North American Freshwater Invertebrates.* Academic Press.

Timms LL, Bowden JJ, Summerville KS, Buddle CM. 2013. Does species-level resolution matter? Taxonomic sufficiency in terrestrial arthropod biodiversity studies. *Insect Conservation & Diversity.* 6: 453–62.

Valdés À, Gosliner TM. 1999. Phylogeny of the radula-less dorids (Mollusca, Nudibranchia), with the description of a new genus and a new family. *Zoologica Scripta.* 28: 315–60.

Vecchione M, Shea E, Bussarawit S, Anderson F, Alexeyev D, Lu C-C, Okutani T, Roeleveld M, Chotiyaputta C, Roper C, Jorgensen E, Sukramongkol N. 2005. Systematics of Indo-West Pacific loliginids. *Phuket Marine Biology Centre Research Bulletin.* 66: 23–6.

Walters SM. 1986. The name of the rose: A review of ideas on the European bias in angiosperm classification. *New Phytologist.* 104: 527–46.

Werner YL. 2006. The case of impact factor versus taxonomy: A proposal. *Journal of Natural History.* 40: 1285–6.

Winston JE. 1999. *Describing Species: Practical Taxonomic Procedure for Biologists.* Columbia.

# 5 Species are Units of Evolution

## VARIATION

Colour, size, physiological tolerances, morphology, genetic sequences, and really anything that could be used to describe species, all have a range of variation among individuals (Chapter 2). Some variation is important and leads to obvious selective advantage, while some is inconsequential. Understanding species really rests on assessing variation from that other species' point of view, to judge whether the variation we perceive is actually important to the experience of that organism.

A deeper exploration of the causes of variation in variability seems worthwhile. If we see a population of organisms with a limited distribution and low variability—like the coelacanth, *Latimeria chalumnae* Smith, 1939—it is easy to accept that this is a single species (Casane & Laurenti, 2013). By contrast, the relationships among the highly mobile, fast swimming, predatory tuna, genus *Thunnus*, have been a longstanding phylogenetic problem. The Atlantic Tuna (*Thunnus thynnus* (Linnaeus, 1758)) and Pacific Tuna (*Thunnus orientalis* (Temminck & Schlegel, 1844)) are so closely related that they have alternatively been considered species, or subspecies, and there is evidence for hybridisation, shown in high levels of mitochondrial introgression, between these and other congeneric species (Kitagawa et al., 2000; Díaz-Arce et al., 2016). In the terrestrial realm, the Venus Flytrap, *Dionaea muscipula* Ellis, has a small native range in the southeastern United States and a distinctive morphology and lifestyle. But another group of North American carnivorous plants is much more widespread and morphologically variable, the Trumpet Pitcher *Sarracenia purpurea* Linnaeus. Although a number of putative subspecies or varieties of *S. purpurea* have been named, the variety of pitcher shape and colour seems to simply reflect the changing environment from Florida to Northern Canada rather than diverging lineages of pitcher plants (Ellison et al., 2004; Ellison & Gotelli, 2009). In a species that is variable and extends over a very large range, does broad variation represent one diverse lineage with individuals that respond to their local conditions, or does that variation mask multiple lineages that have not been recognised as separate species? All species could be placed somewhere on a spectrum of confidence in their species assessment, and neither the clearly defined, nor the very fuzzy, is the exception or the rule.

Colour is an obvious visual feature of many animals, but it is often unimportant. The seastar *Pisaster ochraceus* (Brandt, 1835) in the northeast Pacific comes in two extremely distinctive colours, bright purple and bright orange. Although intermediate colour varieties are rare, the purple and orange forms are unquestionably the same species, and the mechanisms for maintaining both colours in a population remains a mystery (Harley et al., 2006). Many molluscs also come in bright, vivid colours,

**FIGURE 5.1**   These sea slugs illustrate the challenges of identifying cryptic species. The two animals on the left are both *Chromodoris joshi* Gosliner and Behrens, 1998; although the bottom one has blue central stripes and a broad outer orange band, this is representative of variation within the same species (see Layton et al., 2018). The two animals on the right look nearly identical, apart from the ruffled posture of the top one, but the top is *Glossodoris pallida* (Ruppell and Leuckart, 1828) and the bottom right was only very recently recognised as a separate species of *Glossodoris* (Matsuda and Gosliner, 2018). (Images provided by Terry Gosliner.)

sometimes in patterns of colour blocks and stripes (Figure 5.1). Different colour forms have long been considered part of the variation within species, and indeed, some species genuinely have wide variation in colour and pattern (e.g., Layton et al., 2018; Sigwart, 2017). New genetic evidence shows that in some lineages, relatively subtle striping patterns correspond to separate species (e.g., Matsuda & Gosliner, 2018). Whether or not that stripe is important to the slugs for a particular reason or just accidentally reflects genetic diversity is not known, but it does provide a warning that there is more diversity out there than has been discovered.

Within a species, the existence of persistent natural variation is the cornerstone for evolution by natural selection. Variation as a feature itself has a key role in species evolution; variation itself is persistent. The removal or dominance of one character state is usually not enough to redefine a species. If all the orange *Pisaster ochraceus* individuals were somehow removed, that would reduce only a single variable that is (apparently) not particularly important from the seastar's own perspective. Even though 'ochre' in the species name refers to the orange colour-morph, purple-only seastars would still be *Pisaster ochraceus*. In that less colourful alternative future, there would still be many other features, ossicle density, arm length, genetics, that varied among individuals and populations. In addition to that variation within populations, individual organisms are also plastic, moulded by their life experience in their environment.

In exploring and building the case for evolution through natural selection, Darwin (1859) wrote extensively about the morphological variation in domestic breeds. This subject will be familiar to anyone who has read *On the Origin of Species* and noticed that this famous work rapidly launches into what appears to be an exhaustive manual on pigeon breeding. (My own first attempt at reading the book floundered in confusion over the pigeon thing, and it took me years to go back to it.) The first chapter of the *Origin*, 'Variation under domestication', deliberately laid out the most accessible arguments for the mutability of species; we are surrounded by clear and easy to understand evidence of changes in morphology through selection in domesticated species. The genetic loci that control various features of fancy pigeons (head crests, wattles, colour patterns), and the interrelationships of established fancy breeds have now been studied with genomic data (Shapiro et al., 2013), but they could be manipulated long before that. Selective breeding is a form of indirect genetic modification; that this results in measurable changes in morphology of offspring was important evidence that evolutionary selection, independent of human interference, has the innate power to shape species lineages.

Domestic species continue to provide important evidence and arguments about speciation; breeding lines of domestic pigeons (*Columba livia* Gmelin, 1789) rapidly revert to wild-type characters when artificial selection is relaxed (Darwin, 1859; Shapiro et al., 2013). The same is true for other domestic and ornamental species, though genome information can now pinpoint loci that control specific traits (e.g., Blackman et al., 2010; Bombarely et al., 2016), and for some domesticated species, it is difficult to envisage any mixture between modified and wild-type individuals. The informative parallel between artificial selection and natural selection is that features can be shaped from part of the general sprawl of variation into a prominent feature. Some natural systems, particularly freshwater fish, have become models for work on speciation because morphological changes apparently occur relatively rapidly and freshwater systems provide for exaggerated isolation (e.g., Schluter, 1993; Vrijenhoek, 1994). The extension of selection as a process to a more detailed general model of speciation, the actual division into fully segregated species lineages, is not a trivial point of theory and one that Darwin (somewhat ironically) never fully confronted in the *Origin* (Stamos, 2007).

The fundamental idea of natural selection is that variation in a trait or feature provides a spectrum, and selective processes can pick out one focal point on that spectrum. Fixing that one point, stopping or narrowing the variation in one trait, does not stop or narrow the total scope of all variation in a lineage, because the spectra of variation in other morphological or genetic features can remain or even expand. Variation occurs, and evolved, on multiple dimensions. Nuclear, mitochondrial, and chloroplast genomes each evolve separately, and phenotype is non-linearly connected to genotype. The transition from microevolution to macroevolution—or from variation to speciation—is a step change from stochastic fluctuations in traits to permanent changes that correlate with a redirection of the evolutionary trajectory of a lineage. In large part, this transition is an issue of time. Morphological variation occurs over a broad range of timescales, and it is important to clarify the vocabulary for different scales of change (Figure 5.2).

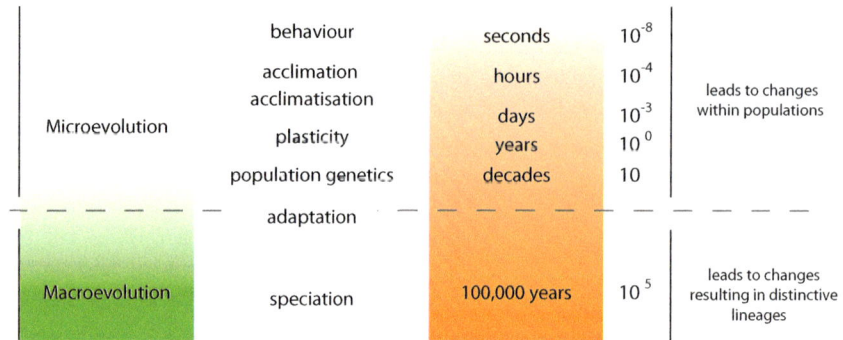

**FIGURE 5.2** Schematic representation of micro- and macroevolutionary processes and indicative timescales across many orders of magnitude.

## PLASTICITY AND ADAPTATION

The word 'adaptation' is often used in an imprecise way, and different authors have divergent preconceptions about what they mean when they say it. Here, we are predominantly thinking about species and their origins and histories, or an explicitly macroevolutionary perspective, which informs what qualifies as 'adaptive' or an 'adaptation'. A stunted tree on a windswept cliff has grown in response to environment, but that is not an adaptation. Giant tails on fancy fantail pigeon varieties are a change in morphology resulting from selection, but they are not adaptation. Lines on nudibranchs are diagnostic features, but they are probably not adaptive. Birds that increase their vocalisations in urban environments to compensate for noise pollution experience higher fitness, but this is probably not adaptive because that is a trait induced by environment, not a permanent segregation of lineages. '**Adaptive**' refers to evolutionarily relevant phenotypic features that provide increased fitness (reproductive potential) for the species as a whole and become a permanent part of that species.

Many authors use adaptation, acclimation, acclimatisation, and phenotypic plasticity interchangeably. This liberal approach is almost guaranteed in popular writing, but also extremely common in the technical scientific literature. Within the experience of a single generation of a single organism, the differences are sometimes imperceptible. From a macroevolutionary or species lineage perspective, they are profoundly different. The use of these terms can be arranged in terms of their temporal scale, from fast-acting and transient to more permanent: acclimate, acclimatise, plasticity, epigenetic change, adapt. This spectrum starts with effects that can manifest in seconds, to lifetimes, to multi-generational, and finally to macroevolutionary changes that affect species (Figure 5.2).

The difference between acclimate and acclimatise in common usage is mostly a difference of English dialects: the older (plesiomorphic) word form, acclimate, is counter-intuitively more common in American English, and the more recently derived usage 'acclimatise' is more common in international English. In technical terms, acclimation refers to acute physiological adjustment, typically in an experimental setting, and acclimatisation is a longer-term adjustment to an adjusted homeostasis, usually in the natural environment (Speakman, 2001).

To **acclimate** is to experience short term, adjustive responses to a rapid change in environment, or getting used to a new setting. Experimental physiology is primarily concerned with measurements of responses to a single variable (temperature, or water or atmospheric chemistry). In comparative physiology experiments, a group of organisms is divided among experimental treatments, and then held in captivity such that their environments are exactly the same, except for one variable of interest, allowing time for the subjects to get used to their experimental environment (Obernier & Baldwin, 2006). Similarly, jumping into a cold swimming pool or a hot shower has an immediate effect on our vascular system, but after time the immediate reaction subsides. The compensatory physiological changes to a new setting produce acclimation (Eckert & Randall, 1983).

To **acclimatise** refers to an overall shift, including phenotypic changes, that come with longer-term exposure to different environments. A classic example is the way many non-human mammals grow additional fur in response to seasonal cooling. Acclimatisation also refers to the summative changes associated with moving to a new habitat, especially in humans. The experience of culture shock has real physiological effects (Oberg, 1960; Winkelman, 1994). These include acute stress responses, potentially from thermal stress in a different climate, temporary disruption of the circadian rhythm from jet lag or a difference in seasonal daylight, and psychological exhaustion from navigating an unfamiliar environment with unfamiliar food, soundscape, and social expectations. These responses, the experience of culture shock, can occur even in a new job or school without moving to a different country or climate. The relatively quick shift in circadian rhythm to overcome jet lag is effectively *acclimation*: disruption, compensation, and a return to homeostasis with a slight shift to one parameter. Deeper acculturation or re-aligning the mental and physiological states to a new environment, is a form of *acclimatisation*. There are both heritable traits and individual variation in physiology that make people more or less tolerant to hot or cold climates, influenced by longer-term life at certain latitudes. This interplay of natural variability and response to environmental pressures has significant overlap with phenotypic plasticity.

**Phenotypic plasticity** refers to the innate spectrum of variability that can be induced over ontogeny through environmental control (West-Eberhard, 1989). These are usually morphological changes that can be expressed during the lifetime of a single individual, shaped by pressure from the environment. This may be due to food or nutritive input, or lack thereof, or physical forces such as wind and waves. Barnacles grow longer cirri (feeding legs) in faster currents, with lengths that are precisely tuned to local flow dynamics (Arsenault et al., 2001). Phenotypic plasticity is beneficial: a tree that adamantly grows upward, in a region of strong directional winds, will be blown over and uprooted. An animal that incorporates pigment from its food plant consequently matches the colour of its local habitat and becomes better camouflaged from potential predators. Phenotypic plasticity can lead to different, apparently quite separate, morphotypes within a species. However, plastic traits are potentially transient and reversible; transplant experiments usually reverse the expression, though it sometimes takes another generation.

Plasticity is itself a trait, and varies within and among species—broadly defined, plasticity includes the malleability of any feature, including physiological responses that control acclimation or acclimatisation. Although flexibility seems advantageous in a changing environment, such responses may come with significant and deleterious energetic costs at the individual level (DeWitt et al., 1998). For a species that occupies a narrow niche, under long-term stable environmental conditions, plasticity may be a trait that is selected against in evolution. In species with more variation, the ability of a genotype to produce a phenotype perfectly matched to the environment is still wonderful and mysterious. Many recent studies focus on phenotypic or genotypic changes that may be associated with or driven by climate change; limited understanding of the possible range of plasticity means that in many cases it is still difficult or impossible to tell what is plasticity and what may be evolutionarily relevant adaptation (Merilä & Hendry, 2014).

This interaction of organism and environment is not a superficial relationship; species may actively respond to environmental change at the genomic level (Cullis, 1999). In this way, the mechanisms of microevolution contribute to macroevolutionary effects, but in unpredictable ways. Environmental influences, usually at early stages of development, can create changes in an organism's genome, referred to as **epigenetic changes**. For example, stress or exposure to toxins can alter gene expression, and this was understood long before genomic studies. This kind of environmentally-induced differential expression is found in all major groups of eukaryotes (Penny et al., 2014). Colour forms of plants and insects or the sex of reptiles is determined by environmental conditions during early development. Furthermore, some environmental cues induce genomic changes that can be passed on when pressure on the parent influences gene expression in their gametes. In such cases, the effects are expressed in their offspring even when the next generation is raised in a very different environment. If a history of stress on a future mother or father could potentially disadvantage their eventual offspring, even if the child is healthy and raised in ideal conditions, this obviously has important implications for human medicine (e.g., Thayer & Kuzawa, 2011). Heritable epigenetic modification is relatively common in plants and was documented by the much-maligned botanist Jean Batiste Lamarck (Weigel & Colot, 2012). His observations can now be understood through epigenetic mechanisms, but Lamarck extended the inheritance of acquired characteristics to speciation, and this is not compatible with a modern, holistic view of evolution. Invoking Lamarck's work is not to say that he or his followers were correct all along (Penny, 2015). Epigenetic variation is another form of plasticity or complexity within lineages, the effects of which have been observed for some time but only recently appreciated or understood.

A question remains whether epigenetic variation can persist, directionally, in such a way that this might be a mechanism for adaptive change to species as well as the observed, microevolutionary effects at individual or population level. Some time ago, experiments on *Drosophila* showed that the heat-shock protein Hsp90 buffers variation in other genetic loci; when this buffering is compromised by temperature or chemical inhibition, additional variation is expressed in other developmental pathways (Rutherford & Lindquist, 1998). Because this was a case where environment has an indirect, but clearly causative impact on morphological development, it was immediately noted as a mechanism for potential epigenetic influences on macroevolutionary processes (Wagner et al., 1999; Erwin, 2000).

Inter-generational epigenetic effects (parent to offspring) have now been recorded in a range of organisms and the molecular mechanisms studied in some detail (Heard & Martienssen, 2014). Parental effects such as added stress or nutritional deficiencies during development, can impact offspring (F1), and weaker offspring may be disadvantaged in terms of their reproductive capacity and thus the initial effects also indirectly impact the next generation (F2). But this is not the same as truly sustained trans-generational effects. More recently, experimental work found that certain epigenetic changes are trans-generational, persisting as alterations to the genome that are present for multiple later generations not exposed to the initial trigger (e.g., Slotkin & Martienssen, 2007). Stable trans-generational effects are extremely rare and remain controversial. Mechanisms for enacting these changes and making them potentially sustainable—persistent on a scale relevant to species evolution—are not really understood, but could also be interacting with cryptic genetic variation, sustained environmental influences, or heritable impacts on the microbiome (Heard & Martienssen, 2014).

Epigenetic modifications are an essential part of the plasticity and variation within species that shapes their evolutionary trajectories. But the responses to a changing environment do not always have a positive or productive effect on the population. Mother stickleback fish that were experimentally exposed to repeated threats of predation gave birth to baby fish that grew up less able to avoid predators, not more agile (McGhee et al., 2012). Stress is generally bad for organisms, it does not make them stronger. The *a priori* ability to be plastic and respond dynamically to fit in with the local environment has clear advantages for the organisms and we can speculate that at the species level there could be selective pressure favouring such flexibility. When epigenetic changes due to stress decrease the potential for plasticity, this response is potentially maladaptive. But adaptive plasticity (like everything) is not linear: constantly changing in response to a changing environment could buffer a lineage from selection pressure and reduce evolutionary rates, or on the other hand, rapid plastic responses in phenotype could increase fitness, hone features through selection and accelerate permanent phenotypic change (West-Eberhard, 2003).

## SELECTION AND SUFFICIENCY IN THE MOMENT

Adaptation is a powerful central structure in evolutionary theory. It is perhaps an unfortunate word choice, though; its colloquial use suggests a pro-active change, where an object moulds itself to accommodate new circumstances. This is the process we observe in phenotypic plasticity, and the physiological responses of acclimitisation (Eckert & Randall, 1983); however, this is not equivalent to adaptation in an evolutionary sense. In terms of species evolution, two tenets must be held in the front of our minds:

Evolution selects *against* less favoured traits.
Species do not evolve with a goal in mind.

In the evolution of species, adaptations are permanent, relevant changes in phenome and genome. Judging something as an adaptation implies an evaluation of data about heritability (permanence), and a sustained impact over the temporal range of a species. Whether or not a feature is an 'adaptation' is a question: Did this trait have some discernable influence on the evolutionary trajectory of this species? This can only be judged *post hoc*, in comparison to the success of related lineages in the present and past. Our judgement of something as an adaptation rapidly becomes teleological—it is almost impossible to talk about adaptations without implying that the feature was a deliberate strategy to gain some kind of advantage. Distinctive features often get called adaptations (the trunk of an elephant, the thorns of a cactus) and functional morphology often offers a straightforward explanation for the 'adaptive significance' of features that are obviously useful. We usually discuss adaptation in terms of the utility of the end product rather than the mechanisms that led to the change, in part because it drives an interesting narrative. Adaptive significance can be used as a rhetorical device to explain evolutionary pathways and lineage diversification.

Projecting unnecessary meaning into 'adaptive' features is a well-known but still dangerous pitfall in macroevolution. These types of over-interpreted traits especially are now often called 'spandrels' after a metaphor proposed by Gould and Lewontin (1979; Gould, 1997). They wrote about an architectural feature common in cathedrals: spandrels, the symmetrical spaces between arches that are often filled with ornamental decoration (Figure 5.3). The question is, are these spaces designed to accommodate the art form, or is the art form a byproduct of the building design? Assumptions about the 'perfecting' influence of selective processes lose sight of the noise introduced by sufficiency and, importantly, opportunism. Features may be inadvertent artefacts of construction, like the spandrels, or retained because there was no particular disadvantage, and these may become useful in later evolutionary circumstances; but this does not mean there was some premeditated plan.

Adaptationist interpretations of macroevolution can be snared into circular logic common to creationist interpretations of the world, satirically exemplified by the fictional character Dr Pangloss, who declared 'Observe, that the nose has been formed to bear spectacles—thus we have spectacles. Legs are visibly designed for stockings—and we have stockings' (Voltaire, 1759; Gould & Lewontin, 1979). Entirely serious evolutionary studies write about inherited features that are 'pre-adapted' to exploit opportunities, just as noses seem to anticipate reading glasses, in that a feature originally evolved in a non-adaptive way in one context but later proves to be particularly useful (or not disadvantageous) to descendents in a different niche. Gould and Vrba (1982) coined the term 'exaptation' to describe such co-opted traits, to emphasise that evolutionary pathways are not driven by anticipating future goals.

While the evolution of some groups seems to provide progressively better and better solutions to a problem, this is not always true. A descendent clade might actually lose an adaptive phenotype (Strathmann, 1978). Many fossil animals had heavy armour, such as placoderm fish, crinoids, or machaeridian worms (Vinther et al., 2008), but these are in the evolutionary stems of groups with unarmoured descendents. Extinct groups include morphological disparity that cannot be predicted from living species alone (Chapter 9); extinct types may represent inferior or obsolete adaptations or they might simply have been replaced by equally effective solutions to the next version of the same evolutionary problem (Vermeij, 1973).

Adaptation in less familiar organisms may be more difficult to understand, and perhaps more prone to misinterpretation. The brittlestar *Ophiocoma wendtii* Müller and Troschel, 1842, has a very strong sensitivity to light, and clear behavioural

**FIGURE 5.3** Spandrels represent an incidental feature of the construction of archways, but these spaces present such a rich artistic opportunity, it could be believed that the arches were created to provide the spandrels. The spandrel painting shown here is a detail from a spectacular set by Gustav Klimt embedded within the architechture of the Kunsthistorisches Museum Vienna. (Old Italian Art, detail, 1891, image reproduced with permission of KHM, Vienna.)

responses to differing light levels, including changing its own colour. This species also has many transparent, round crystalline elements within its arms; for many years these were interpreted as lenses that we thought to focus light onto photoreceptors beneath them (Aizenberg & Hendler, 2004). Further studies examined the photonic properties of these lenses and their usefulness in bio-inspired design of small scale dispersed sensor arrays. Later, though, it became clear that the micro-lenses are not used to focus light; *Ophiocoma* has photoreceptors in its skin on top of the skeletal lenses, not below them (Sumner-Rooney et al., 2018). The optical properties of the 'lenses' were called an exaptation, but this may refer to a structure co-opted for human inspiration rather than utility for the brittlestar.

Adaptations also usually relate to a particular life stage, or at least the context of a focal life stage is essential to developing hypotheses about the evolution of phenotypes (Merilä & Björklund, 2004). Gastropod molluscs all undergo 'torsion', a twisting of the body axes during early larval metamorphosis that positions the anus above the head, which may relate to the ability to withdraw into the shell. This has long been well known as a feature of the group (Spengel, 1881) but there has been a long debate about whether it is primarily adaptive for free-swimming larvae (Garstang, 1928) or for benthic adults (Morton, 1958). Torsion undoubtedly is significant in the broad ecological and morphological diversification of gastropods as a clade.

Some species groups share a particular 'key innovation' or trait in an ancestral lineage, which led to an 'adaptive radiation' or a rapid burst of evolution that expands phenotypic and species diversity. Adaptive radiations have almost always been defined primarily in terms of their ecology, as focussed on the ecological drivers that may have promoted speciation (Schluter, 1996). The species complexes that result from such relatively clear evolutionary patterns are extremely important for understanding evolution and biodiversity. One of the most dramatic adaptive radiations is the fish eating habits of cone snails (family Conidae): δ-conotoxin peptides in cone snail venom enabled crawling marine snails to hunt and kill fish, and this is interpreted as an innovation that kicked off multiple evolutionary radiations totalling over 100 piscivorous snail species (Aman et al., 2015). Another famous example is beak forms in finches of the subfamily Geospizinae found in the Galápagos Islands and made famous by Darwin. It is inferred that the modifications to beak shape enabled incipient finch lineages to better exploit particular food sources, adaptations that drove the evolution of at least 13 species (Schluter, 1996). The parallel example in plants is the 'silversword alliance' a radiation of over 50 endemic plant species in Hawai'i. While this is considered a roughly equivalent case study to the Galápagos finches, the silversword plants are all dramatically different in their morphology and ecology and the 'key innovation' could be described simply their ancestor's ability to arrive and survive in remote Hawai'i. Rapid bursts of speciation in general are not necessarily dependent on a single lucky feature.

Identifying an adaptation *post hoc* does not mean that its future impacts can be predicted. The Galápagos finches are one of the best-studied radiations, including important observations of the role of hybridisation (Farrington et al., 2014) and a detailed understanding of the genetic control over that key feature, beak morphology (Mallarino et al., 2011). Meticulous observations of the animals in the field over several decades found substantial variation, and evidence of natural selection, but the longer-term effects were not predictable based on the starting conditions (Grant & Grant, 2002). Perhaps the most important lesson that this model radiation can offer is the importance of natural history to understanding the underlying complexity of any real species radiation.

Considering biodiversity at a global level, certain groups of organisms take up more than their fair share of species richness: insects are the most species rich animal group, and even within insects several subgroups dwarf the diversity of other animals. (There is a famous story attributed to J.B.S. Haldane that the main thing to be learned about the mind of God from studying species, is 'an inordinate fondness for beetles' (Hutchinson, 1959).) A 'group' or 'clade' may be of arbitrary size or scope, but here we understand groups to have clear morphological diagnoses, which are important for communication about diversity. A lot of research in evolutionary biology is driven by understanding large radiations and traits that seem to correlate with high speciation.

In some cases, it appears that a key morphological or genetic adaptation created opportunities to exploit new niches: in fish-hunting cone snails, the ability to paralyse fish seems to have opened a new frontier in how to make a living as a gastropod, and a single peptide in their ancestors' venom underlies not one but several bursts of speciation. The adaptability of beak morphology in Galápagos finches, coincidentally deployed in an island archipelago with few ecological competitors, is another example in more limited geographical scope. The silversword alliance highlights an important flaw in the narrative of 'key innovations'. Though rapid radiations are widespread in terrestrial and marine environments, very few clades are actually correlated with a relatable and obvious adaptive trait.

In deeper evolutionary time, key innovations are more accurately but prosaically described as an advantageous and successful approach to some of life's necessities. Many challenges in life are shared by disparate groups of organisms—how to eat or obtain nutrients, how to move or defend oneself from predation—and a good tactic can be reinvented many times. The best approach could be found multiple times using the same building blocks, or through completely convergent means. At a large scale, across the whole diversity of animals, consider that animals have evolved vermiform bodies dozens of times independently: multiple times among reptiles (snakes and legless lizards), but also repeatedly across animal phyla including worm-molluscs (aplacophorans), worm-like burrowing sea anemones, and countless more. This is not because of traits inherited from a wormy early animal ancestor, but because many lineages have independently found benefits from their own perspectives in smooth locomotion or the ability to burrow or fit in small holes. Vermiform or serpentine bodies in most cases could be thought of as an 'advanced', adaptive morphology (and a clear example that all living lineages are equally evolved).

Natural selection acts on individuals, populations, and species, by removing the most unsuccessful and thus favouring the least bad. It does not, contrary to popular teaching, necessarily favour the most fecund. So, while adaptation is one essential evolutionary mechanism, it is not the case that species are actually identifying specific problems and then proactively solving them. In some cases, such as examples touched on here, there is strong evidence for a correlation and perhaps even a causative relationship of some kind between adaptive features and speciation. Famous examples of adaptive radiations focus on narrative or descriptive scenarios for how biodiversity came to be as we observe it now. The evidence for these narratives comes from gene expression, biochemistry, and morphology. We use this evidence to reconstruct a parsimonious story about what factors influenced the speciation patterns. Other radiations remain perplexing. New methods including genomic data will not suddenly 'solve' adaptive radiations; these hypotheses necessarily remain largely impossible to test, because there is no control group. This is not to undermine explanations that are entirely plausible and interesting, but the point to be made is that there is no universal magic key that will unlock the secret to adaptive radiations. On the front lines of evolution, all a species needs to do is be good enough to survive, just today. Adaptation is not design, it is a record of sufficiency in the moment. And what was sufficient in one moment, in that niche, that ecosystem and that climate, could easily change in the next moment.

## MICROEVOLUTION AND MACROEVOLUTION

We mostly interact with species in the present, represented by observations of individual organisms. This *synchronic* variation, the range exhibited within the present time, is only a fraction of the species. Each species has a geographic range or distribution, this is its 'horizontal' distribution in the sense that it is restricted to a single time slice (the present, or a slice through the past in the fossil record), spread over a certain geographic range within a small moment in geological time of Earth's long history. Species also have a 'vertical' dimension in time, extending into the past, linked to direct individual ancestors, and into the future through individual children. That whole population extending through space and through time can define the extent of a species. So, the variation in species encompasses several dimensions—morphological and genetic variation among a group of co-occurring individuals, but also variation that can occur diachronically within the lifetime of an individual, or a few generations, or for longer over the whole evolutionary span of a species. Both micro- and macroevolutionary patterns are subject to selective pressure, but microevolution does not extrapolate linearly to speciation (Eldredge & Cracraft, 1980). This variation in all directions puts fuzzy edges on the boundaries that define species (Figure 5.4).

The capacity for variation in organisms and species is necessary, but not sufficient, to explain speciation. But a lineage is not always in the process of speciation, nor extinction. Most species lineages apparently spend most of their time in equilibrium, that is, periods of geological time when an ancestor-descendent population lineage

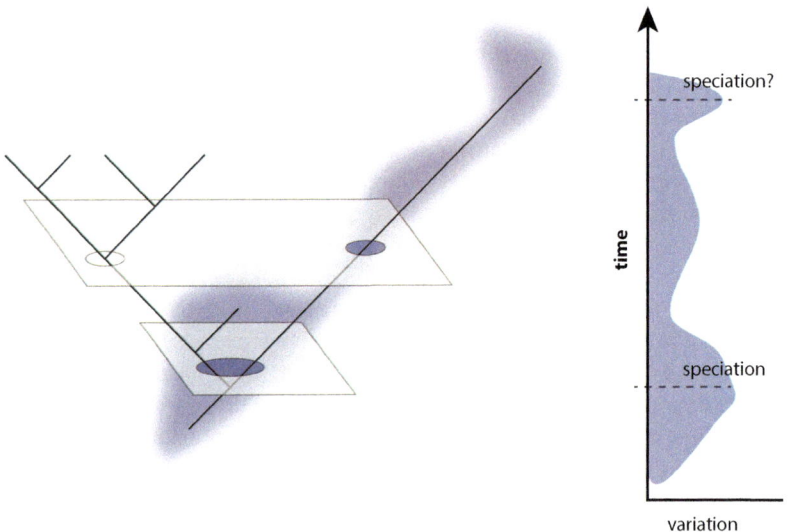

**FIGURE 5.4** Schematic representation of microevolutionary variation overlaid on one lineage in a species-tree (a diagram of macroevolutionary changes). Within-lineage variation changes through time, as shown to right; larger excursions in variation in the earlier time point correlated to a speciation, lineage-splitting event. An observation of the horizontal sample represented by the bottom slice may have been equivocal about whether that represented one species, or two, at the time.

is persisting, neither in process of speciation nor waning toward extinction. Even at equilibrium, variation is present within and among populations of a species. Microevolutionary processes clearly are still continuing during equilibrium phases, and the additive consequences of that variation could result in drift. The concept of anagenesis, where this accumulated gradual change could result in a substantively different population through linear descent, is not usually considered an example of speciation (Bull & Maron, 2016); under anagenesis, species richness has not increased, the lineage is intact and unchanging, the descendants simply expanded the boundaries, or shifted the range of variation observed within the vertical species. Drift remains controversial in population genetics, especially considering whether apparent state shifts attributed to drift may simply be samples from variability among a larger stochastic process (Millstein, 2016).

The process of speciation intrinsically means that all species have, or had, closely-related species to split from; this is the Sarawak Law (Wallace, 1855). Looking around our current random sample, the present moment that is a slice in time through the tree of life, we will encounter species in all phases of speciation: some in equilibrium, and others in transit through the process of splitting into daughter lineages that will in the future be seen as separate species (Figure 5.4). In a single time slice, closely related lineages may overlap; it may be impossible to satisfactorily divide lineages definitively into an absolute or fixed number of species. Microevolutionary variation, like variability within a population, may or may not be directional or related to speciation.

Variability, plasticity, and even epigenetic changes fall within the realm of microevolutionary change. These may create measurable and substantial changes in phenotype, but not necessarily to the extent that it would lead to a new and separate species lineage. Lineage splitting and the formation of new species and species groups (cladogenesis) fall in the realm of macroevolution. Macroevolution considers of biodiversity dynamics over geological time spans, the origination and extinction of species and species groups (Erwin, 2000). It is an open question to what extent microevolutionary processes can be traced or expanded to macroevolutionary processes, that is, whether the same mechanisms are transferable to these very different timescales (Eldredge & Gould, 1972; Gould, 1980; Plotnick & Sepkoski, 2001). Some wonder whether evolution is just 'wheels within wheels'; can microevolutionary processes multiply again and again until there is a new species? There are trends in living and fossil taxa that seem to point to parallels in responses to minor environmental fluctuations or major perturbations in climate (Simons, 2002). But it is apparent from experimental work and from long-term field observations that variability is not linearly related to speciation. Is most microevolution just frittering about with the details, working in the available range of variation in a species, such that even substantial changes could still be reversible?

In ecology, it is not necessary to track or understand the actions of individual organisms in order to measure and predict emergent patterns of the system. Similarly, there are emergent patterns in macroevolutionary trends that overcome the participation of individual species. These include the skew distribution in size-frequency of higher taxa (Chapter 6) and the dynamics of biodiversity across major extinction horizons in the fossil record, when diversity was reduced but later recovered (Chapter 9). Microevolutionary processes do not explain these system-level behaviours.

Species do not have an isolated one-on-one relationship with their geographic environment; interactions with other co-occurring species also control the evolution of lineages, and their extinction. Biotic drivers for evolution are often explained with reference to the Red Queen from *Alice Through the Looking Glass*, who said 'It takes all the running *you* can do, to keep in the same place. If you want to get somewhere else, you must run at least twice as fast as that!' (Carroll, 1872). The idea of a Red Queen process in evolution is based on observations that the probability of extinction in a given environment is not related to the age of the species, but that probability applies equally (a constant) to all co-occurring species, and this was called the 'law of constant extinction' (Van Valen, 1973). The metaphor of the Red Queen describes the survivorship of lineages against stochastic extinction, continuously evolving in order to maintain a competitive advantage in a specific 'adaptive zone' in a constantly changing environment (Liow et al., 2011). Empirical evidence has mostly rejected Van Valen's Red Queen hypothesis as originally intended in terms of constant extinction (Finnegan et al., 2008). Much of the later literature expanding the idea relates to molecular clock patterns and some aspects of microevolutionary phenotypic change. In order to explain larger scale macroevolutionary patterns, Barnosky (2001) extended the metaphor to the 'Court Jester'. In very broad strokes, the Red Queen is dynamic tension, and the Court Jester provides a periodic shake-up that jolts everything into new positions. Diversification of clades results from elements of both the 'Red Queen', and the 'Court Jester' (Barnosky, 2001). Over small scales in time or space, mostly the biotic interactions of the Red Queen type dominate the microevolutionary trajectories of species lineages (Figure 5.5). But the domain of the Court Jester is large-scale abiotic patterns and events, like latitudinal gradients in temperature, or volcanism, that have stronger effects on macroevolutionary diversification and extinction, and emergent global biogeographic patterns (Benton, 2009; Valentine & Jablonski, 2015).

## EXPERIMENTAL APPROACHES

The most productive line of research to understand the transition between these different levels—within-lineage variation (microevolution), and cladogenesis (macroevolution)—probably comes from comparative developmental biology (Gilbert et al., 1996). The evolution of developmental control genes may be the basis for macroevolutionary change (e.g., Jacobs, 1990). Certainly, comparative developmental biology could help disentangle intraspecific adaptive change and change that leads to speciation, since develop represents the linkage from ancestor to descendant and thus the continuity of species lineages.

Conflating micro- and macroevolutionary phenomena is a systemic problem in much experimental evolutionary biology (Erwin, 2000; Wheeler, 2008). The primary impediment is time; a force that is essential in evolution and cannot be substituted in empirical studies. Evolution takes a long time. Consider, for example, some of the species we best understand, mammals, which have the best-studied fossil record of any group and an unparalleled wealth of data on living species. We know that the median species longevity in mammal species is about 1.6 million years (Marshall, 2017). That is, among thousands of extinct mammal species that can be traced in the fossil record, the usual time between when a mammal species first appears and when it goes extinct

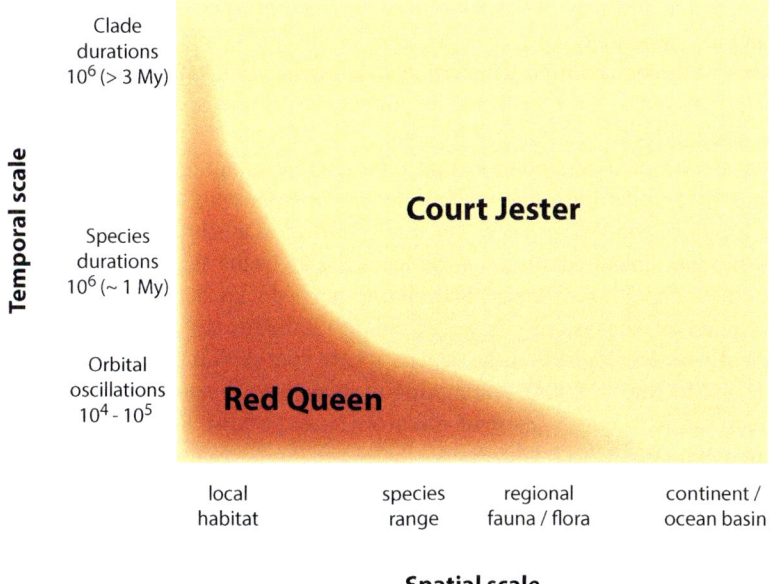

**FIGURE 5.5** The domain of the Red Queen. Biotic interactions under control of the (microevolutionary) Red Queen influence lineage evolution at relatively small spatial and temporal scales, whereas large-scale clade and contintental patterns are controlled by the mercurial Court Jester through extinction and geological perturbations. (Redrawn from Benton, 2009.)

is about 1.6 million years. This is a really very short time in geological time scales. Among living mammals, the median generation time is about 3 years (Pacifici et al., 2013). Generation time and lineage persistence are interconnected, and both relate to body size and biogeographic factors, but these broad approximations are adequate for our thought experiment. A quick bit of arithmetic suggests that the lifetime of a species is around 530,000 generations; that is demonstrably enough time to accumulate real macroevolutionary change. Let's speculate that a span on the order 100,000 years may be sufficient for mammal lineages to split, or around 33,000 generations. Perhaps in another type of species, something that has a 1 day generation time might plausibly speciate within 33,000 days, or certainly by 530,000 days. That would make a better laboratory experiment to determine how microevolutionary manipulations translate to macroevolutionary change, but 33,000 days is more than 90 years, 530,000 days is over 1,400 years. Long-term experiments following this logical model have been attempted with microbes, most importantly with the long-term evolution experiment of *E. coli* bacterial cultures that has been ongoing since 1988 (Lenski, 2004). Indeed, a substantial adaptive innovation was observed in those cultures at around 31,000 generations (Blount et al., 2012). Even so, experimental evolution has an empirical bias for process-driven outcomes, and a designed environment with no ecology may not be broadly comparable to natural systems.

There is no magic number of generations or years that automatically leads to a new species, because it depends on the environmental and ecological context of the lineage. The primacy of variation in understanding species extends to variation in variation—repetitive but true—the level of variability may be different among separate species, or over time. The frequency or tempo with which new lineages arise also clearly is not a constant factor for all species. Some lineages seem to simply persist, phenotypically unchanged for long spans of time.

Early bursts of diversification that characterise adaptive cladogenesis (when a lineage diversifies into new unexploited niche spaces) may induce synergistic diversification effects. When a new 'key innovation' is established, minor modifications rapidly produce many additional varieties, such as in the famous finches. We can imagine how changing diversity, and adding new species and competitors, shapes the evolution of ecosystems; however, adaptive radiations are still occurring at geological—not ecological—time scales, where 'rapid' is a period of 1–2 million years (e.g., McGowan, 2004). Attempting to infer burst dynamics from living biota alone can be misleading, because an unknown spread of diversity is lost through past extinction (Marshall, 2017; Chapter 9).

## MACROEVOLUTION AND SPECIES IDENTITY

Species themselves are ancestor-descendent population lineage segments between terminal events (splitting or extinction). Microevolutionary processes occur within lineage segments, and contribute to speciation, but macroevolution is not just an emergent property of ongoing microevolution. This view of macroevolution does not necessitate saltational speciation, a model with abrupt emergence of new species. After all, most natural 'experiments' in evolution probably fail, in that many incipient species never establish a really independent trajectory and merge with adjacent lineages. A process of incipient speciation, however, may be indistinguishable from periods of expanding variation within a single lineage. There is reticulation, or cross-linking, in the phylogenetic relationships among closely-related living species (Mallet et al., 2016), and even more deeply separated species groups (Rothfels et al., 2015). These are the fuzzy edges around species lineages (Chapter 7), blurred by hybridisation and introgression. The frequency and degree of reticulation may be a future quantitative metric to identify the boundary between increased variation and incipient species in living taxa.

A couple of significant points that have been neglected in this discussion are the many ways organisms reproduce and grow, and the extraordinary impacts that can have on the relationship between genotype and phenotype. Many organisms reproduce asexually—it is common in plants, and some animals, and there are parthenogenic animals. These do not really disrupt the idea of a dichotomy between microevolution and macroevolution, in that asexually reproducing organisms still have variation within time-horizontal species and ancestor-descendant relationships. Among all species, changes in genotype are not linearly correlated with changes in phenotype.

Another important point that deserves mention is the influence of ontogeny. Hennig (1966) coined the term 'semaphoront' to signify that morphological characters have to be assessed on a specific specimen, at a specific life stage, to be meaningful. Animals have a single ontogeny, potentially punctuated by metamorphic changes. Plants and fungi have multiple, or even simultaneous ontogeny, through being reduced and growing back, and more importantly plant parts are created sequentially. The older branches are genetically different than the apical meristem within a single individual in *Arabidopsis* spp. and many other plants with branching patterns, including trees (Alberts et al., 2002). The experience of epigenetic effects, and other factors that influence genotypic change and inheritance, are more complex in other organisms than in animals. Importantly, ontogeny is an intrinsic feature; trees have been growing the same way for many millions of years (Chapter 9).

Micro- and macroevolutionary distinctions account for the difficulty in assessing species, but it does not offer direct solutions for how to make a difficult call about a particular species identification. Everything to do with species is context dependent. Different lineages, and even their individual genetic and morphological features, evolve at different rates and in response to different selective pressures. There are clearly species that stand apart, while other clades form a closer alliance. Overall, there are substantial morphological and genetic gaps that separate lineages, established through dynamic changes in speciation rates (as one lineage accelerates away from its nearest relatives) or through diversity that is lost to extinction. There is not a totally continuous spectrum of diversity, so it is more reasonable to believe that there are identifiable, separate tips in the tree of life that we could give names to and classify relative to each other. Every group of organisms has different typical problems that confound studies of their evolution—working scientists should recognise that solving one problem with their study organism is not necessarily a panacea, and likewise the problems I face are not necessarily a universal indictment of all taxa. Some flowering plants have outrageously high levels of morphological plasticity, genetic variation, and hybridisation that make confidently differentiating individual species lineages problematic (Diggs & Lipscomb, 2002). Most invertebrate animals suffer from large discovery gaps, with high numbers of undescribed species and poor data for much known diversity. Sponges have morphological characters that can differentiate species, but high levels of homoplasy that confound higher-level systematics (Morrow et al., 2013). Birds are taxonomically the most completely described group of any organisms, and the high quality data available lead to ornithologists identifying and discussing finer and finer divisions within species (Barrowclough et al., 2016), to the point where other fields quietly think bird systematists could make better contributions by working on other organisms where there is more need of their skills.

Species are more than their morphology, and more than their DNA. From a single time slice (i.e., the horizontal species we normally interact with), it is nearly impossible to tell whether the whole time-vertical species lineage is at the beginning, close to other sisters, alone in equilibrium, or approaching an end. Previous debates over species 'concepts' have often tried to resolve this conundrum by presenting alternative philosophical approaches or lines of evidence that would provide a single accessible explanation for species essences (Chapter 7). Hanging onto a single concept about how to define species leads to conflict and hanging onto a single view can make variation seem like an unresolvable muddle with species graded one into the next. Species are shaped by their evolutionary history, but as lineages they also extend into future time. This explains why there is, as yet, no unequivocal evidence of observed lineage splitting in the wild (Endler, 1986). Macroevolutionary processes can only be conclusively recognised *post hoc*, when the products of speciation or extinction become conclusively apparent.

## REFERENCES

Aizenberg J, Hendler G. 2004. Designing efficient microlens arrays: Lessons from nature. *Journal of Material Chemistry*. 14: 2066–72.

Alberts B, Johnson A, Lewis J, Raff M, Roberts K, Walter P. 2002. *Molecular Biology of the Cell*, Fourth Edition. Garland Science.

Aman JW, Imperial JS, Ueberheide B, Zhang MM, Aguilar M, Taylor D, Watkins M, Yoshikami D, Showers-Corneli P, Safavi-Hemami H, Biggs J. 2015. Insights into the origins of fish hunting in venomous cone snails from studies of *Conus tessulatus*. *Proceedings of the National Academy of Sciences*. 112: 5087–92.

Arsenault DJ, Marchinko KB, Palmer AR. 2001. Precise tuning of barnacle leg length to coastal wave action. *Proceedings of the Royal Society B*. 268: 2149–54.

Barnosky AD. 2001. Distinguishing the effects of the Red Queen and Court Jester on Miocene mammal evolution in the northern Rocky Mountains. *Journal of Vertebrate Paleontology*. 21: 172–85.

Barrowclough GF, Cracraft J, Klicka J, Zink RM. 2016. How many kinds of birds are there and why does it matter? *PLoS ONE*. 11: e0166307.

Benton MJ. 2009. The Red Queen and the Court Jester: Species diversity and the role of biotic and abiotic factors through time. *Science*. 323: 728–32.

Blackman BK, Strasburg JL, Raduski AR, Michaels SD, Rieseberg LH. 2010. The role of recently derived FT paralogs in sunflower domestication. *Current Biology*. 20: 629–35.

Blount ZD, Barrick JE, Davidson CJ, Lenski RE. 2012. Genomic analysis of a key innovation in an experimental *Escherichia coli* population. *Nature*. 489: 513–18.

Bombarely A, Moser M, Amrad A, Bao M, Bapaume L, Barry CS, Bliek M, Boersma MR, Borghi L, Bruggmann R, Bucher M. 2016. Insight into the evolution of the Solanaceae from the parental genomes of *Petunia hybrida*. *Nature Plants*. 2: 16074.

Bull JW, Maron M. 2016. How humans drive speciation as well as extinction. *Proceedings of the Royal Society B*. 283: 20160600.

Carroll L. 1872. *Through the Looking-Glass, and What Alice Found There*. Project Gutenberg edition 2016.

Casane D, Laurenti P. 2013. Why coelacanths are not 'living fossils'. *BioEssays*. 35: 332–8.

Cullis CA. 1999. The environment as an active generator of adaptive genomic variation. In: Lerner HR (editor). *Plant Responses to Environmental Stresses*, pp. 149–60. Marcel Dekker Inc.

Darwin CR. 1859. *On the Origin of Species by Means of Natural Selection, or the Preservation of Favoured Races in the Struggle for Life*. John Murray, publishers.

DeWitt TJ, Sih A, Wilson DS. 1998. Costs and limits of phenotypic plasticity. *Trends in Ecology & Evolution*. 13: 77–81.

Díaz-Arce N, Arrizabalaga H, Murua H, Irigoien X, Rodríguez-Ezpeleta N. 2016. RAD-seq derived genome-wide nuclear markers resolve the phylogeny of tunas. *Molecular Phylogenetics & Evolution*. 102: 202–7.

Diggs GM Jr, Lipscomb BL. 2002. What is the writer of a flora to do? Evolutionary taxonomy or phylogenetic systematics? *SIDA, Contributions to Botany*. 20: 647–74.

Eckert R, Randall D. 1983. *Animal Physiology: Mechanisms and Adaptations*, Second Edition. Freeman and Company.

Eldredge N, Cracraft J. 1980. *Phylogenetic Patterns and the Evolutionary Process*. Columbia University Press.

Eldredge N, Gould SJ. 1972. Punctuated equilibria: An alternative to phyletic gradualism. In: Schopf TJM (editor). *Models in Paleobiology*, pp. 82–115. Freeman, Cooper & Co.

Endler JA. 1986. *Natural Selection in the Wild*. Princeton University Press.

Ellison AM, Buckley HL, Miller TE, Gotelli NJ. 2004. Morphological variation in *Sarracenia purpurea* (Sarraceniaceae): Geographic, environmental, and taxonomic correlates. *American Journal of Botany*. 91: 1930–5.

Ellison AM, Gotelli NJ. 2009. Energetics and the evolution of carnivorous plants—Darwin's 'most wonderful plants in the world'. *Journal of Experimental Botany.* 60: 19–42.

Erwin DH. 2000. Macroevolution is more than repeated rounds of microevolution. *Evolution & Development.* 2: 78–84.

Farrington HL, Lawson LP, Clark CM, Petren K. 2014. The evolutionary history of Darwin's finches: Speciation, gene flow, and introgression in a fragmented landscape. *Evolution.* 68: 2932–44.

Finnegan S, Payne JL, Wang SC. 2008. The Red Queen revisited: Reevaluating the age selectivity of Phanerozoic marine genus extinctions. *Paleobiology.* 34: 318–41.

Garstang W. 1928. The Ballad of the Veliger or How the Gastropod got its Twist. In: Garstang W. 1951. *Larval Forms and Other Zoological Verses*, pp. 36–7. Blackwell.

Gilbert SG, Opitz JM, Raff RA. 1996. Resynthesizing evolutionary and developmental biology. *Developmental Biology.* 173: 357–72.

Gould SJ. 1980. Is a new and general theory of evolution emerging? *Paleobiology.* 6: 119–30.

Gould SJ. 1997. The exaptive excellence of spandrels as a term and prototype. *Proceedings of the National Academy of Sciences.* 94: 10750–5.

Gould SJ, Lewontin RC. 1979. The spandrels of San Marco and the Panglossian paradigm: A critique of the adaptationist programme. *Proceedings of the Royal Society B.* 205: 581–98.

Gould SJ, Vrba ES. 1982. Exaptation—a missing term in the science of form. *Paleobiology.* 8: 4–15.

Grant PR, Grant BR. 2002. Unpredictable evolution in a 30-year study of Darwin's finches. *Science.* 296: 707–11.

Harley CD, Pankey MS, Wares JP, Grosberg RK, Wonham MJ. 2006. Color polymorphism and genetic structure in the sea star *Pisaster ochraceus. Biological Bulletin.* 211: 248–62.

Heard E, Martienssen RA. 2014. Transgenerational epigenetic inheritance: Myths and mechanisms. *Cell.* 157: 95–109.

Hennig W. 1966. Phylogenetic Systematics. In: Zangerl R (editor) 1999. *Phylogenetic Systematics*, translated by DD Davis and R Zangerl. University of Illinois Press.

Hutchinson GE. 1959. Homage to Santa Rosalia or why are there so many kinds of animals?. *The American Naturalist.* 93: 145–59.

Jacobs DK. 1990. Selector genes and the Cambrian radiation of Bilateria. *Proceedings of the National Academy of Sciences.* 87: 4406–10.

Kitagawa T, Nakata H, Kimura S, Itoh T, Tsuji S, Nitta A. 2000. Effect of ambient temperature on the vertical distribution and movement of Pacific bluefin tuna *Thunnus thynnus orientalis. Marine Ecology Progress Series.* 206: 251–60.

Layton KK, Gosliner TM, Wilson NG. 2018. Flexible colour patterns obscure identification and mimicry in Indo-Pacific *Chromodoris* nudibranchs (Gastropoda: Chromodorididae). *Molecular Phylogenetics & Evolution.* 124: 27–36.

Lenski RE. 2004. Phenotypic and genomic evolution during a 20,000-generation experiment with the bacterium *Escherichia coli. Plant Breeding Reviews.* 24: 225–65.

Liow LH, Van Valen L, Stenseth NC. 2011. Red Queen: From populations to taxa and communities. *Trends in Ecology & Evolution.* 26: 349–58.

Mallarino R, Grant PR, Grant BR, Herrel A, Kuo WP, Abzhanov A. 2011. Two developmental modules establish 3D beak-shape variation in Darwin's finches. *Proceedings of the National Academy of Sciences.* 108: 4057–62.

Mallet J, Besansky N, Hahn MW. 2016. How reticulated are species? *BioEssays.* 38: 140–9.

Marshall CR. 2017. Five palaeobiological laws needed to understand the evolution of the living biota. *Nature Ecology & Evolution.* 1: 00165.

Matsuda SB, Gosliner TM. 2018. Glossing over cryptic species: Descriptions of four new species of *Glossodoris* and three new species of *Doriprismatica* (Nudibranchia: Chromodorididae). *Zootaxa.* 4444: 501–29.

McGhee KE, Pintor LM, Suhr EL, Bell AM. 2012. Maternal exposure to predation risk decreases offspring antipredator behaviour and survival in threespined stickleback. *Functional Ecology*. 26: 932–40.

McGowan AJ. 2004. Ammonoid taxonomic and morphologic recovery patterns after the Permian–Triassic. *Geology*. 32: 665–8.

Merilä J, Björklund M. 2004. Phenotypic integration as a constraint and adaptation. In: Pigliucci M, Preston K (editors). *Phenotypic Integration: Studying the Ecology and Evolution of Complex Phenotypes*, pp. 107–29. Oxford University Press.

Merilä J, Hendry AP. 2014. Climate change, adaptation, and phenotypic plasticity: The problem and the evidence. *Evolutionary Applications*. 7: 1–4.

Millstein RL. 2016. Genetic drift. In: Zalta EN (editor). *Stanford Encyclopedia of Philosophy*. Stanford University, USA.

Morrow CC, Redmond NE, Picton BE, Thacker RW, Collins AG, Maggs CA, Sigwart JD, Allcock AL. 2013. Molecular phylogenies support homoplasy of multiple morphological characters used in the taxonomy of Heteroscleromorpha (Porifera: Demospongiae). *Integrative & Comparative Biology*. 53: 428–46.

Morton JE. 1958. Torsion and the adult snail; a re-evaluation. *Proceedings of the Malacological Society of London*. 33: 2–10.

Oberg, K. 1960. Culture shock: Adjustment to new cultural environments. *Practical Anthropology*. 7: 177–82.

Obernier JA, Baldwin RL. 2006. Establishing an appropriate period of acclimatization following transportation of laboratory animals. *ILAR Journal*. 47: 364–9.

Pacifici M, Santini L, Di Marco M, Baisero D, Francucci L, Marasini GG, Visconti P, Rondinini C. 2013. Generation length for mammals. *Nature Conservation*. 5: 87–94.

Penny D. 2015. Epigenetics, Darwin, and Lamarck. *Genome Biology & Evolution*. 7: 1758–60.

Penny D, Collins LJ, Daly T, Cox SJ. 2014. The relative ages of Eukaryotes and Akaryotes. *Journal of Molecular Evolution*. 79: 228–39.

Plotnick RE, Sepkoski Jr JJ. 2001. A multiplicative multifractal model for originations and extinctions. *Paleobiology*. 27: 126–39.

Rothfels CJ, Johnson AK, Hovenkamp PH, Swofford DL, Roskam HC, Fraser-Jenkins CR, Windham MD, Pryer KM. 2015. Natural hybridization between genera that diverged from each other approximately 60 million years ago. *American Naturalist*. 185: 433–42.

Rutherford SL, Lindquist S. 1998. Hsp90 as a capacitor for morphological evolution. *Nature*. 396: 336–42.

Schluter D. 1993. Adaptive radiation in sticklebacks: Size, shape, and habitat use efficiency. *Ecology*. 74: 699–709.

Schluter D. 1996. Ecological causes of adaptive radiation. *American Naturalist*. 148: S40–64.

Shapiro MD, Kronenberg Z, Li C, Domyan ET, Pan H, Campbell M, Tan H, Huff CD, Hu H, Vickrey AI, Nielsen SC. 2013. Genomic diversity and evolution of the head crest in the rock pigeon. *Science*. 339: 1063–7.

Simons AM. 2002. The continuity of microevolution and macroevolution. *Journal of Evolutionary Biology*. 15: 688–701.

Slotkin RK, Martienssen R. 2007. Transposable elements and the epigenetic regulation of the genome. *Nature Reviews Genetics*. 8: 272–85.

Speakman JR. 2001. Thermoregulation in vertebrates. eLS.

Spengel JW. 1881. Die Geruchsorgane und das Nervensystem der Mollusken. *Zeitschrift für wissenschaftliche Zoologie*. 35: 333–83.

Stamos DN. 2007. *Darwin and the Nature of Species*. State University of New York Press.

Strathmann RR. 1978. Progressive vacating of adaptive types during the Phanerozoic. *Evolution*. 32: 907–14.

Sumner-Rooney L, Rahman IA, Sigwart JD, Ullrich-Lüter E. 2018. Whole-body photoreceptor networks are independent of 'lenses' in brittle stars. *Proceedings of the Royal Society B*. 285: 20172590.

Thayer ZM, Kuzawa CW. 2011. Biological memories of past environments: Epigenetic pathways to health disparities. *Epigenetics*. 6: 798–803.

Valentine JW, Jablonski D. 2015. A twofold role for global energy gradients in marine biodiversity trends. *Journal of Biogeography*. 42: 997–1005.

Van Valen L. 1973. A new evolutionary law. *Evolutionary Theory*. 1: 1–30.

Vermeij GJ. 1973. Adaptation, versatility and evolution. *Systematic Zoology*. 22: 466–77.

Vinther J, Van Roy P, Briggs DE. 2008. Machaeridians are Palaeozoic armoured annelids. *Nature*. 451: 185–8.

Vrijenhoek RC. 1994. Genetic diversity and fitness in small populations. In: Loeschcke V, Jain SK, Tomiuk J (editors). *Conservation Genetics*, pp. 37–5. Birkhäuser.

Voltaire. 1759. *Candide, or, Optimism*. Project Gutenberg edition 2006.

Wagner GP, Chiu CH, Hansen TF. 1999. Is Hsp90 a regulator of evolvability? *Journal of Experimental Zoology*. 285: 116–8.

Wallace AR. 1855. On the law which has regulated the introduction of new species. *Annals and Magazine of Natural History*. 2nd Series. 16: 184–96.

Weigel D, Colot V. 2012. Epialleles in plant evolution. *Genome Biology*. 13: 249.

West-Eberhard MJ. 1989. Phenotypic plasticity and the origins of diversity. *Annual Review of Ecology and Systematics*. 20: 249–78.

West-Eberhard MJ. 2003. *Developmental Plasticity and Evolution*. Oxford University Press.

Wheeler QD. 2008. Introductory: Toward the new taxonomy. In: Wheeler QD (editor). *The New Taxonomy*, pp. 1–19. CRC Press.

Winkelman M. 1994. Cultural shock and adaptation. *Journal of Counseling & Development*. 73: 121–6.

# 6 Natural Patterns in Classification

## FRAMES OF REFERENCE

The standard terminology for species is based on nested groups, starting with the two-part scientific name with a genus name and a species name. A genus can contain multiple species, a family can contain multiple genera, and so on. Users accept this system as intuitive and practical, but it is cause for great concern to certain specialists among the name-makers. Potential problems with ranked taxonomy mainly follow two lines of argumentation: first, that these groupings may or may not reflect evolutionary relatedness. Second, that named, higher-ranked groups do not follow transferable, *a priori* criteria (a 'family' in one group is not defined on the same criteria as a 'family' in another). Here, we will consider the origin and application of ranked groups and to what extent these concerns impact their utility.

Understanding taxonomic ranks depends on a clear understanding of the relevant frame of reference. Species lineages are subject to ongoing evolutionary change but over very long timespans; this change does not alter their classification *per se* but the fact that systematics is trying to put stable boxes around dynamic, hard-to-define things creates an ongoing tension. Species evolve at different rates, under different selective pressures (Chapter 5), so there is a wide variation in what morphological or genetic distinction that diagnoses a species. A 'species' is not a rigidly defined measure of variability that can be uniformly applied across different groups. Therefore, it follows logically that there cannot be simple criteria for groups of species (Chapter 2).

In physics, relativity describes the way interactions of objects depend on a specific reference frame. The speed of a ball bouncing down the aisle of a passenger train is very different when measured from a reference frame inside the train or the ball's absolute speed over ground seen from outside the train. Reference frames profoundly change the interpretation of data. The question for species classification is, are we trying to establish some universal reference frame that defines what ranks mean, the Newtonian rules for the classification of species, or is it of more practical use to have a system with ranks that reflect the reference frame of the species in question? The application of variable reference frames is surprisingly intuitive; to understand the issue of speed with our ball, it is just a matter of visualising yourself inside a train, or watching the train go by, or looking down from a point off-planet. When we use taxonomic ranks, we intuitively accept that the primary frame of interest is within a local part of the Tree of Life, not the global view. The point of taxonomic ranks is to communicate about where a species fits into the pantheon of diversity, relative to other species. In ranked taxonomy, the reference frame is the explicit taxon of interest—for that given species, it is more similar and closely-related to those in its own genus than it is to other genera, more similar to other members of its own family than to other families.

Alternative, unranked classification systems have been proposed, including models where names of species groups depend on phylogeny. In this view, the ideal classification would begin with a phylogenetic tree, a map that shows the inter-relatedness of all (or maybe most) species in a group, and the branching structure of that tree would be used to identify and name groups of species. There are a number of problems with this approach. Key information about inter-relatedness is simply not available for most organisms. When many new species are being discovered every year, a stable phylogeny sounds like a pipe dream. In an earlier version of this text, my editor crossed out 'pipe dream' and suggested 'unlikely in the near future'. This reveals a common point of misunderstanding that warrants reiteration: for researchers who work with understudied groups of organisms, the idea of a stable, well-resolved, usable phylogeny can seem literally hallucinatory. Bouchet and colleagues (2016) estimated that at current rates of productivity it will take another 300 years to describe the species of only marine molluscs that are already known to be new. 'Understudied' taxa represent the vast, overwhelming majority of diversity, where basic documentation is sparse and phylogeny may be a luxury (Wheeler, 2004). Starting with a phylogeny and applying classification afterward is a somewhat backwards approach to these data-limited systems. Basic starting hypotheses about informative comparisons and similarities can be stated without explicit phylogenetic analyses, and ranked taxonomy is one useful tool.

Traditional ranked groups may not be flexible enough to articulate the growing understanding of similarities and inter-relatedness of taxa, and there may be strong evidence for certain evolutionary groupings that do not fit comfortably with ranked names. Many groups of plants above the family level are classified as 'unranked clades' (APG, 2016), and also for various groups of animals. The point of naming clades is to communicate about the shape of the tree. This is incredibly important, as only a phylogentic framework can communicate information about ancestry, diversification, and adaptation. Phylogeny, though, is not the sole interest for users who need names of species and groups of species.

Data taken without their reference frame lose meaning: the ball in the train is travelling at 792,000 km/h. Relative to what? That is the estimated speed of our solar system, including ball, relative to the galactic centre (Fraknoi, 2007). Polypodiopsida and Cyatheales are monophyletic groups—but shared origin is the only thing that statement communicates. The class Polypodiopsida comprises the modern ferns, including the order Cyatheales, the tree ferns (Rothfels et al., 2015). Unranked, unexplained names of clades do not communicate relative similarities or differences among groups of species. As we will explore in this chapter, Linnaean ranks are compatible with phylogeny. These ranked groups indicate relative—not absolute—degrees of similarity or differentness among nested groups of species, and the membership of these species groups can be revised to reflect evolutionary relatedness when it makes sense to do so. The Linnaean System is the source of a lot of confusion, and it has taken a lot of abuse in scientific discourse (Lambertz & Perry, 2015; but see Giribet et al., 2016), but it has an enduring and intuitive simplicity that makes it a powerful tool for communication with a broad audience.

## TAXONOMY AND SYSTEMATICS

Describing biodiversity, and analysing phylogeny, are separate processes and separate fields of science, though both are strongly connected with taxonomy. Some people say that taxonomy is an off-putting word, because it has 'tax' in it. (It also makes people tired, because as soon as you hear 'taxonomy' you might think, 'Save me now, someone is going to start going on about Latin and Greek roots'. And you would be right, but it will only last one paragraph.) The etymology of 'taxes', as in what we pay, comes from the Latin verb *taxare*, to assess or to censure. Taxonomy (and taxidermy, and indeed taxi cabs) is derived from the much nicer Greek root *taxis* (τάξις) referring to direction or arrangement. The suffix -*taxis* is a familiar reference to orientation, such as phototaxis, orienting toward light as plants do, or other stimuli through magnetotaxis (magnetic fields) and thigmotaxis (touch). Likewise, the same root appears in our words for the arrangement of names (*onoma*, in tax*onomy*) or skin (*derma*, in taxi*dermy*). So, try to put tax collectors out of your mind and imagine yourself riding in the taxonomy taxi, confidently navigating through a beautiful leafy park, perhaps homing in on a nice natural history museum.

In recent years, the word 'systematics' has become a more popular alternative to describe the science of giving names to species, perhaps because it sounds more like science and less like finance. Systematics and taxonomy in current usage are not entirely equivalent things, and though the difference is somewhat minimally important to non-specialists, it is worth pausing to clarify what words mean. Phylogeny is concerned with reconstructing lineage inter-relationships. All these endeavours are all related, but not necessarily interdependent:

Taxonomy—naming things
Systematics—classifying things
Phylogenetics—the evolutionary relationships of things, visualised in branching
    patterns

It is entirely possible to revise classification of a species (Chapter 4) without naming any new species (Chapter 8). Specimens of a bizarre marine organism that looked more than anything like a mushroom were discovered in 1986 and finally described in 2014 as a new species and new genus, *Dendrogramma*, as a metazoan of uncertain affinity (Just et al., 2014). The animal was so strange the authors hesitated to assign it to a phylum, but they did narrow it down to being not a bilaterian. Following on from that discovery, only two years later new specimens were recovered that were suitable for DNA analysis and confirmed that *Dendrogramma* is a cnidarian, in the siphonophore family Rhodaliidae, so its classification finally could be filled in 30 years after the first specimens were collected (O'Hara et al., 2016). It is also possible to propose classifications of species or groups of species, through revision to the understanding of relationships among existing units (e.g., Sirenko, 2006; Morrow & Cárdenas 2015; APG, 2016).

In modern biology, we recognise the primacy of shared ancestry and a classification that reflects homologies—characteristics that are shared through evolutionary relationships—not analogies. This is only one approach, but it is instructive that we use it. In a completely deconstructionist view, we could create any sort of arbitrary classification that suits our needs (Dyke & Sigwart, 2005). We could write treatises on the shared features of all the species whose names begin with the same letter, or create subgroups based on adult body mass, so that redwoods, giant squid, and dinosaurs form a group Gigantea. We could even imagine a sound scientific justification for Gigantea: body mass is a very important factor in physiology and ecology, and it is one of the first features you notice about these organisms (Peters, 1986). But we do not do that. While the few outlandishly large organisms might form a natural group of outliers in adult body size, this system rapidly falls down— what do we do with earlier life stages? How do we conscribe other groups of non-gigantic things, which fall on a continuous spectrum of sizes? Is it body mass or dimension that is more important? Evolutionary relatedness is considered the most universally informative basis for comparison of organisms and lineages at a global scale. Furthermore, other sources of data for comparative assessments (morphology, genetics) ultimately arise from evolutionary processes. So evolutionary relationships are the strongest foundation to build a system that has predictive power. If one species has a feature, it is more likely that other closely-related species share the same characteristics. The separate process of analysing phylogeny builds on this further. In practice, taxonomists and systematists have always formed groupings of species based on shared characteristics and based on a consensus approach to identify similarities.

The aspiration of modern systematics and taxonomy is to express these evolutionary relationships in the classification and naming of all organisms. In simple form, this means separating things that look the same, if the balance of evidence suggests that they reached the same end point by convergence—this is homoplasy. The beaks of parrots and squids are obviously products of separate evolutionary pathways, and their owners are in separate phyla, just as plants that mimic the forms of insects are not taxonomically confused with arthropods. The venomous Eastern Coral Snake (*Micrurus fulvius* (Linnaeus, 1766)) is in a separate family from other harmless snakes (*Lampropeltis elapsoides* (Holbrook, 1838)) that mimic their colouration (Slowinski, 1995). Gastropods with cap-like shells in the form of a limpet have evolved at least 54 times (Vermeij, 2016) so there is no single order or family that encompasses 'limpets'. These distinctions are important, and classification (and phylogeny) transmits information about the evolutionary distance separating convergent traits. By contrast, a diagnostic feature that is shared because multiple species inherited it from the same common ancestor is called a synapomorphy. In phylogenetic systematics, higher-ranked taxa are diagnosed by synapomorphies.

In the last few decades, the view that species, and consequently higher taxa, are hypotheses about evolutionary relationships has become increasingly accepted (Chapter 4). The Linnaean system of classification pre-dates tree thinking, and indeed was formulated at a time when species were thought of as having fixed essences, so it is worth serious debate to consider whether the Linnaean System remains fit for purpose (e.g., Schuh, 2003). This has played out in the literature over the last fifteen years and despite some angst in the specialist literature, Linnaean ranks remain robust to criticism.

## THE LINNAEAN SYSTEM

Relatedness among species is conventionally articulated in a seven-level nested hierarchy, which has grown from a five-level approach originally formulated by Linnaeus (1735); Table 6.1. There are a couple points of history that are important here. First, it should be clear that the classification system we use today did not spring forth fully formed in Linnaeus's *Systema Naturae.* As in all science, his work built on the contributions of others, including the use of binomial names, which was established but not yet a universal standard—some binomials documented in Linnaeus were in use as vernacular names in classical Roman times (Stearn, 1959). The five ranks that Linnaeus defined (kingdom, class, order, genus, species) have been refined and expanded by later workers, and much of his classification has been overturned in revision. Linnaeus did also propose various standard rules for assigning ranks, and while the sequential terms for the nested hierarchy remain the same, the original rules are largely obsolete because of later scientific developments (Ereshefsky, 1997).

In Linnaeus's original works, he developed classification schemes and criteria for plants, animals, and also minerals. There are now other, modern standards for classification of rocks, gems, and minerals, which refer to their chemical composition and crystalline form. In the 18th century, the economic development of mining was hampered by limited geological knowledge, so a comprehensive and predictive framework to understand the formation and properties of valuable minerals was sorely needed (Laudan, 1987). But Linnaeus's early efforts at geological classification, based on large crystal forms and chemical behaviour in fire, were unpopular and did not find any practical application. The foundations of modern mineral classification came from contemporary geologists like Friedrich Mohs (b. 1773–d. 1839) who focussed on mineral hardness rather than chemistry, and Abraham Werner (b. 1749–d. 1817), who was an early proponent of ideas about the dynamic Earth system (Rampino, 2017). Linnaeus's approach was rejected for minerals, but for animals and plants it was immediately popular.

Linnaean ranked classification now applies to life forms, but with some distinctions in its application among different branches. In Linnaeus's works in the 18th century, he recognised two types of life: animals and vegetables. Later, Haeckel recognised 'Protista' as a third type of organism, the unicellular life forms (Figure 6.1). In nearly three hundred years of additional research, we now recognise a minimum of seven separate kingdoms of living organisms; there are more, but there is ongoing research about the evolutionary divisions among microscopic eukaryotes and microbes (Table 6.2). Increasing recognition of the staggering diversity of unicellular lifeforms— including prokaryotes, and also many independent lineages of organisms historically called 'protists'—has precipitated dramatic and ongoing revisions to the higher classification of life (e.g., Raoult et al., 2004; Schulz et al., 2017). This is noted for completion, but that diversity is quite firmly outside the scope of this book. The seven traditionally ranked hierarchical groups of the Linnaean system apply within the domain Eukarya (Table 6.2).

Haeckel (1874) introduced the term 'Metazoa' which is synonymous with 'Animalia'. Metazoa was restricted to multicellular animal life with tissues and organs, whereas Animalia had no such restriction and in some uses historically extended to include protozoans (Nielsen, 2011). However, there are unicellular animals: the Myxozoa have

## TABLE 6.1
## Complete Ranked Taxonomy: Three Examples

The Linnean System has seven standard ranks (in bold text here) and a large number of tags for additional intermediate groupings. These can be used in combination with additional clade names assigned to groups without reference to rank. Not all ranks or aspects of classification are used in describing species. Some ranks ('infraorder', 'tribe') are only used regularly for certain groups. In this table, each column gives the complete classification of an example species. All the words in grey text are accurate parts of the full classification for these organisms, but were never mentioned in the original papers that described them as new species (because they are not essential to that context). Even between two closely related organisms, one rank may be informative in one case and redundant in another. Ranks reflect the relative similarities within a group of related species, they do not have absolute meaning that can be transferred to other organisms. Arecaceae, the palm family, contains about 2600 species. The whole class Polyplacophora contains about 1000 living species. But while living chitons are similar to fossil species from the Devonian (ca 390 Mya), the earliest fossil palm is much younger, from the Cretaceous, around 80 Mya.

| | | | |
|---|---|---|---|
| Domain | Eukarya | Eukarya | Eukarya |
| **Kingdom** | **Animalia** | **Animalia** | **Plantae** |
| Subkingdom | — | — | — |
| Superphylum | Lophotrochozoa | Lophotrochozoa | — |
| **Phylum [=Division]** | **Mollusca** | **Mollusca** | — |
| | Aculifera (unranked clade) | Aculifera (unranked clade) | — |
| | | | Angiosperms (unranked clade) |
| Subphylum | — | — | — |
| Superclass | — | — | — |
| **Class** | **Polyplacophora** | **Polyplacophora** | — |
| | | | Monocots (unranked clade) |
| Subclass | Neoloricata | Neoloricata | — |
| Infraclass | — | — | — |
| Superorder | — | — | — |
| | | | Commelinids (unranked clade) |
| **Order** | **Lepidopleurida** | **Chitonida** | **Arecales** |
| Suborder | Lepidopleurina | Acanthochitonina | — |
| Infraorder (mammals) | — | — | — |
| Superfamily | — | Cryptoplacoidea | — |
| **Family** | **Leptochitonidae** | **Acanthochitonidae** | **Arecaceae** |
| | Leptochitonidae 'Clade I' (unranked clade) | | |
| Subfamily | — | — | Coryphoideae |
| Supertribe | — | — | — |
| Tribe (insects, plants) | — | — | Chuniophoeniceae |
| Subtribe | — | — | — |
| **Genus** | ***Leptochiton*** | ***Acanthochitona*** | ***Tahina*** |
| Subgenus | — | — | — |
| **Species** | ***cascadiensis*** | ***lanae*** | ***spectabilis*** |
| Subspecies | — | — | — |
| Authority, date | Sigwart & Chen, 2017 | Sirenko & Saito, 2017 | J. Dransfield & Rakotoarinivo, 2008 |
| Species description includes molecular phylogeny? | Published previously | No | Yes |

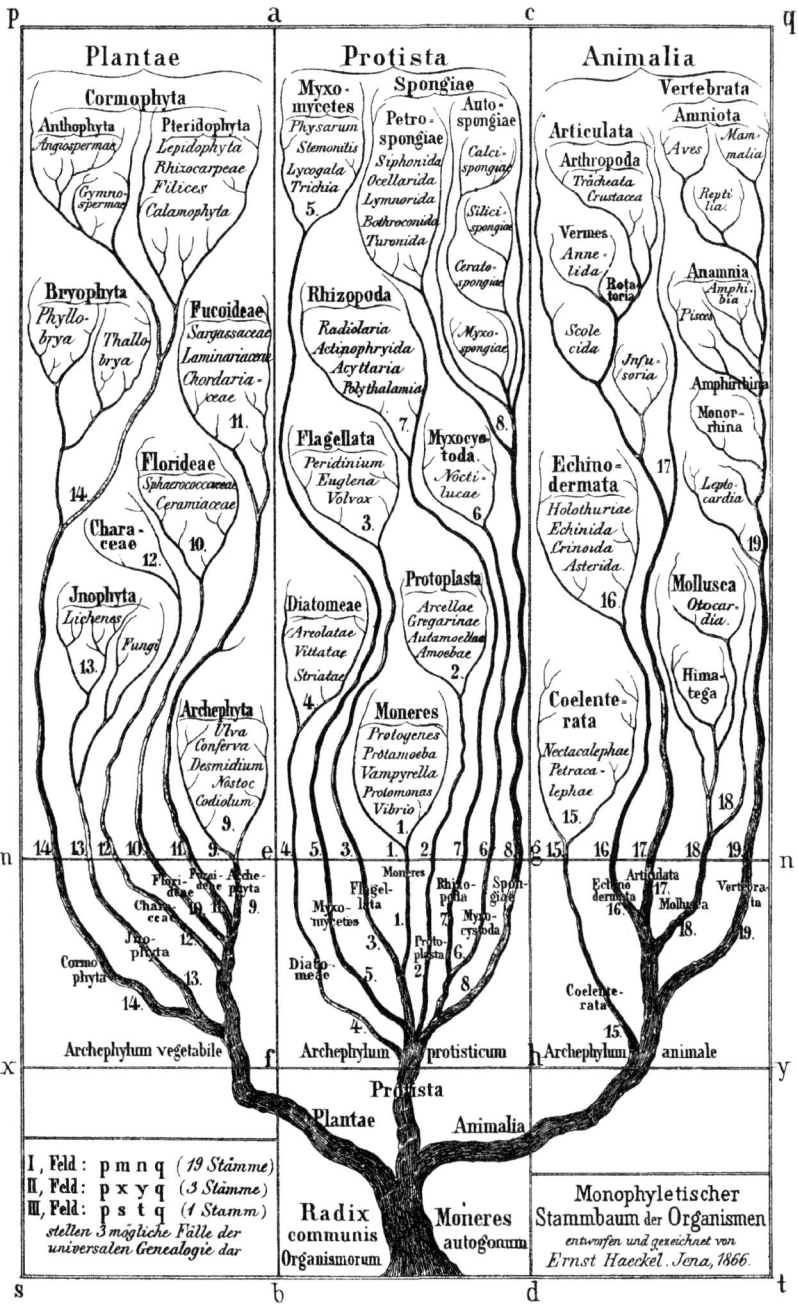

**FIGURE 6.1** Ernst Haeckel (1866) recognised a single origin for all eukaryotic life forms in his *Monophyletic tree of organisms*. However, the vertical placement in the illustration indicates the relative value of 'higher' forms such as angiosperms or vertebrates and 'lower' forms such as bryophytes and echinoderms.

## TABLE 6.2
## An Abbreviated History of Higher Level Taxonomic Concepts

This is not a complete history of the classification and revision of major groups of Life. There are especially many more alternative arrangements that were proposed in the 1990s and 2000s, but this chronology represents some key milestones in 250 years of understanding the organisation of the largest subdivisions of life on Earth. Important introductions of newly-recognised taxa are represented in **bold type**. This book only considers issues of species with regard to macroscopic eukaryotes (animals, fungi, and plants).

| | | Archaea | Bacteria | Eukaryota | | | | |
|---|---|---|---|---|---|---|---|---|
| | | | | Protozoa | Chromista | Animalia | Fungi | Plantae |
| Linnaeus, 1735 | 2 kingdoms | — | — | — | — | Animalia | Vegetabilia (spans Fungi–Plantae) | |
| Haeckel, 1866 | 3 kingdoms | — | Protista (spans Bacteria–Chromista) | | | Animalia | Plantae (spans Fungi–Plantae) | |
| Chatton, 1937 | **2 EMPIRES** | **PROKARYOTIC CELLS** (spans Archaea–Bacteria) | | EUKARYOTIC CELLS (spans eukaryotes) | | | | |
| Copeland, 1956 | 4 kingdoms | Monera (spans Archaea–Bacteria) | | Protista | | Animalia | Plantae (spans Fungi–Plantae) | |
| Whittaker, 1969 | 5 kingdoms | Monera (spans Archaea–Bacteria) | | Protista | | Animalia | **Fungi** | Plantae |
| Woese & Fox, 1977 | **3 URKINGDOMS** | **ARCHAEBACTERIA** | EUBACTERIA | URKARYOTES (spans eukaryotes) | | | | |
| Cavalier-Smith, 1981 | 8–10 kingdoms | Prokaryota (spans Archaea–Bacteria) | | **4–5 kingdoms of 'protists'** | | Animalia | 1–2 kingdoms of fungi | Viridiplantae |
| Woese et al., 1990 | 3 DOMAINS | ARCHAEA | BACTERIA | EUCARYA (spans eukaryotes) | | | | |
| | >6 kingdoms | 1 kingdom | 1 kingdom | 'Protista' (containing multiple, unnamed kingdom-level groups) | | Animalia | Fungi | Plantae |
| Cavalier-Smith, 2004; | 2 DOMAINS | PROKARYOTA (spans Archaea–Bacteria) | | EUKARYOTA (spans eukaryotes) | | | | |
| Cavalier-Smith, 2018 | 6 kingdoms | Archaea | Bacteria | Protozoa | Chromista | Animalia | Fungi | Plantae |
| Raoult et al., 2004 | **4 DOMAINS—including a new domain, giant viruses** | ARCHAEA | **BACTERIA** | EUKARYOTA (spans eukaryotes) | | | | |

a parasitic stage of their life cycle that is unicellular or syncytial (Canning & Okamura, 2004), but these are a bizarre group of derived cnidarians that have been unequivocally shown to be the descendants of multicellular animals (Siddall et al., 1995). We must rely on a phylogenetic definition of the animals—shared descent from the common ancestor of all animals, rather than a count of cells—and the two terms, Metazoa and Animalia, can be used interchangeably. (Confusingly, statutory legal definitions of animals may be restricted to tetrapod vertebrates, or all vertebrates, or extend to some larger non-vertebrates, but that is a separate matter.) Most comprehensive taxonomic treatments use Animalia (e.g., Ruggiero et al., 2015), but metazoan is often used as a vernacular term to emphasise a broader group than familiar vertebrates.

The Linnaean System does not impose intrinsic value judgements on organisms placed in different parts of the classification arrangement. As noted in Chapter 4, this is a remarkably progressive feature of the system, detaching the classification of species from issues of 'higher' or 'lower' placement. In Linnaeus's original work, vertebrate animals are arranged with the same hierarchical ranks, in species, genera, and on to phyla, and so are plants and so are worms (the 'everything else' group of animals he termed *Vermes*). The language some people use to describe nominally phylogenetic relationships, by contrast, often falls into the rhetorical trap of labelling organisms and traits as 'primitive' (Krell & Cranston, 2004).

## RELATIVISM AND REGULATION

A common question is how much difference is 'enough' difference to warrant separation at a given rank, from species, and genera, and so on. Different specialist groups do rely on particular criteria to justify a 'genus-level' or a 'class-level' separation for the taxa they work on. I could tell you that certain features of the radular tooth-structures in chitons are sufficient to separate genera. Why? Often, the explanation for such conventions is abbreviated to 'because we say so', which might give the unfortunate impression that both the characters and the taxonomic ranks are arbitrary. Features of the radula are highly conservative in chitons, and substantial changes in morphology seem to correlate with genetic divergences (Saito, 2004). In that group, we have data about the radula for far more species and individuals than we have DNA (at the time of writing, about 60% of chiton genera—not species—have sequences deposited in GenBank, and chitons are a reasonably well-known group by the standards of marine invertebrates). It is not really the radula itself, more that differences in the radula that are correlated with other important features used in taxonomic determinations. The combination is what is important, but the radula is a bit of a 'tell'. This feature is available for living species, but hardly ever preserved in fossils. Similarly, osteological features of vertebrates (including, by coincidence, dentition of mammals) can be used to identify and distinguish species groups because these features are things that all members of the group have in common, features that change more slowly than species lineages separate. Learning to distinguish features of the radula requires a lot of experience and expertise, and that one feature is useful for a particular class of molluscs does not guarantee its applicability to other molluscs, and other species do not have a radula, a uniquely molluscan dental apparatus. So, this level of technical specificity is naturally hard to articulate in general terms. Extending the concept of context dependence in species, higher ranked groupings, including genera, families, and upward are hypotheses about evolutionary relatedness. The main purposes of ranked taxonomy is to communicate, clearly, to the broadest possible audience, the context-dependent similarity of the species contained in the groups of interest. When we can say that a new discovery is not only a species new to science, but it represents a new genus or a new family, this clearly transmits the intangible idea of its uniqueness.

Species are not all separated by a universal metric; likewise, ranks are relativistic. Following this logic, a genus is a group of phylogenetically related species that are more closely related to each other than to anything outside the group. A family is a group of genera. As the assessment of similarity is relativistic, specific criteria are more or less applicable in different organisms. We acknowledge that different species, indeed different populations, evolve at differing rates and in response to a range of abiotic and biotic pressures. Logically, then, genera and other higher ranked groups are also variable entities, with more or less species that are more or less diverse. What matters is not our perception or value of genetic, morphological or ecological differences, but rather what matters to long term genetic exchange from the species' points of view. So, we have defined species based on relativity—a different reference frame—a comparison with other closely-related species. In fact, this comparative framework begins with a single specimen, the type material or the first individual specimen that was attributed to a particular species (Chapter 4). The whole framework rests on a comparative approach.

One objection raised against the Linnaean system is that ranks fail to transmit absolute meaning or a clear standard of evaluating some sort of distance or heuristic distinctiveness. For example, some researchers are concerned that a rank has no temporal equivalence, in terms of the amount of time that has passed since the existence of the common ancestor of a group (e.g., Avise & Johns, 1999; Hedges et al., 2015). An ambition to such a standard seems odd, because evolutionary rates are variable even within groups, and many groups do not have any fossil record, so time is not a particularly informative metric. The diversification of the brown seaweed family Fucaceae is dated to around 19 Mya (Silberfeld et al., 2010; Cánovas et al., 2011), a global family with about two dozen species. The origin of the antlion family Myrmeleontinae is estimated at around 135 Mya (Michel et al., 2017); it includes about 2000 species. Is it problematic to call both of these groups 'families'?

One alternative proposed that a temporal definition of 'family' would be limited to monophyletic groups that share a most recent common ancestor in the range 33–56 million years ago (Mya) (Avise & Johns, 1999). There is a significant fundamental problem that definitions relative to our current time slice suggest that the present is a *de facto* end point to evolution, ignoring the full, time-vertical extent of lineages that extend into the future (Chapter 5). Applying a set metric definition to taxa would communicate something about evolution, but it could also remove substantial implicit information about whether species of terrestrial insects and marine plants, or more closely comparable groups, are evolving in equivalent contexts, have equivalent generation times, or with similar speciation rates, extinction rates or selective pressures. In ranked taxonomy, and indeed in clades defined by phylogeny, each family is a reference frame based on its own species, not an absolute measure of diversity.

There is no *a priori* reason to expect temporal equivalence of ranked taxa; but it is an important caution about how named groups are combined or compared. Most people are familiar with mammals and birds. These are each considered a class-level rank. The two groups diverged around 340 million years ago in the Carboniferous period (Irisarri et al., 2017), and both comprise large radiations with about 16,000 living species between them. They are colloquially familiar animals that are demonstrably different. There are also two class-level ranked clades of worm molluscs, which diverged perhaps in the Ordovician (another 100 million years earlier than the split between mammals and birds), and each has a few hundred known living species all in marine environments (Sigwart et al., 2014). Solenogastres and Caudofoveata admittedly look nearly the same to anyone but their mothers (and they have no eyes) or a mollusc specialist. But I can explain their significance in 10 seconds to anyone by saying, these two animal groups are different taxonomic classes, so they are as different from each other as mammals are different from birds. This concept would be impossible to explain concisely without access to familiar, ranked classification: geological ages of clades might be useful, except that most people do not memorise divergence dates of even very familiar groups. Fur and feathers are familiar; the importance of a pedal groove or posterior gills to worm molluscs are likely harder to understand. These vertebrate and invertebrate classes have very different histories and contain different numbers of species, comprising different amounts of genetic, morphological or ecological diversity that have accumulated at different rates over their histories. Such disparities might seem to support the view that 'taxa of higher rank than species do not exist in the same sense as do species' (Eldredge & Cracraft, 1980). Each group of organisms seems to have a different standard or set of requirements to qualify for a higher rank, but this is explained by their different reference frames.

Rank definitions are relative to their constituent members. Consider the relative career trajectory and resultant responsibility and power of a flag officer in the US Navy, a war force of over 300,000 personnel, hundreds of ships and most of the world's aircraft carriers, compared to a flag officer of the Irish Naval Service, serving a nation that is constitutionally neutral with a fleet of 8 ships that mostly focusses on fisheries protection. Within a navy, personnel of different ranks have comparatively different leadership responsibility and experience (ensigns are junior to their captain) regardless of the size of the ship. Comparing between the two countries' navies, an American or Irish officer with the same rank would find the fleets under their command may be markedly different. 'Rank' in the military sense of increasing authority is rather different from the taxonomic application, but the analogy could tentatively be extended to note that increasing grade demonstrates increasing independence and decision-making responsibility (NATO, 2017).

Classification is not a unilateral bottom-up sorting of species into groups. This is both the power and the weakness of the Linnaean System; it allows us to communicate both the very broad distinctions (kingdom, phylum), and very fine (species and subspecies). In practice, the kingdoms and phyla are established in recognition of major distinctions and evolutionary separations, including unknown gaps caused by extinctions (Marshall, 2017). Classes represent subdivisions within a phylum. Thus, the lower three ranks (species, genus, family) are generally applied bottom-up, or by grouping species by similarities; the top three ranks (kingdom, phylum, class) are top-down, identified by the gaps between subgroups. And the middle level, the order, floats in the middle and its use is probably mainly as a bridge between 'higher' and 'lower' classification strata.

The seven classical ranks of the Linnaean System may or may not be sufficient to encapsulate the evolutionary diversity of any particular major branch in the tree of life. The problem with classification is not the system, but misunderstanding the most important audiences for the results of classification: non-specialists. Some phyla or divisions may be neatly expressed in seven levels, others may seem to need less. There are ample options for additional structure, or groupings can be named as unranked clades (Table 6.1). There are two practical corollaries: first, not all identifiable clades or aspects of a phylogenetic topology need to be named. A phylogenetic reconstruction is also a hypothesis and can change with additional data or analyses; naming should be preferentially restricted to stable, well established nodes. In a perfectly resolved tree there are many nodes that are probably not interesting to most people. Second, the primary ranks of the Linnaean System have universal acceptance in science and beyond, and so they should be applied to groupings of taxa that are not only stable, but significant to a broad audience.

## TREE THINKING

We all have an intuitive expectation that nested groupings reflect relatedness. In 'phylogenetic systematics' this is formalised and depends on a criterion of monophlyly to define species groups. When this system was first proposed by Hennig in the 1960s, it was radical and controversial. Now many aspects of this system have become normalised in systematics, in particular the basic use of coherent evolutionary lineages as the main criterion for defining species, and the primacy of monophyletic groups in defining higher ranked taxa.

Classification based on monophyletic groups becomes problematic when there is substantial uncertainty around the tree, the evolutionary relationships among the species or groups of interest. In most groups of organisms, this is chronic, for the simple reason that species outnumber the scientists who work on them (Clark & May, 2002). The notion that all relationships within a clade need to be perfectly resolved before any taxonomic changes are made is a matter of personal taste, but a handicap that can be avoided by accepting that science is incremental and names of species are working hypotheses.

The criteria required for naming a group of species at higher rank (above the genus level) have been articulated as (1) monophyly of the species group in phylogenetic analysis, (2) stability, or strong support for phylogenetic coherence of a group, through multiple lines of evidence, and (3) 'phenotypic diagnosability' (Vences et al., 2013). All three of these pillars are important and were applied well before the advent of phylogenetics; the idea of common origins, or homology, have always been fundamental to classification (Pachen, 1999). These criteria are ideals rather than prerequisites, but they involve assessing ranked taxa in a phylogenetic framework, and explicitly considering how a species are connected to their ancestors.

In the *Origin of Species*, Darwin included only a single figure, a branching diagram illustrating ancestor-descendant relationships (Figure 6.2). Such

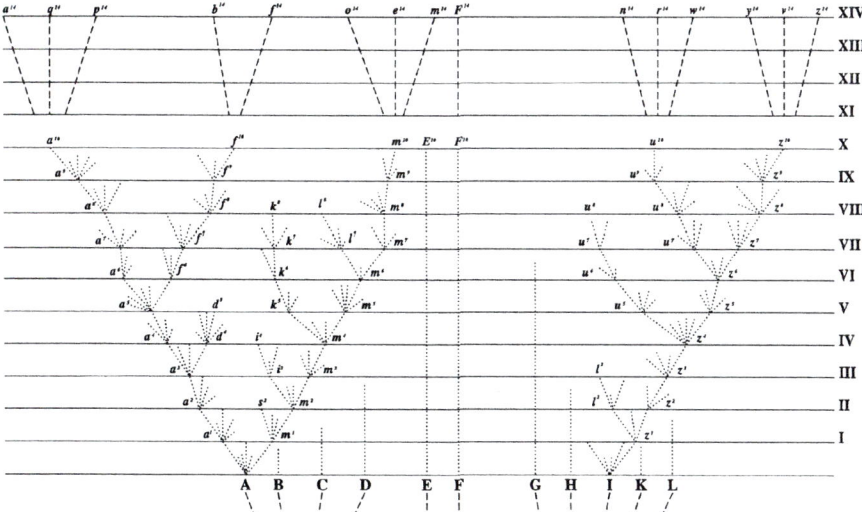

**FIGURE 6.2** The only figure in the *Origin of Species* (Darwin, 1859). In this tree, the vertical axis is a time axis and the majority of taxa (or individuals, it is never specified what the branches represent) go extinct.

branching diagrams are fundamental to the way that we understand species and, now, systematics. Much has been written about the importance of 'tree thinking' and also the importance of basic instruction about how to read and understand phylogenetic diagrams (Baum et al., 2005; see especially the instructive 'Thinking Quizzes' published as supplementary data to that paper). The lines and nodes in those diagrams represent species—population lineages—and speciation events (Hennig, 1966). While tree diagrams are familiar to most students of evolutionary biology, there are a few very basic points that bear repeating, as the structure of tree diagrams lends them to easy misinterpretation (Figure 6.3). First, this is a simplification, a schematic diagram, and we know that speciation processes are

**FIGURE 6.3** Two identical phylogenetic trees. The tree on the left is presented with the vertebrate groups in the traditional linear arrangement, from left to right. The tree on the right is identical in topology, the branches are rotated at the two nodes marked on both trees with black dots. No information aspect of the tree has changed.

not instantaneous or dichotomous (Chapter 5; Figure 5.4). Second, the structural arrangement of branches is important, not their order on the page. Just because a species is at the left or right side, or the top or the bottom of the page does *not* mean that it evolved first or last. Sets of branches can be 'rotated', or their order flipped, around any node without changing the relationships among the branches (Figure 6.3). Finally, there are many different ways to illustrate evolutionary trees; the branches could be calibrated to time, as implied in Darwin's tree, or scaled to show genetic changes, or simplified to show only the topology without indicating any other measure of separation among species (as in Figure 6.3). The phylogeny, or inter-relatedness, is the same in all cases, but like any form of scientific illustration, the style of diagram can be chosen to emphasise a particular message (Archibald, 2014).

Trees are composed of sets of branches, called clades, not just the species labelled on their tips. Relatedness of taxa is a matter of shared common ancestry; on phylogenetic diagrams this is indicated by the depth of their common ancestor (and not by the adjacency of tips on the tree). The branches that have the closest relationship to the root of the tree are called the 'basal branching' lineages or the 'earliest diverging' lineages. A clade is a group of branches that share a common ancestor, a monophyletic group. The idea of a 'clade' is deliberately ambiguous and has no taxonomic level or rank, it is only a statement about shared ancestry. So, Eukarya is a clade, and populations within a species may be clades. As discussed above, unranked clade names provide no additional information about the size of the group or evolutionary distance, and require reference to a tree diagram or phylogenetic explanation. And because the concept of a clade is intrinsically linked to phylogeny, for the most part, multi-species clades cannot be properly understood until a sufficiently complete phylogenetic analysis is available.

The core concept of phylogenetic systematics is that only clades, monophyletic groups, should be allowable as group names under formal classification. In the strictest interpretation of the principles of phylogenetic systematics, the criterion of 'phenotypic diagnosability' is limited to synapomorphies, homologous features inherited from a most recent common ancestor that unite the descendants. This all seems eminently reasonable. We want classification that is grounded in an evolutionary framework, and to find diagnostic characters that the shared evolutionary history of groups of species—the things they have in common based on ancestry, not accident or environment. Yet many specialists have highly emotional reactions to this idea. Not because they are rejecting a modern approach, but because, as usual in biology, things are more complicated than we might prefer. Monophyly as a good criterion seems uncontroversial; the problem is what to do with established taxonomic groups that are not monophyletic, and the eternal conflict between perfectionism and pragmatism.

## THE PROBLEM WITH PARAPHYLY (PART 1: GENERA)

If you put a perfectly resolved species tree in front of a systematist, it is not obvious how the species on that tree should be partitioned into genera. In part, there is the weight of history. Most people do not want to make unnecessary or unwarranted taxonomic revisions. Everyone wants these ranks to be biologically meaningful, that is, to have some character, morphology, behaviour, or even a gene that is a unique synapomorphy. This is the criterion of phenotypic diagnosability (Vences et al., 2013). To be *useful*—relevant to others, outside of the study of phylogeny—a species group should be identifiable, and there should be some feature or set of characters that can diagnose a new species as a member of that group. These diagnostic features in fact are more important to many people than phylogenetic accuracy, but the phylogenetic history of the group confers a predictive power based on shared evolutionary history.

The taxonomic group of our closest relatives, including our own species, the taxonomic tribe Hominini within the mammal family Hominidae, contains an instructive example of a taxonomic conundrum over genus names; this is the taxonomic group that all of us take personally, so it may also illustrate some of the heightened emotion that can accompany opinions about taxonomic revision. *Homo* is a small genus that contains one living species (us, *Homo sapiens* Linnaeus, 1758) and about six extinct species known only from fossils, and this group forms a stand-alone monophyletic clade of all the species descended from a common ancestor. Our other nearest relatives, slightly more distant fossil species of early humans, are classified in the extinct genus *Australopithecus* (Figure 6.4). There are important skeletal

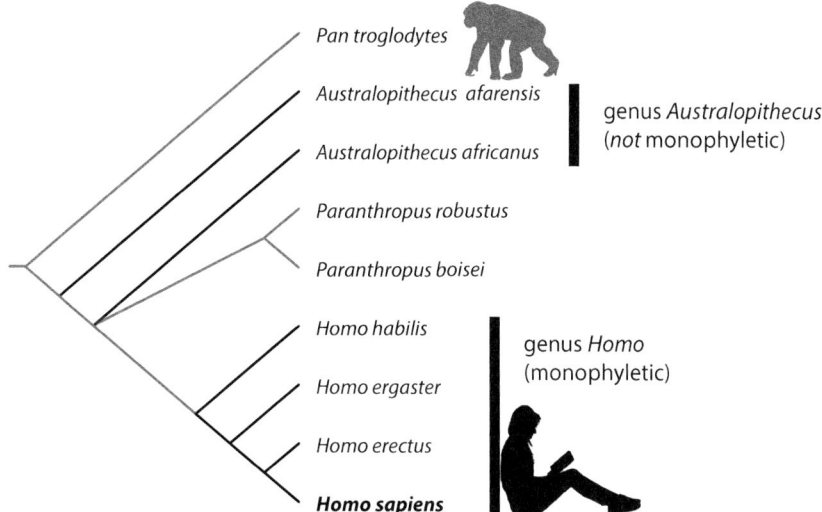

**FIGURE 6.4** Simplified tree of hominid evolution, in a selection of species in the fossil and living genera *Australopithecus, Paranthropus*, and *Homo*. (Phylogenetic reconstruction analysed under parsimony, matrix reanalysed from Dembo et al. (2015a, 2015b) using a subset of their taxa and only parsimony-informative characters.) Phylogenetic trees, like any form of scientific illustration, can be presented in various forms in order to emphasise a particular message or finding. This cladogram highlights the topology or branching patterns; there is no scale to indicate the timing of speciation or extinction events. Only *Homo sapiens* Linnaeus, 1758 and the chimpanzee *Pan troglodytes* (Oken, 1816) are living species, the rest are extinct early hominin species that had different temporal ranges. Species in the genus *Homo* form a **monophyletic** group (four species shown here, black tips that link to a common ancestor); the genus *Australopithecus* is **paraphyletic** (represented by only two of its species here).

characteristics that unite the species within *Australopithecus*, and these features also separate that genus from species in the genus *Homo* (Cela-Conde & Ayala, 2003). In formal parlance, *Australopithecus* spp. are united by hominin symplesiomorphies. That is, all the species in that genus retain primitive features of their common ancestor—that ancestor is also part of the more distant heritage of the genus *Homo*, but our genus has evolved new distinct features. The members of *Australopithecus* are all similar to each other and clearly different to us. The problem with this is that *Australopithecus* is not monophyletic, it is paraphyletic (Figure 6.4). Each species branches off from the main stem of early hominid evolution, there is no single evolutionary origin of the *Australopithecus* species group; *Australopithecus* is not a clade.

Paraphyletic or polyphyletic taxa (like the genus *Australopithecus*) are groups that cannot form an exclusive and complete monophyletic group on a phylogenetic tree. These non-monophyletic groups are rejected under the strict criteria for phylogenetic systematics. There are only two phylogenetically 'correct' solutions to resolve the genus *Australopithecus* so that its species are in monophyletic higher taxa: The first option is to name all lineages of *Australopithecus* each a separate genus, multiple genera with one species each. That seems like a lot of genera (there are several more species of *Australopithecus* not shown in Figure 6.4), but is there a good reason not to do that? We do not want to fall prey to a biased, emotional reaction, thinking those primitive failures in *Australopithecus* do not deserve to get a separate genus each to themselves (Skelton and McHenry, 1998). The second option is to extend the genus and include all the species in *Australopithecus* + *Homo* + *Paranthropus* in a single genus. That expanded version of the genus *Homo* would contain one living species (*Homo sapiens*) and about 16 fossil species (Villmoare, 2018). This is also considered unacceptable by experts.

Palaeoanthropologists present an entirely rational argument for maintaining *Australopithecus* as a paraphyletic genus: all these *Australopithecus* species are morphologically similar. It is a group with clear diagnosis based on evolutionarily relevant features. Both 'correct' alternative classifications would decrease the information content of the names. If all of the *Australopithecus* species were classified in new genera, the additional names would imply that there is as much dissimilarity separating those ex-*Australopithecus* genera as there is from any one of them compared to all the species in *Homo*. This is not true, they share inherited similarities, so there is valid reason for concern about confusion. *Australopithecus* and *Homo* and *Paranthropus* each have distinctive morphological features; any newly discovered fossil remains could be diagnosed based on those features and put into one of those three genera. There is an unusually high level of interest in new fossil hominids, but there is nonetheless clear morphological separation among different lineages within the clade. The purpose in giving names to species groups is to communicate their evolutionary similarity.

For the time being, the genus name *Australopithecus* remains paraphyletic. Additional data, such as a future discovery of a new fossil lineage might prompt reconsideration, perhaps a distinct cluster of lineages among the *Australopithecus* species could be elevated to a new genus. Future revisions could solve the 'problem' of a paraphyletic genus, but importantly that would be driven by a need to communicate new information about the relative evolutionary and biological similarities among the relevant species.

Paraphyletic groups or phylogenetic 'grades' do share common ancestry, so there is a common evolutionary origin to the contained lineages and an evolutionary explanation for their shared similar features. The problem with paraphyly is that it reflects the real, untidy nature of species evolution (Hörandl & Stuessy, 2010; Seifert et al., 2016). And here is another important point about paraphyly: time. *Australopithecus* was a perfectly fine, monophyletic clade until the ancestor of *Homo* evolved into something quite different, splitting into several different species on the way. Should the shared ancestry and homologous features of *Australopithecus* become an inadequate diagnosis, just because one of their descendent lineages went off in a very different direction? Although all but one of the species in the Hominini are now extinct, the same patterns play out in many groups of living as well as fossil species.

Palaeoanthropologists have a wealth of data about species in Hominini; despite relatively little specimen material of the fossil species, the members of this clade are far better studied than any other living group of organisms. *Australopithecus* remains paraphyletic for a specific reason, whether or not you agree, it is a well-reasoned argument about communicating accurate relevant information through nomenclature. Evolution in different species lineages within a clade may be progressing at very different rates, especially where these species are dispersed into separate environments. There is strong evidence that the rate of evolution in *Homo sapiens* was accelerated synergistically with our development of tools and technology, which enabled our ancestors to colonise the whole planet (Zimmer, 2010). With or without technology, change in one lineage in a clade does not cause every other related species around the world to change too.

There are several informative points to be learned from this example. Foremost among these is that phylogenetic systematics—a system where higher ranked names are restricted to monophyletic clades—is optional (Nordal & Stedje, 2005). We consider it ideal, and especially where phylogeny is well resolved, evolutionary biologists want to rationalise the *tree* with the *names*. We do this because we think it helps the names to carry additional information and increase their predictive power; this is the primary reason to avoid non-monophyletic names (Ward et al., 2016). Nomenclatural rules long pre-date any concept of phylogeny, so a rule about phylogenetic systematics has never been written into the codes of nomenclature; but invoking the long history of taxonomy would be an overly simplistic dismissal of the issue. The rules are constantly updated, and any revision after 1966 might have included guidance in the relevant codes of nomenclature about resolving names as clades (see Chapter 4; ICZN, 1999; ICN: McNeill et al., 2012). It is not always desirable to demand absolute lockstep of phylogeny and nomenclature.

To reiterate these points in a more abstract example, the species within a paraphyletic genus are very likely not to have any good characters to support separating them, because they are more similar than they are different (Figure 6.5). Paraphyly is equally common in living species (or species co-occurring in any time slice) as it is among an assemblage of fossil and extant species such as hominid taxa. That overwhelming similarity among a paraphyletic assemblage is itself an interesting point about evolution; the grouping has retained symplesiomorphies while the descendent clade that 'breaks' its monophyly has some other additional synapomorphies. The cartoon example illustrates omniscient knowledge of the phylogeny of cartoon shrubbery; in

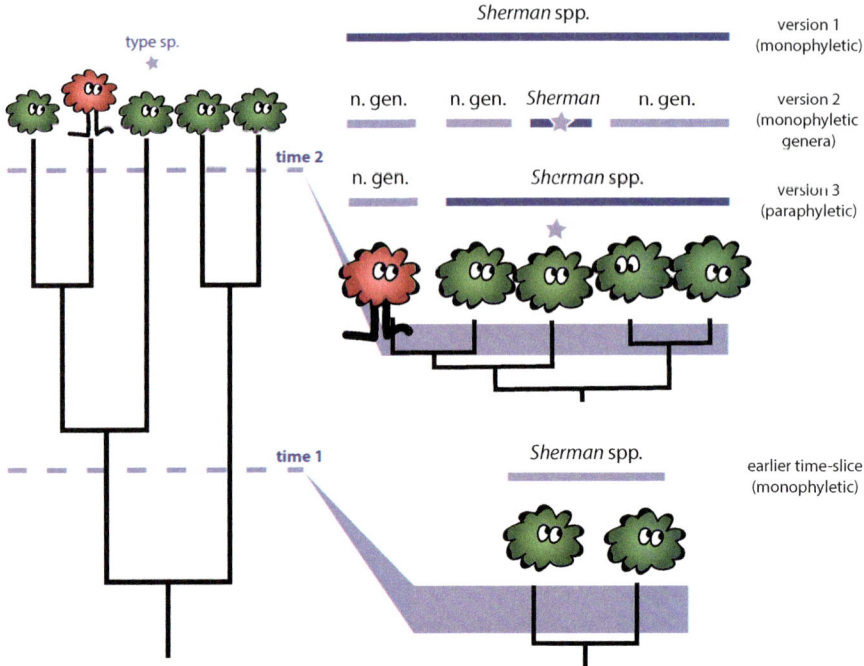

**FIGURE 6.5** Paraphyly in a cartoon phylogeny of seemingly similar shrubbery, *Sherman* spp. The complete tree at top is redrawn as sampled at two different time points (time 1, bottom right; time 2, top right). Divergence of new additional species lineages at time 2 renders the genus paraphyletic, though all species but one are united by symplesiomorphies (they share the ancestral morphology seen in species that were extant at time 1). There are three possible approaches to revising the taxon *Sherman*, shown in the vertical bars—(1) the original (time 1) diagnosis of *Sherman*, excluding the derived form; (2) splitting the genus into monophyletic units, erecting two new genera (n. gen.) for those other than the type species; and (3) extending the diagnosis of an otherwise morphologically conservative *Sherman* to include shrubbery with legs. As a thought experiment, consider how this classification should be approached if species at time 1 represent fossils, and time 2 are extant species, or by contrast, if time 1 is the present, and time 2 are future descendants.

an earlier time point, there are two species in a genus, run the clock forward through some additional diversification, and at a later time there are more species that look superficially similar but one has adapted appendages. Lumping all species into an expanded definition of the genus fails to communicate that one of these is really very different; splitting would exaggerate the differences among most of them.

Some taxon names are explicitly and horrendously non-phylogenetic. 'Wastebucket' taxa are sometimes works in progress (Chapter 4), although this is not obvious to non-experts, and some are accidents of history while others are deliberate. The terrestrial flatworm 'collective genus' *Australopacifica* was established as a genus to signify inadequate descriptions, where specimens can be identified to family but there is insufficient data to assign it to a genus (Ogren & Kawakatsu, 1991). Thus, it is a genus name that has some meaning but cannot be revised to form a monophyletic clade.

Decisions about classification are left to taxonomic experts to argue about their own organisms; perhaps future workers will find it useful to include phylogenetic data in systematic descriptions. In the planarian literature, taxonomists sometimes refer to 'genus *Australopacifica* (collective group)'. Likewise, paraphyletic-by-choice taxa could be noted, for example, 'genus *Leptochiton* (paraphyletic)'. While phylogenetics is important, the practical application of taxonomy usually takes precedence: we need to be able to identify things. Among well studied groups—human fossils, for example, or many other vertebrate groups—the rate of discovery of new species is relatively low, and research focusses on inter-relationships more than species discovery. In most organisms—invertebrate animals, plants, seaweeds—the rate of discovery remains very high. To those who are just trying to figure out the correct order of magnitude for how many species there are, perfecting the organisation of nomenclature sounds like a 'first-world problem' (Chapter 12). At the genus level, the criterion to be 'phylogenetically meaningful' (Vences et al., 2013) is not synonymous with monophyly, and the priorty for nomenclature should be heuristic diagnosis of relevant groups.

## THE APPLICATION OF RANKS

Hennig (1966) argued that allowing paraphyletic groups in classification would undermine synthetic analysis of biodiversity patterns. Including an unrestricted combination of monophyletic and non-monophyletic taxa in analysis means we may not be comparing like with like. The counts of genera or families in different regions, or at different time points, carry an implicit assumption that a 'genus' or a 'family' has transferable biological or phylogenetic meaning. To date, the influence of paraphyly, or any non-monophyletic assemblages in that type of analysis has never been directly tested. But this argument is based on an explicit motivation that taxonomy should create comparable groups, which returns to the idea that species groups at a particular rank should have an absolute definition or criteria, which is not true in practice or in theory. The good news is, phylogeny can be relied on to bring signal out of the chaos even when we are not explicitly using it.

In the early 20th century, Yule (1925) reported that the size-frequency distribution of genera was similar in many different groups. In groups from plants to beetles, small genera with few species are common, while large genera are comparatively rare (Figure 6.6). The hollow curve or right-skew distribution follows the same type of natural patterns as body size (small organisms are common, large organisms are less abundant), and economics (the skewed distribution city sizes, with a few large cities and many small towns, corporation sizes, or personal wealth). Many other people have rediscovered this pattern in genus sizes, but the pattern was frequently dismissed as some kind of human-induced phenomenon—though it was never clear how different tendencies for lumping or splitting by individual taxonomists could produce a consistent pattern across all organismal groups (Williams & Gaston, 1994; Strand & Panova, 2015). Simulation approaches have now shown that idealised or omniscient, rule-based systematic classifications applied to simulated phylogenetic trees consistently produce the same hollow curve—the skew distribution in genus size is an emergent mathematical property of species phylogeny (Sigwart et al., 2018).

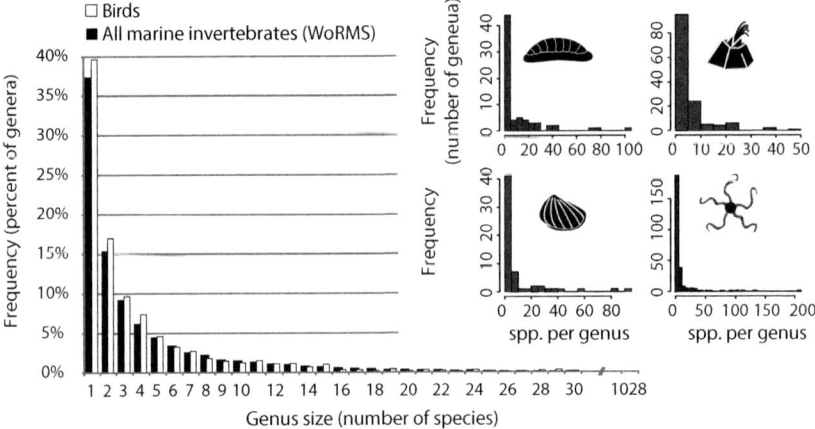

**FIGURE 6.6** Around one-third of genera are monotypic (containing only one species), in all animal groups. But monotypic genera represent only a small fraction of species richness in each clade. This pattern is similar to skew distributions from economic theory. Average 'species per genus' has been used as an estimator of total diversity, or to compare the diversity of different environments. But the central tendency (mean) is always lower than would be intuitively expected.

We are intellectually fascinated with large genera and the processes that lend meaning and order to large radiations (Rabosky et al., 2013), but small genera are the normal product of speciation processes. There are many evolutionary pathways to a small genus; it may have split off from a larger clade or represent a lineage that persisted unchanged for a long time or be the last standing member when all related lineages succumb to extinction. In simulations, and in real-world taxonomic data, genera with only one species are the largest fraction of genera (Figure 6.6). Large genera, by contrast, can only arise through rapid, sustained bursts of speciation. If too much time passes between speciation bursts, new evolutionary distance pushes subgroups apart, decreasing the maximum size of the genus group. So, there are many ways to become a small genus, which are consequently common, and the circumstances required to produce a large genus are comparatively rare. This skew distribution is also a property of higher ranked taxa and larger more inclusive clades (Gaston & Williams, 1999; Ricotta et al., 2002; Chapter 10; Figure 10.4). The overall global patterns in the size of species groups (genera, families) as currently known actually reflect patterns that would be expected from phylogeny, even though in many groups the taxonomy is incomplete, contains unresolved paraphyly, and may be limited by a significant discovery gap.

## THE PROBLEM WITH PARAPHYLY (PART 2: A HIGHER PROBLEM)

The next rank above the genus, the family, represents a group of genera that are closer to each other than to anything outside the group. A family-level group *de facto* comprises additional, accumulated, evolutionary change, more than in the individual genera it contains. During the course of this accumulated change, there will have been extinction and speciation events that are not uniformly distributed across the taxa in the family. The scope of this evolutionary separation should create separate, diagnosable groups. Thus, while paraphyletic genera are broadly unavoidable, and not necessarily problematic, the same exemption is not easily extended to families or higher groups.

While paraphyly may be acceptable in some cases at the genus level, there is a widespread consensus that non-monophyletic taxa are absolutely not acceptable at higher ranks. At the top end of the phylogenetic spectrum, the idea of paraphyletic kingdoms or phyla is actually repulsive to most biologists. (This has created chaos among microscopic life forms, including the 'protists', which fall outside the scope of this book, but see some of the references cited in Table 6.2.) When large unnatural groups are labelled under a name that has no evolutionary cohesion, this leads to scientific ideas that are uninformative, not predictive, and can reinforce biased views about superficial similarity.

The majority of metazoan diversity is colloquially referred to as the 'invertebrates', a gigantic paraphyletic assemblage of millions of species defined only by the common ancestor of all animals on the one hand, and on the other hand the exclusion of one objectively minor clade, the vertebrates. This has never been used as a formal taxonomic classification—even Linnaeus (1758) separated the Insecta from the Vermes. Gould (2000) traced the early history of formal invertebrate classification in the 50 years after *Systema Naturae*. In 1793, Jean Baptiste Lamark separated vertebrates from non-vertebrates, uniting Linnaeus's separate classes of chordates as distinct from animals without backbones (Satoh et al., 2014). By 1809, Lamarck had formally recognised ten invertebrate animal phyla, but these were interpreted as progressive series of forms from least to most vertebrate-like (Gould, 2000). Although that legacy of 'progressive' evolution still haunts biology (Chapter 5), 200 years ago, Lamarck (1815), no doubt influenced by the comparative anatomical approaches of Frédéric Cuvier (b. 1773–d. 1838), began to realise that the classification of invertebrates could only be managed through sprawling arrangements of nested, branching patterns (Kutschera, 2011). Invertebrates comprise some 97% or more of animal species, so it should be unsurprising that their morphological, ecological, developmental, genetic, and physiological diversity outstrip the capacity of vertebrates. Within vertebrates, 'fish' also represents a number of major clades, both extinct and extant, that form a paraphyletic grade within the phylogeny of vertebrate animals. Above the genus level (and also for many paraphyletic genera), naming these groups creates an inaccurate impression that the group is homogeneous. Naming paraphyletic higher taxa underestimates diversity and reinforces preconceptions about the organisms as 'primitive'.

Fish, invertebrates, bryophytes, and many others, are examples of informal groupings that are no longer formally used or acknowledged by taxonomists, because they are not monophyletic clades. But these names remain in the names of scholarly societies, journals, and field guides. Some of these names persistently appear in the literature, used inadvertently when an author follows the names as used in older textbooks. Naming these groups implies that they have significant unifying characteristics, and they do. However, it is inherited symplesiomorphies—shared ancestral characters—that define them. These groups each contain multiple radiations in their own right, with interesting adaptations that often push the boundaries of the capability of life itself. Defining such groups based on shared ancient characteristics puts the emphasis on their supposedly primitive natures. It is a potential way to belittle less-valued lineages in comparison to more 'advanced' groups. The critical question is: Are we keeping these groups together because it is biologically relevant and informative, or because of a historical bias that they are less diverse, or more economically relevant or familiar than a later-derived clade (e.g., tetrapods over fish, vertebrates over invertebrates, or angiosperms over bryophytes)?

## RANKS AND COMMUNICATION

Evolutionary trees are complex, and the number of internal nodes in a phylogeny is directly dependent on the resolution of the tree and the number of taxa included. Can this structure be resolved with traditional ranks from the Linnaean System? In the phylogeny that separates a species from the common ancestor of its kingdom, there are many more than seven nodes or branching-points, and probably more than seven nested groupings with diagnostic features. This is equally true in the phylogenetic history that separates an anemone, *Anemonia viridis* (Forsskål, 1775), from the common ancestor of all animals, or an anemone, *Anemone coronaria* Linnaeus, 1753, from the common ancestor of all plants. Indeed, the more species that are discovered, the more structure we understand in the tree of life. Not all of this structure is useful to a broad audience, and not all nodes require names. The standard Linnaean hierarchy has an admittedly arbitrary number of levels, and they are not a full reflection of phylogeny; there is no biological magic to seven ranks but these familiar and flexible group names are a crucial communication tool.

The deepest and most transferable information about organisms is an encoded record of evolution, and relative relationships to other similar things. We want to know about kinship for many reasons. Kinship carries the implication that an observation on one specimen or one species can be applied with some confidence to the rest of the group because they are inter-related. The fact that many different groups of fishes have their own independent evolutionary histories is relevant to appreciating fish diversity, and also understanding that, for example, we cannot make sweeping generalisations about fictional idealised fish features as relevant to the primitive condition of early tetrapod vertebrates. In more proximate groupings, like genera, accurate and meaningful information about kinship may not require such absolute adherence to monophyly.

Expressing phylogeny is not the sole purpose of taxonomy, and ranked groups are useful. We often cannot identify things to species level in the field. Identifying material in marine plankton samples, 'I got it to *phylum*' is tongue-in-cheek triumph. But, usually, with some practice and access to reference materials, you can confidently get something to family or genus level. There may be a fixed ceiling to how much identification can be refined by technology; it is the impractical sequence every individual of every species that you observe. (Even with a future technique for rapid sampling of environmental DNA, is there ever a scenario that would accommodate every birder zapping a rare bird, when hundreds of people flock to see a rarity?) Data limitation is an intrinsic fact of biological observation. This is equally true for fossils, where species level characters are sometimes simply not preserved. Nested additional levels—groups of species, groups of genera, and so on, with suites of morphological or genetic characters that distinguish them—are additional levels of resolution that are necessary to make sense of biodiversity.

## REFERENCES

[APG] Angiosperm Phylogeny Group. 2016. An update of the Angiosperm Phylogeny Group classification for the orders and families of flowering plants: APG IV. *Botanical Journal of the Linnean Society*. 181: 1–20.

Archibald DJ. 2014. *Aristotle's Ladder, Darwin's Tree: The Evolution of Visual Metaphors for Biological Order*. Columbia University Press.

Avise JC, Johns GC. 1999. Proposal for a standardized temporal scheme of biological classification for extant species. *Proceedings of the National Academy of Sciences*. 96: 7358–63.

Baum DA, Smith SD, Donovan SS. 2005. The tree-thinking challenge. *Science*. 310: 979–80.

Bouchet P, Bary S, Héros V, Marani G. 2016. How many species of molluscs are there in the world's oceans, and who is going to describe them? *Tropical Deep-sea Benthos*. 29: 9–24.

Canning EU, Okamura B. 2004. Biodiversity and evolution of the Myxozoa. *Advances in Parasitology*. 56: 43–133.

Cánovas FG, Mota CF, Serrão EA, Pearson GA. 2011. Driving south: A multi-gene phylogeny of the brown algal family Fucaceae reveals relationships and recent drivers of a marine radiation. *BMC Evolutionary Biology*. 11: 371.

Cavalier-Smith T. 1981. Eukaryote kingdoms: Seven or nine? *Biosystems*. 14: 461–81.

Cavalier-Smith T. 2004. Only six kingdoms of life. *Proceedings of the Royal Society B*. 271: 1251–62.

Cavalier-Smith T. 2018. Kingdom Chromista and its eight phyla: A new synthesis emphasising periplastid protein targeting, cytoskeletal and periplastid evolution, and ancient divergences. *Protoplasma*. 255: 297–357.

Cela-Conde CJ, Ayala FJ. 2003. Genera of the human lineage. *Proceedings of the National Academy of Sciences*. 100: 7684–9.

Chatton E. 1937. *Titres et travaux scientifiques*. Sottano.

Clark JA, May RM. 2002. Taxonomic bias in conservation research. *Science*. 297: 191–2.

Copeland HF. 1956. *The Classification of Lower Organisms*. Pacific Books.

Darwin CR. 1859. On the Origin of Species by Means of Natural Selection, or the Preservation of Favoured Races in the Struggle for Life. John Murray, publishers.

Dembo M, Matzke NJ, Mooers AØ, Collard M. 2015a. Bayesian analysis of a morphological supermatrix sheds light on controversial fossil hominin relationships. *Proceedings of the Royal Society B*. 282: 20150943.

Dembo M, Matzke NJ, Mooers AØ, Collard M. 2015b. Data from: Bayesian analysis of a morphological supermatrix sheds light on controversial fossil hominin relationships. *Dryad Digital Repository*. https://doi.org/10.5061/dryad.5025v.2

Dyke GJ, Sigwart JD. 2005. The search for a 'smoking gun': No need for an alternative to the Linnaean System of classification. In: Minelli A (editor). *Animal Names*, pp. 49–65. Istituto Veneto di Scienze, Lettere ed Arti.

Eldredge N, Cracraft J. 1980. *Phylogenetic Patterns and the Evolutionary Process. Method and Theory in Comparative Biology*. Columbia University Press.

Ereshefsky M. 1997. The evolution of the Linnaean hierarchy. *Biology & Philosophy*. 12: 493–519.

Fraknoi A. 2007. How fast are you moving when you are sitting still? *The Universe in the Classroom*. 71: 1–5.

Gaston KJ, Williams PH. 1999. Mapping the world's species—the higher taxon approach. *Biodiversity Letters*. 1: 2–8.

Giribet G, Hormiga G, Edgecombe GD. 2016. The meaning of categorical ranks in evolutionary biology. *Organisms Diversity & Evolution*. 16: 427–30.

Gould SJ. 2000. A tree grows in Paris: Lamarck's division of worms and revision of nature. In: Gould SJ (editors). *The Lying Stones of Marrakech*, pp. 115–43. Harvard University Press.

Haeckel EHPA. 1866. *Generelle Morphologie der Organismen*. G. Reimer.

Haeckel EHPA. 1874. *Anthropogenie; oder, Entwickelungsgeschichte des Menschen*. Eilhelm Engelmann.

Hedges SB, Marin J, Suleski M, Paymer M, Kumar S. 2015. Tree of life reveals clock-like speciation and diversification. *Molecular Biology & Evolution*. 32: 835–45.

Hennig W. 1966. *Phylogenetic Systematics*. In Zangerl R (editor). 1999. *Phylogenetic Systematics*, translated by Davis DD and Zangerl R. University of Illinois Press.

Hörandl E, Stuessy TF. 2010. Paraphyletic groups as natural units of biological classification. *Taxon*. 59: 1641–53.

[ICZN] International Commission on Zoological Nomenclature. 1999. *International Code of Zoological Nomenclature*, Fourth Edition. International Trust for Zoological Nomenclature, London.

Irisarri I, Baurain D, Brinkmann H, Delsuc F, Sire JY, Kupfer A, Petersen J, Jarek M, Meyer A, Vences M, Philippe H. 2017. Phylotranscriptomic consolidation of the jawed vertebrate timetree. *Nature Ecology & Evolution*. 1: 001370.

Just J, Kristensen RM, Olesen J. 2014. *Dendrogramma*, new genus, with two new non-bilaterian species from the marine bathyal of southeastern Australia (Animalia, Metazoa incertae sedis)–with similarities to some medusoids from the Precambrian Ediacara. *PLoS ONE*. 9: e102976.

Krell FT, Cranston PS. 2004. Which side of the tree is more basal? *Systematic Entomology*. 29: 279–81.

Kutschera U. 2011. From the scala naturae to the symbiogenetic and dynamic tree of life. *BMC Biology Direct*. 6: 33.

Lamarck JBABM de. 1815. *Histoire Naturelle des Animaux sans Vertèbres*. Verdière.

Lambertz M, Perry SF. 2015. Chordate phylogeny and the meaning of categorial ranks in modern evolutionary biology. *Proceedings of the Royal Society B*. 282: 20142327.

Laudan R. 1987. *From Mineralogy to Geology: The Foundations of a Science, 1650–1830*. University of Chicago Press.

Linnaeus, C. 1735. *Systema Naturae: Sive regna tria naturæ systematice proposita per classes, ordines, genera, & species*. Johann Wilhelm de Groot for Theodor Haak.

Linnaeus, C. 1758. *Systema Naturae. per regna tria naturæ, secundum classes, ordines, genera, species, cum characteribus, differentiis, synonymis, locis. Tomus I. Editio decima*. Salvius.

Marshall CR. 2017. Five palaeobiological laws needed to understand the evolution of the living biota. *Nature Ecology & Evolution*. 1: 00165.

McNeill J, Barrie FR, Buck WR, Demoulin V, Greuter W, Hawksworth DL, Herendeen PS, Knapp S, Marhold K, Prado J, Prud'homme Van Reine WF (editors). 2012. *International Code of Nomenclature for algae, fungi and plants (Melbourne Code) adopted by the Eighteenth International Botanical Congress Melbourne, Australia, July 2011*. Regnum Vegetabile 154.

Michel B, Clamens AL, Béthoux O, Kergoat GJ, Condamine FL. 2017. A first higher-level time-calibrated phylogeny of antlions (Neuroptera: Myrmeleontidae). *Molecular Phylogenetics & Evolution*. 107: 103–16.

Morrow C, Cárdenas P. 2015. Proposal for a revised classification of the Demospongiae (Porifera). *Frontiers in Zoology*. 12: 7.

[NATO] North Atlantic Treaty Organization. 2017. Annex A. In: Scaparotti CM, Mercier D (editors). *NATO Non-Commissioned Officer (NCO) Bi-Strategic Command Strategy and NCO Guidelines. Memorandum SH/CSEL/DP/17- 317925.* Supreme Headquarters Allied Powers Europe.

Nielsen C. 2011. *Animal Evolution: Interrelationships of the Living Phyla.* Oxford University Press.

Nordal I, Stedje B. 2005. Paraphyletic taxa should be accepted. *Taxon.* 54: 5–6.

Ogren RE, Kawakatsu M. 1991. Index to species of the family Geoplanidae (Turbellaria, Tricladida, Terricola). Part II: Caenoplaninae and Pelmatoplaninae. *Bulletin of Fuji Women's College.* 29: 25–102.

O'Hara TD, Hugall AF, MacIntosh H, Naughton KM, Williams A, Moussalli A. 2016. *Dendrogramma* is a siphonophore. *Current Biology.* 26: R457–8.

Pachen AL. 1999. Homology—History of a concept. In: Bock GR, Cardew G (editors). *Homology. Proceedings of the Symposium on Homology*, pp. 5–23. Novartis Foundation, London, 1998. Wiley.

Peters RH. 1986. *The Ecological Implications of Body Size.* Cambridge University Press.

Rabosky DL, Santini F, Eastman J, Smith SA, Sidlauskas B, Chang J, Alfaro ME. 2013. Rates of speciation and morphological evolution are correlated across the largest vertebrate radiation. *Nature Communications.* 4: 001958.

Rampino RM. 2017. *Cataclysms: A New Geology for the Twenty-First Century.* Columbia University Press.

Raoult D, Audic S, Robert C, Abergel C, Renesto P, Ogata H, La Scola B, Suzan M, Claverie JM. 2004. The 1.2-megabase genome sequence of *Mimivirus. Science.* 306: 1344–50.

Ricotta C, Ferrari M, Avena G. 2002. Using the scaling behaviour of higher taxa for the assessment of species richness. *Biological Conservation.* 107: 131–3.

Rothfels CJ, Li FW, Sigel EM, Huiet L, Larsson A, Burge DO, Ruhsam M, Deyholos M, Soltis DE, Stewart CN Jr, Shaw SW, Pokorny L, Chen T, dePamphilis C, DeGironimo L, Chen L, Wei X, Sun X, Korall P, Stevenson DW, Graham SW, Wong GK, Pryer KM. 2015. The evolutionary history of ferns inferred from 25 low-copy nuclear genes. *American Journal of Botany.* 102: 1089–107.

Ruggiero MA, Gordon DP, Orrell TM, Bailly N, Bourgoin T, Brusca RC, Cavalier-Smith T, Guiry MD, Kirk PM. 2015. A higher level classification of all living organisms. *PLoS ONE.* 10: e0119248.

Saito H. 2004. Phylogenetic significance of the radula in chitons, with special reference to the Cryptoplacoidea (Mollusca: Polyplacophora). *Bollettino Malacologico.* 39(suppl 5): 83–104.

Satoh N, Rokhsar D, Nishikawa T. 2014. Chordate evolution and the three-phylum system. *Proceedings of the Royal Society B.* 281: 20141729.

Schuh RT. 2003. The Linnaean system and its 250-year persistence. *The Botanical Review.* 69: 59–78.

Schulz F, Yutin N, Ivanova NN, Ortega DR, Lee TK, Vierheilig J, Daims H, Horn M, Wagner M, Jensen GJ, Kyrpides NC. 2017. Giant viruses with an expanded complement of translation system components. *Science.* 356: 82–5.

Seifert B, Buschinger A, Aldawood A, Antonova V, Bharti H, Borowiec L, Dekoninck W, Dubovikoff D, Espadaler X, Flegr J, Georgiadis C. 2016. Banning paraphylies and executing Linnaean taxonomy is discordant and reduces the evolutionary and semantic information content of biological nomenclature. *Insectes Sociaux.* 63: 237–42.

Siddall ME, Martin DS, Bridge D, Desser SS, Cone DK. 1995. The demise of a phylum of protists: Phylogeny of Myxozoa and other parasitic Cnidaria. *Journal of Parasitology.* 1: 961–7.

Sigwart JD, Sutton MD, Bennett KD. 2018. How big is a genus? Towards a nomothetic systematics. *Zoological Journal of the Linnean Society.* 183: 237–52.

Sigwart JD, Todt C, Scheltema AH. 2014. Who are the 'Aculifera'? *Journal of Natural History.* 48: 2733–7.

Silberfeld T, Leigh JW, Verbruggen H, Cruaud C, De Reviers B, Rousseau F. 2010. A multi-locus time-calibrated phylogeny of the brown algae (Heterokonta, Ochrophyta, Phaeophyceae): Investigating the evolutionary nature of the 'brown algal crown radiation'. *Molecular Phylogenetics & Evolution.* 56: 659–74.

Sirenko B. 2006. New outlook on the system of chitons (Mollusca: Polyplacophora). *Venus.* 65: 27–49.

Skelton RR, McHenry HM. 1998. Trait list bias and a reappraisal of early hominid phylogeny. *Journal of Human Evolution.* 34: 109–13.

Slowinski JB. 1995. A phylogenetic analysis of the New World coral snakes (Elapidae: *Leptomicrurus, Micruroides,* and *Micrurus*) based on allozymic and morphological characters. *Journal of Herpetology.* 29: 325–38.

Stearn WT. 1959. The background of Linnaeus's contributions to the nomenclature and methods of systematic biology. *Systematic Zoology.* 8: 4–22.

Strand M, Panova M. 2015. Size of genera–biology or taxonomy? *Zoologica Scripta.* 44: 106–16.

Vences M, Guayasamin JM, Miralles A, De La Riva I. 2013. To name or not to name: Criteria to promote economy of change in Linnaean classification schemes. *Zootaxa.* 3636: 201–44.

Vermeij GJ. 2016. The limpet form in gastropods: evolution, distribution, and implications for the comparative study of history. *Biological Journal of the Linnean Society.* 120: 22–37.

Villmoare B. 2018. Early *Homo* and the role of the genus in paleoanthropology. *American Journal of Physical Anthropology.* 165: 72–89.

Ward PS, Brady SG, Fisher BL, Schultz TR. 2016. Phylogenetic classifications are informative, stable, and pragmatic: The case for monophyletic taxa. *Insectes Sociaux.* 63: 489–92.

Wheeler QD. 2004. Taxonomic triage and the poverty of phylogeny. *Philosophical Transactions of the Royal Society B.* 359: 571–83.

Whittaker RH. 1969. New concepts of kingdoms of organisms. *Science.* 163: 150–60.

Williams PH, Gaston KJ. 1994. Measuring more of biodiversity: Can higher-taxon richness predict wholesale species richness? *Biological Conservation.* 67: 211–7.

Woese CR, Fox GE. 1977. Phylogenetic structure of the prokaryotic domain: The primary kingdoms. *Proceedings of the National Academy of Sciences.* 74: 5088–90.

Woese CR, Kandler O, Wheelis ML. 1990. Towards a natural system of organisms: Proposal for the domains Archaea, Bacteria, and Eucarya. *Proceedings of the National Academy of Sciences.* 87: 4576–9.

Yule GU. 1925. A mathematical theory of evolution, based on the conclusions of Dr. J. C. Willis, F.R.S. *Philosophical Transactions of the Royal Society B.* 213: 21–87.

Zimmer C. 2010. *The Tangled Bank.* Roberts and Company.

# 7 Are Species Real?

## BACKGROUND

This is the essence of the species problem: first, I will tell you with absolute certainty that there are millions of species on Earth right now that are undiscovered, undescribed, and unknown. I will also tell you with equal confidence that species are not fixed entities made by an omniscient creator and put on the Earth, each in its correct place. So how did they get there, and what makes them different? Also, are they all really different? In some groups, one thing fades into the next, while other species seem to be starkly different from their nearest relatives. How much of that blurriness is our inability to identify things, and how much is an incomplete boundary between lineages? Most species 'users' want to get on with other questions and we all want a simple *answer*. Unfortunately, this chapter only provides some explanation about the problem and the reasons the answer cannot be simple.

The 'species problem', in brief, is the idea that using different lines of evidence to identify species can lead to different conclusions about how many species there are. That different approaches to classification can produce different results extends to a more profound issue. These apparently irresolvable arguments over how to assign individuals to species seems to undermine the idea of species as real entities. However, this confuses criteria for identifying species with the meaning of species themselves. The idea of an ultimate authority, deciding the correct number of species and their identities, is antithetical to the iterative, cyclical nature of scientific hypothesis testing.

There are utilitarian and also epistemological problems with conscribing species units. Another embedded point in this debate that can cause some anxiety is the nature of what qualifies as 'real'. Fundamentally, reality comprises things that actually exist, as opposed to idealised constructs or imagined relationships. So, there is no question that individual organisms exist and are real, but one philosophical concern is whether the patterns that group these individuals into units that we call 'species' represent a phenomenon that comes from the natural world or an imagined order that we invented to impose onto the chaos of nature. A large part of this problem comes from assumptions about what species 'should' be, inherited from a long history of this scientific discipline.

Addressing any complex problem first requires acknowledging and understanding why it is complicated. To provide some context, this chapter includes a very brief sketch of the history and basic issues around both the utilitarian and philosophical aspects of species. This should help the users of species names to understand the sources of controversy, and where they manifest in broader issues outside of taxonomy.

## SPECIES METAPHYSICS

This question of how we explain species is, more than anything, a great example of a divide between scientific disciplines (Chapter 2). For example, if I mention that I regularly deliver an undergraduate lecture on species concepts, an evolutionary biologist would groan sympathetically or offer war stories; but a colleague from any other branch of science, from ecology to engineering, will most often look blank and confused ('how could you talk about that for *an hour*?'). We will come to species 'concepts' shortly, but first it is necessary to establish what the broader issues are in this debate.

The 'problem' or potential ambiguity with species passes several hierarchical levels of logic which have been articulated as three discrete questions (Slater, 2013). Additional questions or issues come into play, but these provide a useful framework:

> The *taxon question* concerns what features are required to identify a specific individual as a member of a given species.

> The *category question* is what sort of differentness or characteristics are required to distinguish a group of individuals as a *species*, as distinct from a genus, or a subspecies.

> The *metaphysical question* asks what kind of thing do we mean when we talk about 'species' in general terms.

These three questions are different issues with different applications and audiences (Devitt, 2008; Mayr, 1982), and there is widespread disagreement about all three. Anyone who observes actual organisms is at work at the coalface of the first level, including anyone who identifies organisms, as well as those who name them (e.g., What are the diagnostic features that allow me to conclude my cat Leonora is a member of *Felis catus* Linnaeus, 1758?). Work on species 'concepts' and speciation usually addresses the second level, the category question, although the aspiration of species concepts approaches the lattermost issue, the metaphysical or philosophical nature of species. (Is the lineage *Felis catus* a species, or a subspecies, *Felis silvestris catus* [e.g., Cameron-Beaumont et al., 2002]?) All three of these levels of enquiry are often confusingly inter-related and sometimes contradictory. A solution to the metaphysical question does not automatically provide a full answer to the category question, or any help with identification.

Metaphysics refers to understanding the nature of the world that is independent from us as observers. Kant (1787) concluded that metaphysics is inherently dialectical; thinking about the world independent of our cognition is a contradiction in terms, and our understanding is limited to the natural world that we can interact with. (A conclusion that has been controversial for centuries.) Most scientists accept that there is a single reality, which is both parsimonious to our experience and much easier to cope with. The 'metaphysical question' about species has two main answers in very simplified terms: species as *sets*, or species as *individuals* (Ghiselin, 1997).

'Sets' of objects are a group of things, inter-relatedness or common origin is not necessarily a requirement to group things into a set. I can group all individual organisms with particular diagnostic features into a species, or define a set of all the objects on my desk. We could add a criterion of monophyly and shared ancestry to define a set and avoid spurious comparisons. Being in a set conveys no inherent synergy—there is no expectation of emergent properties for the set itself, beyond the shared features that we already used to define membership—and thus if species are sets, there should be no expectation that species, as a category, have special properties

(Kitcher, 1984). As sets are defined via abstraction, they lack *a priori* spatiotemporal restriction; there are features that define membership and anything anywhere with those features can be included regardless of potential gaps in temporal or spatial continuity. We could have start and end points in time for a species, but in this view it is awkward for the lineage, as a whole, to experience change through time. If the descendent individuals have a different morphological or genetic range than their ancestors, it becomes difficult to maintain the abstract definition of the set that includes them all as a species. If a species is a set defined by certain criteria, then the variation we see as the 'fuzzy boundaries' of a species requires extra rules and exceptions, and this is a big problem (Chapter 5).

The conceptual framework of 'species as individuals' instead proposed that species can be seen as having an origin (speciation event), a lifetime (lineage persistence), and death (extinction). Ancestor-descendent linkages restrict the occurrence of the species in space and in time, because individuals have a concrete identity (Ghiselin, 1974; Hull, 1978). This is concordant with macroevolution. In practice, some species names refer to lineages (species as individuals) and other refer to diagnosable groupings (species as sets) (Reydon, 2003). But it is one objective of phylogenetic systematics to refine taxonomic names so that they all have evolutionary meaning (Chapter 6).

In general, most people have an intuitive idea about what species mean. In this book, we have considered how species have 'fuzzy' edges or 'vague' definitions (Chapter 2), in the formal sense of fuzzy logic and the technical, philosophical meaning of 'vague'. A good example of other real objects with fuzzy boundaries is with geographical features; a plateau or mesa in the American Southwest has a clear boundary, rising out of the desert as an isolated structure (Figure 7.1). At close range, the millimetre scale

**FIGURE 7.1** Geological features are unambiguously real, and unambiguously separated from the surrounding landscape. But at finer scale, looking closely at the edge of a mesa or similar feature, the assignment of every grain of sand as either 'mesa' or 'non-mesa' would be highly controversial. Image of the Tepees area, Petrified Forest National Park, Arizona, USA; note palaeontologists at far right for scale.

definition of where desert ends and mesa begins rests on an idealised line that is not respected by individual sand grains, yet the mesa is undeniably a real thing.

There is an opposing point of view, that species are not 'real', but that does not suggest that biodiversity is some sort of collective hallucination; it is part of the metaphysical question, and issues about language, epistemology, and the nature of collective objects. A contrarian view is important to progress serious debate (Mishler, 1999). Parts of the debate over how we define species both in terms of criteria (category question) and ontology (metaphysical question) erratically filter into general discussion outside of science (e.g., Barrowclough et al., 2016, and media coverage about that publication). Confusion over taxonomic instability may be undermining confidence in the very basis for understanding global biodiversity. The purpose here is to confront the reasons why this question is so persistent and so difficult, and to suggest a more comprehensive way of thinking about species that may ease some of the tension.

## FROM ARISTOTLE TO THE ORIGIN

The standard narrative of the history of thought on species is a shift from fixed species to dynamic species. During the 19th century, science shifted from an 'essentialist' conception of species, which is basically a classical or creationist worldview that species are fixed and well-defined, and moved toward accepting the views articulated by Darwin (1859) that species are 'mutable' products of long-term natural evolution with feedback relationships between lineage and environment. The work of Charles Darwin (b. 1809–d. 1882) is the touchstone for this shift in perspective, but in fact it occurred much earlier and much more gradually (Richards, 2010), and the idea of dynamic species is indeed still some distance from being truly, deeply accepted into everyday scientific thought. Darwin was not the only figure in the history of thinking about species, which follows the whole history of science more than any other idea. Comprehensive and insightful histories of the science of species and speciation occupy whole books (e.g., Wilkins, 2009a,b); here, we will superficially touch on a few names and events that contributed to the development of modern scientific thought in this field.

The issue of how to define, delimit or understand species has persisted since at least Classical Greek scholarship and probably earlier. Most histories of species concepts place the start of serious study in taxonomy with Aristotle, over 2000 years ago, in the 4th century BCE. The focus of Aristotle's work was not to catalogue the diversity of life as observed in Greece, but he used the exercise of identifying species as part of developing more general principles of formalised classification. Classification of course is a much broader field, encompassing many different types of data, such as library science or cryptography as well as taxonomy (Bowker & Star, 1999). The classical terms for specific kinds, '*eidos*', and general kinds, '*genos*', come from Aristotelean logic and are not directly equivalent to species and genera as we understand them today (Wilkins, 2009a). Careful study of Aristotle's own work and its interpretation by later scholars has shown that Aristotle's views of species were not strictly essentialist in the way that term is usually understood (Winsor, 2006; Richards, 2010). Classical scholars observed and understood variations in form within species, and that ancestor-descendent genealogical relationships were important to defining biological species (Winsor, 2003).

One backward idea that still insidiously persists from Classical times is the Great Chain of Being, the organisation of especially animal life from 'higher' (humans) to successively 'lower' forms of animals, then lower still to plants, and even rocks (Lovejoy, 1965). Variations on this worldview appeared in many other world cultures and indeed still echo today. There is an explicit statement of relative value within this hierarchy. The 'Great Chain of Being' is contradicted by the modern understanding of biodiversity, but its philosophy reappears persistently in teleological assumptions about the 'progressive' nature of evolution, as if all forms move uniformly from simple to more complex (Nee, 2005). Textbooks and museum exhibits are typically organised in a sequence from more 'primitive' to more 'advanced' species that reinforces this unconscious bias about the value of diversity in different types of organisms.

After the Classial period, later Christian scholars transposed the biodiversity categories of Classical scholars into essences or 'ideas in the mind of God' (Richards, 2010). In the Middle Ages in Europe, the closest equivalents to taxonomic works were bestiaries, illustrated tales featuring the characteristics of animals as moral lessons, and herbals, which documented the identification of plant species and their uses (Wilkins, 2009a). Bestiaries have limited relevance to modern science, though there are a few instances of historical importance in the first studies of some species that would have been exotic to European readers (Baxter, 1998). The quasi-theological role of bestiaries, and, to a lesser extent, herbals, has interesting ripples that reach into the start of modern taxonomy. In the Middle Ages (5th to the 15th century CE) and through the Renaissance (14th to the 17th century CE) the prevailing attitude was that the study of natural history, or biology, was not an interesting end to itself but an extension of theology motivated by a desire to understand the works of a supernatural creator. There is a period of over 2000 years between the time of Aristotle and the work of Linnaeus, and it would be incorrect to conclude that nothing of intellectual merit was happening during that time.

Carl Linnaeus (b. 1707–d. 1770) was a trained botanist and proposed a system for formalising names of species and providing descriptions (Chapter 6). This filled a significant gap in the science of the day as the interest in documenting biological diversity was increasing, but there was no standard approach for authors to describe their contributions (Wilkins, 2009a). His twinned approach of classification into hierarchical categories, and the general idea of concise descriptions for species was the foundation for modern taxonomy. A contemporary colleague famously said Linnaeus thought of himself as the 'second Adam', referring to the Christian creation myth, where Adam gave names to all the beasts. This was a bitter remark, as apparently von Haller was piqued that species he had described were 'demoted' to varieties in a subsequent work by Linnaeus (Harrison, 2009); proof that rivalries over taxonomic interpretations have existed for exactly as long as taxonomy itself.

Apparently, Linnaeus also saw his own taxonomic projects as a product of divine inspiration (Harrison, 2009). Linnaeus was certainly a species essentialist and creationist and believed that the species on Earth had been created specifically by God. Interestingly, even Linnaeus later conceded that new species could arise through hybridisation (at least in plants; Gray, 1821; Wilkins, 2009a), but he maintained that genera were fixed elements of divine creation. There is a continuation, or a strong parallel, of this attitude with the idea of 'baramin', a term invented by Christian creationists in the 20th century. 'Baramin' are a generalised organismal kind (e.g., dog-like things), representing the descendants of organisms that survived the Great Flood in Christian mythology. A 'baramin' today may comprise several species, varieties or hybrids. Despite their supposedly divine origin there are, of course, arguments about 'baramin concepts' (Wood et al., 2003; Lightner, 2009). The conflation of baramin or genera as fixed essences is quite the opposite of modern thinking about species groups.

Twenty-first century secular rationalists baulk at the idea of creationist concepts at the foundation of taxonomy as a discipline. The apparent counterpoint to this historical baggage is the work of Darwin, a century later, which offended contemporary scholars with its apparent departure from accepted Christian tradition (Stamos, 2007). Long before Darwin, the idea of the mutability of species was acknowledged by Aristotle, and by Linnaeus, and gained ground with the works of Erasmus Darwin (b. 1731– d. 1802) and others. Variation among forms within a species is undeniable, but it is not a simple logical transition to conclude that such variation could ultimately lead to new species. The work of Charles Lyell (b. 1797–d. 1875) in understanding geological time was the real key to species evolution; the origin of new species through accumulated change only became plausible through envisioning life over sufficiently vast spans of time (Raup, 1994). Nonetheless, the marine invertebrate biologist, Charles Darwin's (1809–1882), elegant arguments that variation could be honed into new forms through the accumulated action of natural selection were a substantial departure from the popular thinking of his own day (Stamos, 2007). In parallel with Darwin, Alfred Russel Wallace (b. 1823–d. 1913) also showed that any adequate explanation for living biodiversity required that species have an inter-related history (Wallace, 1855). Wallace was working in Sarawak, a part of the island of Borneo that is now a state in Malaysia, which has some of the most extravagant biodiversity in the world. Both Wallace and Darwin gained a very different perspective from their European contemporaries through first-hand observations of tropical diversity (Chapter 11). The proposition that all present species are descended from other previous, different species, conflicted with the established contemporary belief in divinely created kinds.

Some of Darwin's writing has been interpreted to imply that he did not believe species to be 'real'. One repeatedly quoted passage in the *Origin of Species* reads 'I look at the term species, as one arbitrarily given for the sake of convenience to a set of individuals closely resembling each other...' (Darwin, 1859). But such statements explicitly refer to what we earlier called the 'taxon question', not the 'metaphysical question', two points which Darwin also evidentially considered to be separate issues (Wilkins, 2009a). As Stamos (2007) has shown, Darwin's work was written with his contemporary audience in mind. That audience had a fixed, essentialist idea of what species mean; Darwin's rhetorical goal was to challenge his 19th century readers' preconceptions. His own views on species also developed over time (Kottler, 1978). From a 21st century perspective we have Darwinian thought as our starting point, and we dabble in the idea that evolution is too complex to produce definable entities called species (Mishler, 1999); however, Darwin apparently had what we would *now* consider a relatively conventional (or pragmatic) view on species.

## THE MODERN SYNTHESIS AND BEYOND

Another century later, scientific discussions in the mid-20th century that comprised the Modern Synthesis also wrestled with the question of how to discuss and explain species. The Modern Synthesis movement successfully united Mendelian genetics (in an era before the discovery of the structure of DNA) with the Darwinian concepts of variation and natural selection. It has taken nearly another century to assess the positive and negative consequences of the Modern Synthesis on how we think about species today (Wheeler, 2008).

The early geneticist Theodosius Dobzhansky (b. 1900–d. 1975), was the first to articulate a comprehensive understanding of time-dynamic evolutionary species. And, importantly, Dobzhansky recognised that, given the ongoing processes of species evolution, it is nearly impossible to judge whether or not a species is genuinely separate from its nearest sibling-lineage in the present moment (Dobzhansky, 1935, 1940; Chapter 5). A few years later, the ornithologist Ernst Mayr rejected Dobzhansky's approach, writing 'A species is not a stage of a process, but the result of a process' (Mayr, 1942). This is fundamentally problematic, because it requires evolution to have an end point. But Mayr (1940, 1942) proposed an influential model that he explicitly called the 'biological species concept' positing that species are defined as 'actually or potentially' interbreeding populations.

The oversimplification of this idea has badly damaged much of the understanding of species. Mayr (1940) immediately acknowledged the difficulty and limitations of interbreeding as a criterion: the inclusion of the word 'potentially' interbreeding in his original work was a critically important caveat. Mayr's concerns were initially pragmatic, and in fact he explicitly differentiated species 'criteria' from species 'concepts' (Mayr, 1942). Others raised, but dismissed, the problem that interbreeding is inapplicable to any species that uses asexual reproduction. Later critics have pointed out that data do not exist to even theoretically test interbreeding in a large proportion of species where life cycles have never been documented. Mayr and his contemporaries considered the benefits and drawbacks of the interbreeding line of evidence in detail and promoted a more nuanced 'polytypic' species concept incorporating multiple lines of evidence (e.g., Mayr, 1942; Camp & Gilly, 1943; Dobzhansky, 1944). Dobzhansky (1944) used this polytypic approach to conclusively demonstrate the commonality among living *Homo sapiens* Linnaeus, 1758, including all our minimal and fluid morphological differences, as proof that all living humans are unequivocally a single species, a topic that had some urgency in that decade.

Notwithstanding any intention of nuance, the original phrase 'biological species concept' (Mayr, 1940) represented a branding masterstroke. Any other idea in opposition to the 'biological species concept' starts from a disadvantaged position. The choice of label allowed the relatively narrow criterion of interbreeding populations to be positioned in later works as an apparently complete solution. It seems elementary that as biologists we would all want to use the *biological* species concept for general purposes, and all the other species concepts sound like special cases. There are many other species 'concepts', several of them including criteria defined by Mayr (1942) for comparative purposes (Wilkins, 2009b).

The intellectual foundation of the Modern Synthesis was built on population genetics, and the focus on quantitative empirical approaches coloured all contributions to the movement. So while this period in the history of science is generally described as a unification of disciplines (Smocovitis, 1992; Cain, 1993), there was also a wealth of important descriptive data that became dangerously marginalised (Wheeler, 2008). Palaeontologists, in particular, struggled to gain recognition among the broadly developing ideas of the Modern Synthesis (Simpson, 1951). Remineralised tissues in fossils do not preserve molecular genetic data, and most fossil species are not preserved as populations, but palaeontologists changed their approach to attempt to fit population-genetic-centric science (Sepkoski, 2016). The 'biological species concept' likewise cannot be tested in fossils. Many saw this as evidence that fossils themselves were inadequate evidence to assess species, though it would seem clear that Mayr's model was inadequate, since it could not comfortably accommodate the majority of species that have ever lived.

Later in the 20th century, with the ongoing development of the field of phylogenetics, further contributions established a new theoretical framework for separating process (the population genetics-driven ideas at the core of the Modern Synthesis) from the observed patterns of species (Wheeler, 2008). The most important contribution in this movement came from the work of entomologist Willi Hennig (b. 1913–d. 1976), who proposed an approach where Linnaean names and classifications were used to reflect evolutionary (phylogenetic) relationships (Hennig, 1966). This helped to make taxonomy and systematics far more rigorous and paved the way for the incorporation of additional later technology, such as DNA sequencing to the assessment of species and their inter-relationships, and these methods could also be applied to all taxa, including fossils.

The works of most of the scholars mentioned above are rightly beloved by the scientific community, but Hennig is seen as a polarising figure (Platnick, 2016; Schmitt, 2016). Hennig's ideas have been quietly incorporated into the mainstream in the fields of phylogenetics and systematics, yet there is a persistent anecdotal belief that Hennigian approaches are a kind of scientific extremism.

The transition from microevolutionary processes (the variation within and among species that can be manipulated experimentally) to macroevolution (lineage change) is the lasting conundrum of both Darwin and the Modern Synthesis (Chapter 5). The core of the Modern Synthesis was to suggest that these observable processes, controlled by genes, explain the principles of variation and selection as laid out by Darwin: genetics underpins the mutability of species, and it is the nuts and bolts of descent with modification. Experimental evidence moves us closer to a comprehensive understanding of genetic dynamics within species. Yet the focus on experimental processes has moved the debate away from the fundamental goal of synthesis, a universal explanation of the origin, persistence, and extinction of species as units themselves (Wheeler, 2008). The self-styled successors to the Modern Synthesis have a new approach they call the 'Extended Evolutionary Synthesis' (Laland et al., 2015). This is contentious at present, and probably not the same sort of watershed in scientific thought as the Modern Synthesis, but the recent body of work represents an important acknowledgement that more data than ever are now available to bear on the questions of how evolution works. More data are available, but they do not bridge the micro- / macroevolutionary divide. Interestingly, unlike contributions that orbited around the Modern Synthesis, new work on evolutionary theory often glosses over the question of species altogether. It may be that some experts have never considered the problems and limitations of the 'biological species concept'. The idea of interbreeding as a primary line of evidence is inadequate to serious questions of species, but the legacy of the Modern Synthesis has left it deeply embedded in evolutionary teaching.

## SPECIES 'CONCEPTS' ARE LINES OF EVIDENCE

In recent history, various scholars have proposed around 30 different supposedly universal approaches to delineate species, or 'species concepts' (Wilkins, 2009b; Zachos, 2016). These descriptions of 'species concepts' generally do not offer a metaphysical explanation of the nature of species, but rather they are categorical criteria used to determine whether a group of organisms qualifies as a species (de Queiroz, 1999). Many of the various lines of evidence that have been proposed as species 'concepts' were shaped by the experience of the proposers—usually research on terrestrial vertebrates. It is far more difficult to understand the features that are relevant to organisms where the nature of individual identity is less clearcut—colonial organisms, encrusting organisms, syncytial cells—but any robust consideration of species should apply to all (Ghiselin, 1997). Most of these species concept metrics do not apply to all organisms, and the different methods still might come up with contradictory assessments for the same organisms, so we are back to the nub of the species problem.

The terminology 'species concepts' unfortunately invokes an impression that these 'concepts' are more philosophically complete than the authors probably intended. The existence of multiple 'concepts' also implies that these various ideas are mutually contradictory worldviews in competition with each other. To a certain extent, the view of agonistic concepts in direct opposition has been actively promoted in the specialist literature, arguing for or against a particular 'concept' (Wheeler & Meier, 2000). And the 'flock of species concepts' includes a number of approaches that are so different that they really are not comparable (Sterelny & Griffiths, 2012). Various species concepts include morphological diagnosis, gene flow, ecological roles, adaptive niches or phylogeny. The 'recognition species concept' is based on mate compatibility as a criterion (Paterson, 1985); this has real biological relevance in some organisms but it is also remarkably easy to think of counter-examples. This is part of the taxon question; the recognition of individuals as 'self' or 'other' is clearly not limited to human preferences, as most other animals can identify species, and they use species-level recognition in selecting preferred food types. My pet tortoise taught me some botany, as he has strong opinions about differences among *Taraxacum* species (dandelions).

Some species concepts focus more on philosophy, or methods, others on data. The 'palaeontological species concept' was articulated (Simpson, 1951, 1961) as an extension and counterpart to the 'biological species concept' (Mayr, 1942). This does not mean that species are not comparable between fossil and living organisms. Fossils have intrinsically different data limitations than specimens of living species. 'Different data' does not always mean less data. Genetic data, natural history, or population-level comparisons can all be equally limited for the many living species that are only known from a few specimens. It is entirely possible that many of the taxa named from fossil evidence may represent cryptic species complexes. This is also true of living species, the only difference being that with living species we hypothetically have a future opportunity to refine the assessment with additional data sources like genetics or field observations that will more than likely never be available for fossils. If we instead consider that morphology, genetics, and population assessments all represent contributing sources of data, not absolute requirements, this is far less problematic.

The appearance of conflict among most species concepts is a false construct. Species are best identified using a total evidence approach, including all available methods and relevant data. The requirement of total evidence, simply stated, is that all relevant data must be given due consideration (Fitzhugh, 2006). The categorical assessments that feed into all of the various proposed 'concepts' (naturally interbreeding populations, genetic distance, ecological niches) all represent biologically interesting data that mainly relate to the 'taxon question' rather than 'category question' and as such, all of them can and should be used in seeking a consensus. That is not to say that all lines of evidence must agree, but rather a conclusion drawn from the balance of available evidence. This total evidence approach is, more or less, how the practical aspects of taxonomy often actually work; however, it is not usually formalised as such, and individual taxonomists have strong views about whether certain forms of data are valid or informative to their consideration of species in their specialist organisms. In order to really address the 'category question' or the nature of species, we have to go another step beyond assembling evidence for the 'taxon question' and embrace both the time-horizontal variation in species and their time-vertical dimension.

## SPECIES IN TIME

We understand that species have evolved and are constantly evolving. But this latter part, ongoing dynamic evolution, is a significant complication to how we would prefer to carve up the world. How do we know that a species we see today is different from, or the same as its ancestor lineage from 1000 years ago, or 100,000 years ago, or 100 million years ago? How does this work? Is it a process of gradual incremental change, with the ancestral form slowly, almost imperceptibly morphing into the modern form over millions of generations? In fact, the fossil record, and molecular phylogenies show that most species have long periods without much change, interrupted by speciation events (Venditti et al., 2010). 'Lineage segments', the lines in phylogenetic trees that represent ancestor-descendent populations, are the best way to visualise the periods between these changes at speciation events.

In a common biology lab exercise, students remove the surface skin from a layer of an onion near its root and look at the cells through a microscope. Among the cells in this growing tissue, you will see cells in all stages of mitosis (Figure 7.2). Some cells are large and mature interphase with round nuclei, some are midway through division, some are small newly-split daughter cells, and everything in between. Counting the individual cells requires some judgement calls to say if some nearly-split entities are still one cell, or really now two cells. The beauty of this as a lab exercise is that

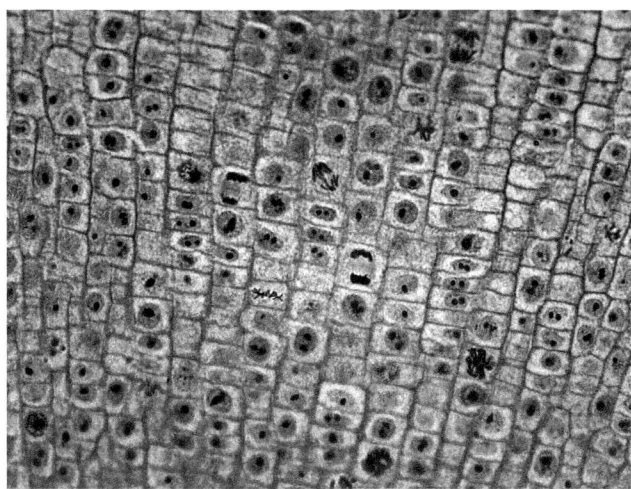

**FIGURE 7.2** Cells in the root of an onion, *Allium cepa* Linnaeus, 1753. Analagous to species lineages, cells have long spans of stable equilibrium (interphase) but that is part of the continuous cycle of cell division, and a single slice captures cells in all stages.

you are guaranteed to see all phases of mitosis at once, because the cycles of all the individual cells are not identical, nor closely coordinated. Beyond our onion's root, the cells in different kinds of tissues in a whole organism take dramatically different times to divide, from minutes in embryogenesis to months or years in living organ tissues. However, no one would take this dynamic diversity as evidence that cells are not real. We can extend this elegant metaphor, proposed by Hennig (1966), to note the important extra challenges presented by species. Unlike speciation, the phases of mitosis are well documented, and mitosis is the only mechanism for somatic cell division. When we glance across an onion tissue though a microscope and see cells in different stages of division, we have a growth chart of the standard phases of the cell cycle to contextualise our immediate observation and infer the past, present, and future trajectory of an individual cell relative to the 'equilibrium' mitotic interphase. We generally do not have such context for species, and it seems increasingly likely that the variety and complexity of speciation processes defy any realistic hope of a universal framework or identifiable phases for species lineages.

Hennig's (1966) metaphor of species as cells experiencing mitosis follows on from that comment of Dobzhansky's (1940), mentioned above, that was rejected by Mayr (1942). Mayr's objection was that he saw each identifiable species as an end point, not a midway point, of evolution. This seems to make sense, in that our scientific goal is to characterise observable differences among the species of the world around us. However, we must accept the slightly uncomfortable fact that evolution is continuing and will continue to shape species. Lineages will divide or persist or go extinct into the future, and they are in the midst of doing so right now. Indeed, Mayr (1940) called these 'species in the making'. Evolution does not have an end point, nor a goal, and there is no magical significance to the specific moment in time we are here to study them. To assume that the present day is a *de facto* end point for every species we study, is to revert to the creationist perspectives of previous centuries. Without access to the distant future, and imperfect access to the past, we are left with an intrinsically data-limited basis to judge what stage of division a species is in at the present. The unavoidable conundrum is, if the organisms I want to identify are members of species in the making, are they species or not? And, to a more practical point, what are we supposed to do about naming them? While every species on Earth at present is indeed the product of its past evolutionary history, some species are closer to their nearest relatives than others, and some are in that awkward almost-but-not-quite or just-barely split stage that leads to differences of opinion about whether it is one species or more.

Modern species are unavoidably the products and the agents of those ongoing evolutionary processes, and thus we find that some species are more clearly separated than others, and many are caught in a moment, midway through a process that may lead to a new species. The pace or frequency of speciation processes interact with body size, generation time, metabolic rates, and many other basic biological factors. Biological rates all vary wildly comparing different organisms, from the explosive development of swarms of tiny plankton to decades-long generation times of trees. Evolutionary rates are not fixed, species evolve faster or slower in response to a wide range of biotic and abiotic influences (Webster et al., 2003; Rabosky et al., 2013). The 'distance' or accumulated changes in the genome that help to define one species as separate from another is not a constant measure but part of a package of evidence to assess whether a lineage segment is separate from other adjacent lineages.

If you had a perfectly resolved family tree of the living world today, this one snapshot in time, showing the relationships among every individual in every population, every species in every part of the world, would it all just blend together into an impenetrable mass without distinctions that define the boundaries between species? Some things would certainly stand out, such as faster evolving lineages, very isolated species, and species in clades where most of their forebears are extinct. Extinction of earlier lineages creates gaps in an expanding phylogenetic tree that mean some lineages are naturally more clearly separated from their relations than others (Figure 7.3). This is true with fixed rates of evolution and is exacerbated if speciation and extinction rates vary over time. A further important point to this thought experiment is that it is absolutely impossible. We will never have *all* the data, so we must necessarily make inferences based on a consensus of the best available data.

Across geological time, species spend long periods of time in an equilibrium state, which is punctuated by changes that lead to new species through splitting or through changes we interpret as adaptations (Eldredge & Gould, 1972). This stop and go of species evolution was a conundrum that worried Darwin and his contemporaries (Bennett, 1997), because it seemed to be contradictory to the continuous application of natural selection. Lindberg (1998) showed that a population-lineage model was embedded in the species descriptive work of William Healey Dall (b. 1845–d. 1927) before the start of the 20th century (see also Dall, 1877), but issues of species dynamics continue to confound modern evolutionary biology. Most species spend most of the time between speciation events, in equilibrium, without substantive change; yet, looking across the vast diversity of life on Earth, it follows naturally that we find a lot of populations or forms that are equivocal in terms of whether they could be called one or more species.

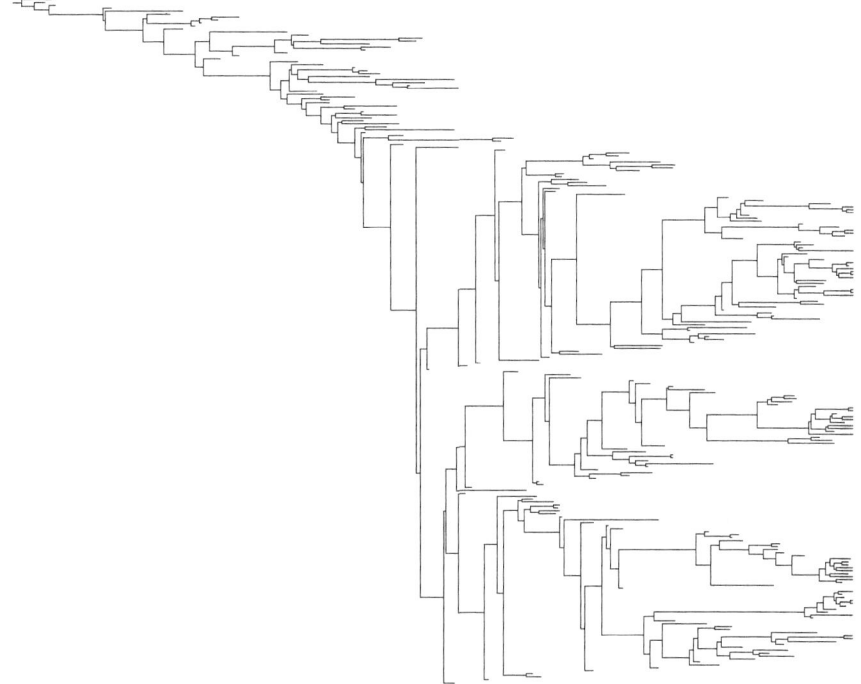

**FIGURE 7.3** A simulated phylogenetic species tree, expanding over time from left to right with fixed probabilities of speciation or extinction at each time slice, showing the 'gaps' and natural groups of lineages that are formed through a simple stochastic process.

Species were not made by a creator to populate the Earth, each with its assigned place and function. Species are constantly evolving in messy, dynamic, beautiful chaos. This long term dynamic process and the history of major evolutionary changes are collectively referred to as macroevolution (Chapter 5). For the most part we can only observe species in a single point in time. A distinction could be made between the apparently fixed, horizontal species samples that are the units of biodiversity, and the dynamic species units that are participants in macroevolutionary processes, but this is a false dichotomy. In a photograph of breaking waves, every speck of seafoam has its perfect place in the composition, but we still have an impression that the image represents something dynamic.

## TOTAL EVIDENCE AND THE NEW PLURALISM

The frustrating lack of a clear universal litmus test ('are you a species yet?'), among other concerns, has led some to conclude that a metaphysical or even a categorical definition of species represents an unsolvable problem (Hey, 2001). Sceptical responses to this conundrum divide between species 'nominalism' or 'pluralism'. Species nominalism is the viewpoint that species are real in name only; the names we apply to natural kinds are a tool for our own communication rather than a reflection of any deeper universal reality. Species pluralism holds that there is no single universally applicable way to describe the nature species, and different species concepts suit different investigator's motivations (ecologists are more interested in ecology; morphologists are more interested in morphology). Pluralism could be interpreted to mean that species are not real, because they are arbitrarily defined by the interests of the observer (Kicher, 1984). However, more recently, Richards (2010) showed that pluralism can lead to an internally consistent and complete solution to the 'species problem', embracing both categorical and metaphysical fuzziness: a solution we call the total-evidence approach.

Now, working within a macroevolutionary framework, the emerging consensus is that species represent 'lineage segments' within a phylogenetic (evolutionary) context (Mayden, 1997; de Queiroz, 1999). This is not entirely the same thing as some previous applications of a 'phylogenetic species concept' but builds on that approach. The 'definitional core' of species is their place within a larger macroevolutionary process (Richards, 2010). Other species 'concepts' actually reflect lines of evidence that describe processes relevant to identifying species (the category question) and the ecological and selective pressures that shape species. The fundamental shift here is that species conceived under any static criteria—including phylogenetic or evolutionary species concepts—were previously often interpreted as products rather than participants in the process of speciation.

All fields treat species as important pivots. Indeed, many other fields of biology contribute to understanding species, so a singular focus on phylogenetic criteria is not necessarily constructive or sustainable (Dupré, 1999). The 'lineage-segment' conceptualisation does not mean it is necessary to have a perfectly resolved phylogenetic analysis before making decisions on the identity of local species. This is entirely impractical. Indeed, the business of identifying species long predates any understanding of evolutionary framework, and the ongoing daily work of taxonomists naming newly discovered species frequently does not include any quantitative phylogenetic analysis, nor should it (Chapter 8). The business of taxonomy has always been integrating multiple lines of evidence into a cohesive, total-evidence approach. This is not to say that it happens by magic. It happens by training, comparison, and critical reflection that provide a basis for sound judgement about how to manage intermediate cases and the blurred edges of species.

The population lineage that forms a species is a real thing, but it does not have a hard, containing boundary that can fence in every individual organism in the species. A paradigm of species as lineages with fuzzy edges can be applied synchronically, comparing all the (horizontal) species that co-occur in a given moment in geological time, and diachronically, assessing the persistence of those lineages vertically through time (Figure 7.4; Figure 5.4). In this framework, the New Pluralism, all of the dozens of species 'concepts' are not competing ideas generating confusion and conflict, but useful contributions to help us identify taxa from different lines of evidence (Richards, 2010). A total-evidence approach provides ample room to understand species with further new kinds of data, which promise to develop quickly with advances in genetic data and gene editing technology.

The time dynamic nature of species precludes a simple 'silver bullet' solution to either the taxon question or the category question. The nuts and bolts identification

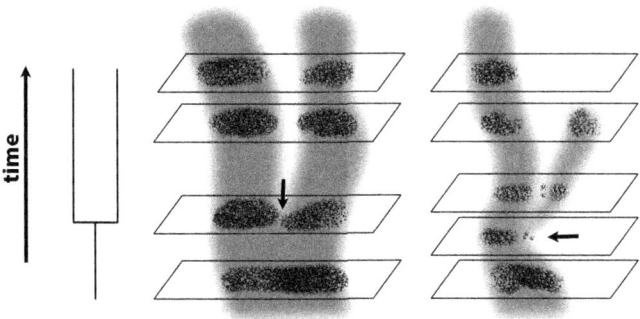

**FIGURE 7.4** Population lineages through time and space (modified from Benton and Harper 2009). The wire diagram to the left is a schematic phylogeny, indicating the sister relationship between two species. Two scenarios to the right show how species as population-lineages would appear as sampled in different 'slices' or moments in geological time.

of individual organisms and the description of novel species remains in the realm of the 'taxon question' and total evidence; that the 'category question' is answered by dynamic, fuzzy boundaries is mostly useful in that it might remove some anxiety over why species are so difficult to put in a box. Recognising species as lineage segments in a large-scale global evolutionary process, and taking a total evidence approach to recognising those units, is the only framework that can incorporate all of the many biological properties of species through time.

## THE SPECIES CATEGORY

The nub of the category question is whether the rank of species has a particular character, emergent properties that are unique to species-level groupings. Is there something about species, independent of us as observers, that determines how we will partition them (Devitt, 2011)? The conception of species as individuals was an important step in embracing the time-dynamic nature of species and species as units that are coherent with evolutionary processes (Hull, 1978). Further developments in understanding molecular mechanisms of evolution, including introgression and horizontal gene transfer, have complicated the picture, with evidence that, at microevolutionary scales of resolution, there is no one single true tree of life (e.g., Morrison, 2014; Soucy et al., 2015); 'species as individuals' may be insufficient to accommodate reticulate phylogenies. That magic moment when the descendants take up their identity as a fully-fledged new species does not happen as an instant birth. These comparisons, like the mesa in the desert, are a matter of scale, and accepting fuzzy boundaries (Figure 7.1).

In Chapter 5, we noted relevant criteria for higher-level groupings of species: monophyly, multiple lines of evidence for a phylogenetic hypothesis, and phenotypic diagnosability. These also apply to species as well as higher groups. A species is delimited by certain identifiable characteristics, and usually also by its geographical range. A continuous geographic range, for example, can be evidence of present interconnectivity of populations. The range of a species is usually larger than the distance any single individual could ever travel in its lifetime. Humans and our domestic animals have a global range, thanks to the intervention of technology. So, ranges are delimited in part by movement, in the case of migratory species, or by dispersal. Many plants and many marine organisms have a biphasic lifestyle, where there is a spore or larval dispersal phase followed by a relatively sessile adult life style. This mode of life allows the species as a whole to maintain a relatively large range, while the gene flow between individual generations may only cross a tiny fraction of the whole available range. Distribution, and the capacity for dispersal, is a critically important trait to understand species (Jablonski, 1987), because it is both heritable and directly interacts with abiotic factors that could cause speciation (e.g., geographical barriers) or extinction (localised or general destruction).

The unique properties of species are what Richards (2010) called 'species category essentialism', that is, that the species-category has a particular essence or characteristic nature. Higher groups, genera and families and so on, do not have a cohesive identity because, if they comprise more than one species, then by definition there is some barrier among those constituent species. Higher ranked groups may be more taxonomically stable in some applications, but they are less epistemologically real (Chapter 6). All members of a species can—hypothetically—interact with each other. They can 'communicate', in the broadest possible sense, in that conspecific organisms can exchange genetic information, share resources, and that activity does not change the fundamental composition of the community or subsequent generations. This has been observed before, in that the 'ecological species concept' approaches this from the converse, noting that conspecifics are necessarily competing for identical resources (Sterelny & Griffiths, 2012), but this is not contradictory. Conspecifics can (hypothetically) survive in the rest of the range of their species (synchronically and diachronically). The long-term trend among descendants of conspecifics is for the same general variability, perhaps expanding and contracting in disparity, but not fragmenting into multiple, co-existing different forms. Fragmentation into independent subgroups each with evolutionary cohesion is evidence of speciation. Many of these features are also true for subspecies, or populations, so species by definition are the broadest group at which this applies.

## SPECIES ARE REAL; SPECIES DESCRIPTIONS ARE HYPOTHESES

All natural systems are dynamic, and species evolution occurs over timescales that are generally too long to be directly observed. Evolutionary biologists are fond of quoting the title of an essay by Theodosius Dobzhansky (1973): 'Nothing in biology makes sense except in the light of evolution'. To other fields of science, this could seem a little dogmatic ('my science is more important than your science'), and one might imagine substituting most other disciplines in place of evolution—geology, chemistry, physiology, and so on. Good arguments could be constructed for the importance of the physical context, chemical make-up, or energetic budget of any biological system. But in talking about species, this sentiment is relevant because evolutionary context may explain most of our difficulties in articulating definitions of species. The complexities of quantifying biodiversity come from the unavoidable fact that species are participants in ongoing evolutionary processes: processes we cannot observe directly and do not yet fully understand. To a working biologist, a time-dynamic view of species as units changing through time as well as space is both the problem, and the answer, for the 'species problem'.

Species are units of evolutionary processes, which we endeavour to understand (the 'metaphysical question'), through investigating the properties of lineages (the 'category question'), and species names and descriptions are our expression of hypotheses about lineage membership (the 'taxon question'). The 'category question' about whether a group of individuals constitutes a species is a hypothesis proposed on the basis of available evidence. Likewise, assigning individuals to a particular species is an instance of a hypothesis. Taxonomists have to weigh up all available evidence to propose a line that separates taxa, and then identify the features that can be used to identify individuals as members of a given species and that will allow future refinements to understand the species and their biology.

This is not easy. The type of analysis required to identify and describe a species would be classified as 'NP-hard' in mathematics (the P is for 'polynomial time', the N is for 'nondeterministic'), meaning that it is at least as hard as the hardest problems that have proofs that can definitively be computed in a finite amount of time. The mathematics of phylogenetic inference is another additional step, and there are formal proofs that determining species trees from genetic sequences was NP-hard, even before we had access to whole genome data (Hamel & Steel, 1996; Ma et al., 2000). There is the further complication that different fields of science fundamentally disagree about what is worthy of investigation (Hull, 2010). In many disciplines within life sciences, particularly ecology, there is a strong trend to atomise functional parameters—isolating influencing factors that can be reduced but reduced no further. What are these separable explanatory variables that provide some measure of differentness for species? Morphology, environment, physiology, and genome are all synchronously interacting. The solutions require a high level of human ingenuity and deduction.

Societies without scientific thought *per se* still have accurate ways to identify and communicate about species or species groups of large organisms (Berlin, 1973; Begossi et al., 2008). Primitive recognition of species has also been used as a counter-argument to the reality of species; if folklore has such strong correlation to scientific knowledge, could it be evidence of the cognitive tendencies of humans rather than any natural pattern in the larger world? Early humans developed biological and technological innovations that allowed us to expand to every continent (Beall et al., 2012). Neolithic Europeans built astonishing monuments that align with astronomical phenomena (Penprase, 2017). (It is interesting that no one, so far, has suggested that megalithic structures are evidence that stars and other celestial bodies or the existence of weather are human cognitive constructs). The fact that a problem has a very long history and universal interest should not suggest that the problem is simple.

Consideration of what species mean has a long history. Over several thousand years, we have transitioned from species as 'essences', or types put in place by a divine force, to realising mutability of species through evolution, to species as entities with dynamic boundaries that shift through time as well as space. The holistic point of view that species can be thought of as hypotheses (real things with fuzzy boundaries), is very modern, but not all that new, as it was probably first articulated explicitly in the 1970s (Bonde, 1977; Valdecasas et al., 2008). The view of species names as hypotheses has gained acceptance among part of the phylogenetics community, and in species discovery, but it is difficult to translate to other fields (Pante et al., 2015).

In species, lineage-splitting events occur throughout geological time, spans of millions of years, a scale that nearly defies human comprehension (Chapter 9). Our observations of species are mostly restricted to a single horizontal slice at the surface of the rock record that represents the present day. Every lineage is changing through time at a different pace and may be persisting, or splitting, or facing extinction, depending on the environment. Large-scale events like mass extinctions and global climate change (in the past and the present) have a disproportionate effect, leading more lineages to extinction or bursts of speciation, just as a sudden change in environment could prompt a burst of growth in our onion cells.

The study of species has a long history and has absorbed many revolutions, and historical context is important to understand what previous authors meant in their scientific interpretations. Darwin (1859) had to push back against the contemporary assumption that species were fixed essences; Mayr (1942) and Hennig (1966) and Ghiselin (1974) pushed back against entrenched practices that were fixated on results (morphological variation) rather than causes (shared ancestry). History shows the ongoing development of our evolutionary or phylogenetic view of species does not mean rejecting taxonomy or the Linnean system of classification. The confusion about defining species is not evidence that species themselves are subjective, rather that our judgement of complex systems should be based on a total-evidence assessment.

## REFERENCES

Barrowclough GF, Cracraft J, Klicka J, Zink RM. 2016. How many kinds of birds are there and why does it matter? *PLoS ONE*. 11: e0166307.

Baxter R. 1998. *Bestiaries and Their Users in the Middle Ages*. Sutton Publishing.

Beall CM, Jablonski NG, Steegmann AT. 2012. Human adaptation to climate: Temperature, ultraviolet radiation, and altitude. In: Stinson S, Bogin B, O'Rourke D (editors). *Human Biology: An Evolutionary and Biocultural Perspective*, Second Edition, pp. 175–250. Wiley.

Begossi A, Clauzet M, Figueiredo JL, Garuana L, Lima RV, Lopes PF, Ramires M, Silva AL, Silvano RA. 2008. Are biological species and higher-ranking categories real? Fish folk taxonomy on Brazil's Atlantic forest coast and in the Amazon. *Current Anthropology*. 49: 291–306.

Bennett KD. 1997. *Evolution and Ecology: The Pace of Life*. Cambridge University Press.

Benton MJ, Harper DAT. 2009. *Introduction to Paleobiology and the Fossil Record*. Wiley.

Berlin B. 1973. Folk systematics in relation to biological classification and nomenclature. *Annual Review of Ecology & Systematics*. 4: 259–71.

Bonde N. 1977. Cladistic classification as applied to vertebrates. In: Hecht MK, Goody PC, Hecht BM (editors). *Major Patterns in Vertebrate Evolution*, pp. 741–804. Plenum Press.

Bowker GC, Star SL. 1999. *Sorting Things Out: Classification and Its Consequences*. MIT Press.

Cain AJ. 1993. *Animal Species and Their Evolution*. Princeton University Press.

Cameron-Beaumont CH, Lowe SE, Bradshaw JW. 2002. Evidence suggesting preadaptation to domestication throughout the small Felidae. *Biological Journal of the Linnean Society*. 75: 361–6.

Camp WH, Gilly CL. 1943. The structure and origin of species, with a discussion of intraspecific variability and related nomenclatural problems. *Brittonia.* 4: 324–85.

Dall WH. 1877. On a provisional hypothesis of salutatory evolution. *American Naturalist.* 11: 135–37.

Darwin CR. 1859. *On the Origin of Species by Means of Natural Selection, or the Preservation of Favoured Races in the Struggle for Life.* John Murray, publishers.

de Queiroz K. 1999. The general lineage concept of species and the defining properties of the species category. In: Wilson RA (editor). *Species: New Interdisciplinary Essays,* pp. 49–89. MIT Press.

Devitt M. 2008. Resurrecting biological essentialism. *Philosophy of Science.* 75: 344–82.

Devitt M. 2011. Natural kinds and biological realisms. In: Campbell JK, O'Rourke M, Slater MH (editors). *Carving Nature at Its Joints: Natural Kinds in Metaphysics and Science,* pp. 155–74. MIT Press.

Dobzhansky T. 1935. A critique of the species concept in biology. *Philosophy of Science.* 2: 344–55.

Dobzhansky T. 1940. Speciation as a stage in evolutionary divergence. *American Naturalist.* 74: 312–21.

Dobzhansky T. 1944. On species and races of living and fossil man. *American Journal of Physical Anthropology.* 2: 251–65.

Dobzhansky T. 1973. Nothing in biology makes sense except in the light of evolution. *The American Biology Teacher.* 35: 125–29.

Dupré J. 1999 On the impossibility of a monistic account of species. In: Wilson RA (editor) *Species: New Interdisciplinary Essays,* pp. 3–22. MIT Press.

Eldredge N, Gould SJ. 1972. Punctuated equilibria: An alternative to phyletic gradualism. In: Schopf TJM (editor). *Models in Paleobiology,* pp. 82–115. Freeman, Cooper & Company.

Fitzhugh K. 2006. The 'requirement of total evidence' and its role in phylogenetic systematics. *Biology & Philosophy.* 21: 309–51.

Ghiselin MT. 1974. A radical solution to the species problem. *Systematic Biology.* 23: 536–44.

Ghiselin MT. 1997. *Metaphysics and the Origin of Species.* State University of New York Press.

Gray SF. 1821. *A Natural Arrangement of British Plants: According to Their Relation to Each Other, as Pointed Out by Jussieu, De Candolle, Brown, &c. Including Those Cultivated for Use: With an Introduction to Botany in Which the Terms Newly Introduced are Explained.* Baldwin, Cradock, and Joy, publishers.

Hamel AM, Steel MA. 1996. Finding a maximum compatible tree is NP-hard for sequences and trees. *Applied Mathematics Letters.* 9: 55–9.

Harrison P. 2009. Linnaeus as a second Adam? Taxonomy and the religious vocation. *Zygon.* 44: 879–93.

Hennig W. 1966. *Phylogenetic Systematics.* In: Zangerl R (editor). 1999. Phylogenetic Systematics, translated by Davis DD and Zangerl R. University of Illinois Press.

Hey J. 2001. *Genes, Categories, and Species: The Evolutionary and Cognitive Cause of the Species Problem.* Oxford University Press.

Hull DL. 1978. A matter of individuality. *Philosophy of Science.* 45: 335–60.

Hull DL. 2010. *Science as a Process: An Evolutionary Account of the Social and Conceptual Development of Science.* University of Chicago Press.

Jablonski D. 1987. Heritability at the species level: Analysis of geographic ranges of Cretaceous mollusks. *Science.* 238: 360–4.

Kant I. 1787. *Critique of Pure Reason.* In: Weigelt M (editor). 2006. Critique of Pure Reason/ Immanuel Kant. Penguin.

Kitcher P. 1984. Species. *Philosophy of Science.* 51: 308–33.

Kottler MJ. 1978. Charles Darwin's biological species concept and theory of geographic speciation: The Transmutation Notebooks. *Annals of Science*. 35: 275–97.

Laland KN, Uller T, Feldman MW, Sterelny K, Müller GB, Moczek A, Jablonka E, Odling-Smee J. 2015. The extended evolutionary synthesis: Its structure, assumptions and predictions. *Proceedings of the Royal Society B*. 282: 20151019.

Lightner JK. 2009. Karyotypic and allelic diversity within the canid baramin (Canidae). *Journal of Creation*. 23: 94–98.

Lindberg DR. 1998. William Healey Dall: A neo-Lamarckian view of molluscan evolution. *Veliger*. 41: 227–38.

Lovejoy AO. 1965. *The Great Chain of Being*. Harper and Row.

Ma B, Li M, Zhang L. 2000. From gene trees to species trees. *SIAM Journal on Computing*. 30: 729–52.

Mayden RL. 1997. A hierarchy of species concepts: The denouement in the saga of the species problem. In: Claridge MF, Dawah HA, Wilson MR (editors). *Species: The Units of Biodiversity*, pp. 381–423. Chapman and Hall.

Mayr E. 1940. Speciation phenomena in birds. *American Naturalist*. 74: 249–78.

Mayr E. 1942. *Systematics and the Origin of Species from the Viewpoint of a Zoologist*. Columbia University Press.

Mayr E. 1982. *The Growth of Biological Thought*. Harvard University Press.

Mishler B. 1999. Getting rid of species? In: Wilson RA (editor). *Species: New Interdisciplinary Essays*, pp. 307–15. MIT Press.

Morrison DA. 2014. Is the tree of life the best metaphor, model, or heuristic for phylogenetics? *Systematic Biology*. 63: 628–38.

Nee S. 2005. The great chain of being. *Nature*. 435: 429.

Pante E, Puillandre N, Viricel A, Arnaud-Haond S, Aurelle D, Castelin M, Chenuil A, Destombe C, Forcioli D, Valero M, Viard F. 2015. Species are hypotheses: Avoid connectivity assessments based on pillars of sand. *Molecular Ecology*. 24: 525–44.

Paterson HEH. 1985. The recognition concept of species. In: Vrba E (editor). *Species and Speciation*. Transvaal Museum Monographs. 4: 21–9.

Penprase BE. 2017. *The Power of Stars*. Springer.

Platnick N. 2016. Willi Hennig and systematics: A personal view. In: Williams D, Schmitt M, Wheeler QD (editors). *The Future of Phylogenetic Systematics: The Legacy of Willi Hennig*, pp. xi–xvi. Cambridge University Press.

Rabosky DL, Santini F, Eastman J, Smith SA, Sidlauskas B, Chang J, Alfaro ME. 2013. Rates of speciation and morphological evolution are correlated across the largest vertebrate radiation. *Nature Communications*. 4: 1958.

Raup DM. 1994. The role of extinction in evolution. *Proceedings of the National Academy of Sciences*. 91: 6758–63.

Reydon TA. 2003. Discussion: Species are individuals—or are they? *Philosophy of Science*. 70: 49–56.

Richards RA. 2010. *The Species Problem: A Philosophical Analysis*. Cambridge University Press.

Schmitt M. 2016. Hennig, Ax, and present-day mainstream cladistics, on polarising characters. *Peckiana*. 11: 35–42.

Sepkoski D. 2016. The 'species concept' and the beginnings of paleobiology. In: Allmon WD, Yacobucci MM (editors). *Species and Speciation in the Fossil Record*, pp. 9–27. University of Chicago Press.

Simpson GG. 1951. The species concept. *Evolution*. 5: 285–98.

Simpson GG. 1961. *Principles of Animal Taxonomy*. Columbia University Press.

Slater MH. 2013. *Are Species Real? An Essay on the Metaphysics of Species*. Palgrave Macmillan.

Smocovitis VB. 1992. Unifying biology: The evolutionary synthesis and evolutionary biology. *Journal of the History of Biology*. 25: 1–65.

Soucy SM, Huang J, Gogarten JP. 2015. Horizontal gene transfer: Building the web of life. *Nature Reviews Genetics*. 16: 472–82.

Stamos DN. 2007. *Darwin and the Nature of Species*. State University of New York Press.

Sterelny K, Griffiths PE. 2012. *Sex and Death: An Introduction to Philosophy of Biology*. University of Chicago Press.

Valdecasas AG, Williams D, Wheeler QD. 2008. 'Integrative taxonomy' then and now: A response to Dayrat (2005). *Biological Journal of the Linnean Society*. 93: 211–6.

Venditti C, Meade A, Pagel M. 2010. Phylogenies reveal new interpretation of speciation and the Red Queen. *Nature*. 463: 349–52.

Wallace AR. 1855. On the law which has regulated the introduction of new species. *Annals and Magazine of Natural History*. 2nd Series. 16: 184–96.

Webster AJ, Payne RJ, Pagel M. 2003. Molecular phylogenies link rates of evolution and speciation. *Science*. 301: 478.

Wheeler QD. 2008. Introductory: Toward the new taxonomy. In: Wheeler QD (editor). *The New Taxonomy*, pp. 1–19. CRC Press.

Wheeler QD, Meier R. 2000. *Species Concepts and Phylogenetic Theory: A Debate*. Columbia University Press.

Wilkins JS. 2009a. *Species: A History of the Idea*. University of California Press.

Wilkins JS. 2009b. *Defining Species: A Sourcebook from Antiquity to Today*. Peter Lang Publishers.

Winsor MP. 2003. Non-essentialist methods in pre-Darwininan taxonomy. *Biology and Philosophy*. 18: 387–400.

Winsor MP. 2006. The creation of the essentialism story: An exercise in metahistory. *History and Philosophy of the Life Sciences*. 1: 149–74.

Wood TC, Wise KP, Sander R, Doran N. 2003. A refined baramin concept. *Occasional Papers of the Baramin Study Group*. 3: 1–14.

Zachos FE. 2016. *Species Concepts in Biology: Historical Development, Theoretical Foundations and Practical Relevance*. Springer.

# 8 How to Name a Species

## WHO ARE THE MAKERS OF NAMES?

The vast majority of species on Earth remain undiscovered, undescribed, and un-named (Chapter 10; Figure 10.3). Inadequate availability of taxonomic expertise, and infrastructure, have prompted concern about the 'extinction' of taxonomists. Isley (1972) envisioned a dystopian future without taxonomists almost 50 years ago. New species continue to be described; in fact, more new species are described every year and certainly more are described now than ever before (Tancoigne & Dubois, 2013). A few better known taxonomic groups do have overall declining rates of discovery (birds, possibly mammals; Figure 10.3), but this is not a general trend. Who is doing this work, and where is it happening?

The rates of discovery and the development of technologies that facilitate our identification of new species are accelerating, and more different kinds of science are incorporated into the grand project to describe Earth's biodiversity (Wheeler et al., 2012). These tools include rapid DNA sequencing (Butcher et al., 2012) as well as technology for understanding morphology, from digital cameras to high throughput three-dimensional imaging (Sumner-Rooney & Sigwart, 2017). Massive digitisation projects have put troves of historical literature into the public domain. Modern travel infrastructure, cultures of open international communication, and electronic communication by email and video enable sharing intellectual infrastructure and building strong bonds between colleagues even when they may have never met face to face.

The question for taxonomy (the science of describing taxa and creating, maintaining, and updating names of species and species groups) is whether enough people have enough training and resources to meet the informatics demand for global efforts to document and describe biodiversity (Peterson et al., 2015). There is a growing community of scientists who contribute to scientific projects that include publication of new species names, as tangibly measured by authorship on scientific papers. But the rate of expanding authorship is much higher than the increase in the number of species described. A study in 2013 demonstrated that more people in more countries are participating in the global project to name species now than ever before (Costello et al., 2013). The authors of that study controversially interpreted their findings as evidence of a *decrease* in the rate of species discovery. Realistically, though, this only reflects broadening participation by researchers whose primary output is not taxonomic (Mora et al., 2013; Poulin, 2014), coupled with the significant and widespread trend in increasing co-authorship in scientific publishing (The Economist, 2016).

It should be noted that an arithmetic average of species-per-author is not a good descriptor of this very skewed distribution—a few authors describe a lot of species, and many contributors are co-authors on one paper (Bouchet et al., 2016)—averages are used to describe normal bell curves (Chapter 2). The 'long tail' of authors who make a single co-authored contribution in a taxonomic publication is probably continuing to grow, as multi-authorship is increasingly common. In a more detailed study of the relationship between authorship and taxonomic descriptive output, Poulin (2014) showed that while the total field of authors participating in taxonomy is growing, and this is a Good Thing, most of those people's participation is transient. The salient point here is that there is a difference between taxonomic authorship, and taxonomic *expertise.*

In taxonomic practice specifically, expertise is recognised by attaching the taxon authority to every name, the author, and date that come after the species name (Chapter 4). In taxonomic descriptions there is the option to specify the species or taxon authority as a subset of authors on the paper, though this is more common in botanical than zoological nomenclature (e.g., ICZN 1999: recommendation 50A). Such detail is a level of technical finesse that may be obscure to non-taxonomists and it is not clear how such authors are counted in larger meta-data analyses or whether that option is popular enough to have any significant effect on broader trends. In modern scientific publishing, more individuals are listed as formal authors who would historically have been included in the acknowledgements (or not at all); this could be seen as 'inflation' or correcting historical injustice. Regardless, all the authors listed on a paper that names a new species have demonstrably participated in at least some aspect of that scientific process. The concern should be whether each person in that team has the skills and confidence to do it again, independently or as part of a different team. Anecdotally, the answer is probably not.

There has also been a recent trend in the globalisation of taxonomic authorship, and scientific publications in general coming from authors in countries outside the traditional centres of North American and Europe. The number of professional scientists who regularly describe species has demonstrably declined in the UK (Boxshall & Self, 2010), and anecdotally in other European and North American countries, but the global number of scientific name-makers may be level or even increasing. Authorship trends show the proportion of papers with contributors from Asia and Latin America have increased over several decades (Costello et al., 2013). The future for taxonomy is not to worry about recalculating authorship data for a more just reckoning of how many people should be counted as 'taxonomists', but to move forward to support the existing workforce and ensure the rate of activity continues to increase.

One of the most common questions people ask of taxonomists is 'how many species have you named?' The long tail of the skewed distribution of authorship is growing, but this is nonetheless a general pattern that has always been true. Most people (including the co-authors on many multi-author papers) who participate in describing species have only named one species; a few have named hundreds. The prolific name-makers are rare but have a large impact. A more recent study, Bouchet and colleagues (2016), showed that in the last 15 years of marine mollusc taxonomy, 6% of the authors contributed 50% of new species. Prolific taxonomists may be beloved, like Linnaeus, who named over 10,000 species (Stearn, 1959; Farber, 2000). Others may be actually quietly resented for their idiosyncratic approaches to systematics in a particular group of organisms (see also Chapters 4 and 6).

Names are familiar to all users, but of course it is not reasonable to start publishing in oceanography because you have been to the beach—new contributions must build on a thorough knowledge of what is already established. In the very permanent world of taxonomy, errors from carelessness create problems that can last literally for centuries. However, the balance of necessity to caution is shifting (Riedel et al., 2013). Enabling taxonomy to thrive as a discipline depends on broad participation. There are too many species that are undiscovered, and limited time to document them as anthropogenic destruction expands (Maddison et al., 2012; Wheeler & Hamilton, 2014). Everyone engaged in organismal biology should consider themselves as potential taxonomists and participants in the global endeavour to document our biodiversity.

The fundamental data for a taxonomic finding are often based on specimens that were collected by others, and always considered in light of descriptions and evidence previously published by others. Blackwelder (1967: p. 312) wrote beautifully about the four pillars of ethics in taxonomic publication: '(1) integrity, (2) the giving credit for the help of others, (3) courtesy and forbearance in language, and (4) humility'. Taxonomic publication, perhaps more than other areas of science, progresses with small increments of new knowledge. Yet taxonomic descriptive papers have a much greater longevity than other areas of science. We can read a species description published 100 years ago, and all of it is still relevant and in fact may remain the primary reference point for all other work on that species.

There are formal, established methods for naming species, codified in rules that must be followed for animals (International Code of Zoological Nomenclature, ICZN, 1999), and plants, fungi and seaweed (the Botanical Code or International Code of Nomenclature, McNeill et al., 2012). The aims of this chapter do not extend to providing a definitive guide to taxonomic practice, or a sufficient manual to help someone directly engaged in preparing a species description for publication. For that, readers are directed to Judith Winston's (1999) excellent desk reference *Describing Species*, though taxonomic practice is developing continuously (see also Quicke, 2013).

This book is primarily intended for users of species names. Knowing the rules of taxonomy is like driving. If I visit a new country and rent a car, it is my responsibility to understand which side of the road to drive on and what traffic signals mean. If you cause an accident, saying 'I never read the Rules of the Road' is unlikely to help your case. The International Codes of Nomenclature, like the rules of the road, are dry and impenetrable, and it helps enormously to have them demonstrated to you by an instructor or at least a more experienced driver. Some knowledge of the rules can help taxonomic passengers avoid doing anything ignorant, and to understand their journey through the world of scientific names.

## PRONUNCIATION

Names in formal scientific nomenclature are colloquially called the 'Latin' names, though the names are really a pot luck of classical Latin and Greek roots, place names, honorifics, and taxonomic in-jokes. Some scholars have strong opinions about using pronunciation that is accurate to the etymology; the argument is that, for scientific names to be effectively universal, there should be a universal sound as well as spelling. Realistically, there are regional differences in agreed pronunciation of scientific names, much as for many English words. A reassuring rule of thumb, is this—Latin is a dead language, so you can say it however you want. This is well-intentioned but imprecise: Greek is very much alive, and in Latin, many expert disagreements about the spoken words spring from differences between classical versus ecclesiastical Latin and the language's long history (Pyles, 1939). Say a name with confidence, and listeners will simply assume you were trained 'somewhere else', and they may self-consciously wonder whether they are saying it wrong themselves.

Jumbles of letters are even more intimidating in italics. It is crucial to say names out loud, and repeat them frequently, to make them familiar and to build associations between the name and the object. Some names are notably, even deliberately, difficult. *Phthipodochiton* Sutton and Sigwart (2012) is the absurd name for a genus of fossil chiton. *Phthi-* is a Greek root ($\varphi\theta\iota$-) pronounced more or less 'fee' with a lisp (*fthee*) and means 'consumed' or 'waned away'. A witty Classics scholar, the wife of a friend of mine, suggested this gem (but my co-author has never forgiven me). The name evocatively describes the organism, a chiton lineage that has lost the ancestral molluscan foot (*Phthi-*, waned away, *podo-*, foot), and it is fun to say.

## WRITING AND SYNTAX

The convention of writing species names in italics or underlined, a typeface to differentiate them from the surrounding text, is the same as the convention for any words that are in a different language from the main text. A few Latin phrases are integrated into English and typically not italicised—e.g., (*exempli gratia*, 'for example'), ad hoc, i.e., (abbreviation for *id est*, 'that is')—but purist editors prefer italics even for those. Keeping species names in italics has a minor practical benefit in scientific writing in that it makes the species names stand out such that they can be found rapidly by eye when skimming a page of text. Vigilance about italics is mainly important because regular text would be taken as evidence of ignorance.

The genus name always has the first letter capitalised, the species name is always lower case. This convention is inherited from the first uses by Linnaeus and has to do with the general grammatical construction of the scientific name as a noun (the genus) and adjective (the species). In other languages, though not in English, this arrangement of having the noun first and adjective following is common. Folk taxonomies also follow this two-part general type + specific descriptor, or noun + adjective format for organism names (Stevens, 2002). In English as well as other languages, it was historically conventional to liberally capitalise nouns in written text, but not adjectives, and that approach is the origin of the capital letter on the genus but not the species name. Capital letters were historically used in species epithets derived from the proper nouns naming places or people (this was true until the late 20th century in botanical nomenclature, but this is no longer done), such as *Caloplaca obamae* Knudsen, 2009, a lichen named after US President Barack Obama.

Taxonomic binomials* are names that identify species as much as your own name identifies you. The correct spelling and use of capital letters are genuinely important, much more than the typeface. Modern word processors can be aggressive

---

* This word is one of the differences between the two major codes of nomenclature, ICN and the ICZN: in the ICN and earlier versions of the ICZN, two-part names were called 'binomial' names, but in the most recent version of the ICZN, the term used is 'binominal'. The two words are used interchangeably in this book.

## TABLE 8.1
### Word Endings for Taxonomic Names

Suffixes in bold are dictated by the relevant nomenclatural code; others are recommended, or where blank here, are variable.

| Rank | Zoological Names | Botanical Names |
| --- | --- | --- |
| Phylum/Division | | –phyta/–mycota |
| Class | | |
| Subclass | | –idae |
| Order | | **–ales** |
| Suborder | | **–ineae** |
| Superfamily | –oidea | n/a |
| Family | **–idae** | **–aceae** |
| Subfamily | **–inae** | **–oideae** |
| Tribe | **–ini** | **–eae** |
| Subtribe | **–ina** | **–inae** |
| Genus | | |

about correcting spelling, which is not helpful when including words in a different language. Many Latin names are similar enough to English words that auto-correcting algorithms can wreck havoc (*muscosa* means mossy, but a computer spell check insists it should be 'mucosal').

The proper taxon names of higher-ranked groups are always capitalised, but the name of the rank is not. We write about kingdom Plantae, family Asteraceae. In English, this is never reversed—you can say orchids are 'in the family Orchidaceae' or simply 'in the Orchidaceae', but not 'the Orchidaceae family' because that sounds repetitious. The suffix –aceae already indicated a family taxon. The endings of major ranked clades are standardised and intended to transmit information about relatedness in shorthand in this way (Table 8.1), but take care, these endings are very different for plants and for animals.

The stem of group names can also be used in the vernacular. Among animals, the family name always ends in –idae, such as Felidae (with a capital letter), and the standard form of a vernacular name for a family is to drop the 'ae', and we can describe a cat as a felid (lower case f). A common name ending in –id is understood among zoological taxonomists to mean a family-level group. Without these signifiers, or explicit mention of rank, you need prior knowledge about the organisms to have any idea what level of diversity might be included. Higher-ranked groups and unranked clade names have less rigidly defined conventions for their English names. Protostomia and Deuterostomia are protostomes and deuterostomes. The phylum Mollusca contains the molluscs, while members of Bryozoa are bryozoans.

Common names, confusingly, may or may not be capitalised, and the spelling may vary regionally. Most species do not have agreed common names (because they have scientific names); threatened or commercially important species may have officially legislated common names (e.g., Page et al., 2013). Officially recognised common names are treated as grammatical proper nouns and therefore capitalised.

This has been formalised in bird common names since the 1970s (Parks, 1978), and it is a broader trend in many organisms but there is no universal consensus (Nelson et al., 2002). It is informative to use capital letters for names that could be misunderstood: a Lemon-scented Gum is a tree, *Corymbia citriodora* (Hooker) Hill and Johnson, 1995, not confectionery.

## ENDINGS AND NAME CHANGES

Many people are intimidated by unfamiliar Latin declensions, including most taxonomists. Learning Latin or Greek is certainly not prerequisite to the science of taxonomy. Even people who learned Latin in school, as I did, often rely on consultation with other colleagues to select names and check grammatical details before they are committed to final publication. The grammatical genders of the species name (basically an adjective) and the genus name (a noun) should agree in number, gender, and case. To be clear, this description is deliberately abbreviated; not all species names are declined adjectives, but most behave that way. In Latin words, the ending of the word usually is a giveaway for the gender, and basically things that end in '–a' are usually feminine, but not always. *Eucalyptus*, for example, ends in an apparently masculine '–us' but actually it is feminine. Honorific names retain the gender of the subject, not their apparent gender as a grammatical word, even if they have a Latin-ish looking ending. (So the gender of a species named after Barack Obama is masculine: the genitive, or possessive form is *obamae*.)

In revisionary taxonomy (Chapter 4), the species epithet stays with the taxon while the genus name can change but the species name may then require a little grammatical adjustment to match its new genus name. The Botanical Code actually recommends that when a genus is split, newly erected genus names should maintain the same grammatical gender to minimise disruption (McNeill et al., 2012: ICN Art. 62). Two points to take away from this are—first, do not panic if the last couple letters of the species name change, and second, copy names *exactly* as they are presented in authoritative sources (Chapter 4).

## PLURAL AND SINGULAR

When referring to a species group, it should be considered a singular entity, the group is an object with its own defining characteristics. The family Tridacnidae is a distinctive group, and *members* of Tridacnidae are giant clams. Similarly, a species is a unit itself, so that *Tridacna gigas* (Linnaeus, 1758) has achieved sizes of over 1 m, and the largest individual *T. gigas* recorded had a shell length of 1.36 m (Hutsell et al., 1999). This is not to be confused with a philosophical view of species-as-individuals (Chapter 7); however, writers should be conscious about which characteristics can be ascribed to a species as a general statement or to individual organisms.

The word 'species' is the same in singular or plural: one species, or many species. The plural form of genus is genera, and the plural of phylum is phyla; this is the same ending as in bacterium (plural bacteria) or vomitorium (plural vomitoria). There is one species in the algal genus, *Johansenia*; it is in the family Corallinaceae, with 67 other genera containing 479 species (Hind & Saunders, 2013; Guiry & Guiry, 2017).

Selecting the names of species is a daunting responsibility, but a great pleasure. Species (and higher groups) can be named after descriptive characteristics, or places, or to honour people (Chapter 4). Honorific names were common in the 19th century, as acknowledgement to patrons who supported scientific work and collecting expeditions. At the end of the 20th century, some institutions experimented with selling species names to raise research funds. Critics worried that this could cast doubt on the taxonomic validity of names, if there was a temptation for deliberate over-splitting to provide names to patrons (Minelli et al., 2000). Philanthropic sponsorship has not become widespread, though it does happen regularly, and despite a few exceptions the financial return on most species names sold is not particularly high (Trivedi, 2005). A charity based in Germany, BIOPAT (*Patenschaften für bioloische Vielfalt*, Sponsorships for Biodiversity), will facilitate a commemorative name in exchange for a standard donation of €2,600, which is split between the research funding for the taxonomic author and other conservation activities (Steghaus-Kovac, 2000).

It has always been common to name a species after the person who first collected it, if they are not in a position to describe the species themselves, as a gesture of acknowledgement and commemoration. In modern science, it is increasingly common to invite colleagues to join a project as a collaborating author if they contributed important specimen material. However, because you cannot name a species after yourself[*], adding someone as a co-author in publication precludes naming a species after her or him. This choice is a matter of personal taste and circumstance: would you rather be an author on the paper or have a species named after you? Scientists building their research reputation may strongly prefer authorship. Established senior scientists, or those without ambitions in research publication, may prefer the species honorific. The balance could rest on what sort of organism is being named.

## HISTORY AND THE ROLE OF COLLECTIONS

Museums of all kinds serve a dual function in society, as they have since the 18th century and continue to today (Barber, 1980). In the case of natural history museums in particular, those functions are to educate the public on issues of biodiversity and the natural world on the one hand; on the other hand, natural history museums are guardians of vast stores of scientific specimens, millions of objects never intended for display, maintained and preserved for the present and future use of researchers (Suarez & Tsutsui, 2004). The common thread that links both purposes is a sense of permanence and stewardship; museum holdings will be preserved effectively forever.

Specimens collected for one purpose find utility for others (McLean et al., 2016; Webster, 2017). Importantly, the contents of scientific collections are expanded, not replaced, with additional ongoing fieldwork. Sometimes multidisciplinary studies get new data out of old specimens, using technology that could not have been imagined when the organisms were alive (Sumner-Rooney & Sigwart, 2017).

---

[*] As a technical point, it is not forbidden, but it would be the height of poor taste.

Modern museums are direct descendants from the 18th and 19th century fashion among educated gentry for a 'cabinet of curiosities' in the home (Sheets-Pyenson, 1987). These individual cabinets were often contained in a physical piece of furniture, but quickly expanded to rooms in the homes of enthusiasts, and gradually expanded to occupy separate, purpose-built buildings for housing and displaying large collections to the interested public. These same cabinets form the earliest collections of major museums and remain important specimens in their care (e.g., O'Riordan, 1976). Early material (and more recent acquisitions) can include type specimens, the most important specimens to taxonomic names (Chapter 4). During the mid-19th century, a huge number of public museums opened in many cities to serve the desire of an increasingly affluent middle class. This 'Museum Movement' developed in parallel in different parts of the world, in North America, Europe, and beyond (e.g., in India: Leviton & Aldrich, 2000). This cultural movement was driven both by an appreciation for aesthetic spaces and for the middle-class interest in improving minds.

Aesthetics is only one minor aspect of the value of museum collections, and the notion of a piece being 'museum quality' is misleading for a biological specimen. In museum parlance, a natural history 'specimen' is any object from the natural world, critically maintained together with information about where, when, how, and by whom it was collected. The difference between a seashell and a museum specimen of a mollusc is not its beauty, but in the archived documentation about the shell's species identification, and where it lived its life as a snail. A 'museum quality' specimen in scientific collections is valued by the sample preparation and having beautiful data. Collections of these objects held in a natural history museum form a physical database, used for biological and environmental sciences. Because of the continuous and permanent nature of museum collections, this database is an ongoing amalgamation of old and new specimens and their related data (Figure 8.1).

The combination of collections of physical specimens with published written and illustrated descriptions of species is the foundation of taxonomy as a scientific discipline. Physical specimens, permanently preserved, coupled with published descriptions that refer to the particular individual example or examples examined, allow these species descriptions to be validated at any point in the future by other scientists. That is, the species description proposed by the author, based on the specimens examined, can be experimentally repeated by any competent observer with access to the same specimen objects, or other comparative examples of the species.

The most important specimens to museum science are the individual examples that were used as the basis for published descriptions of new species, referred to as 'type' specimens (Chapter 4; Winston, 1999). Museums may hold type specimens for species elemental to the local fauna or through the work abroad of scientists connected to the institution or by historical accident. Large museums employ staff scientists whose research programmes are based in the collections, both using the available specimen resources and actively building and expanding the collections. But the collections are available for any researcher to access, either by visiting or by the museum staff sending specimens on loan.

Taxonomic work is impossible without access to the preceding centuries of accumulated data, specimens, and published records. Yet taxonomy, and studies of biodiversity, are perhaps more important in 21st century science than in the

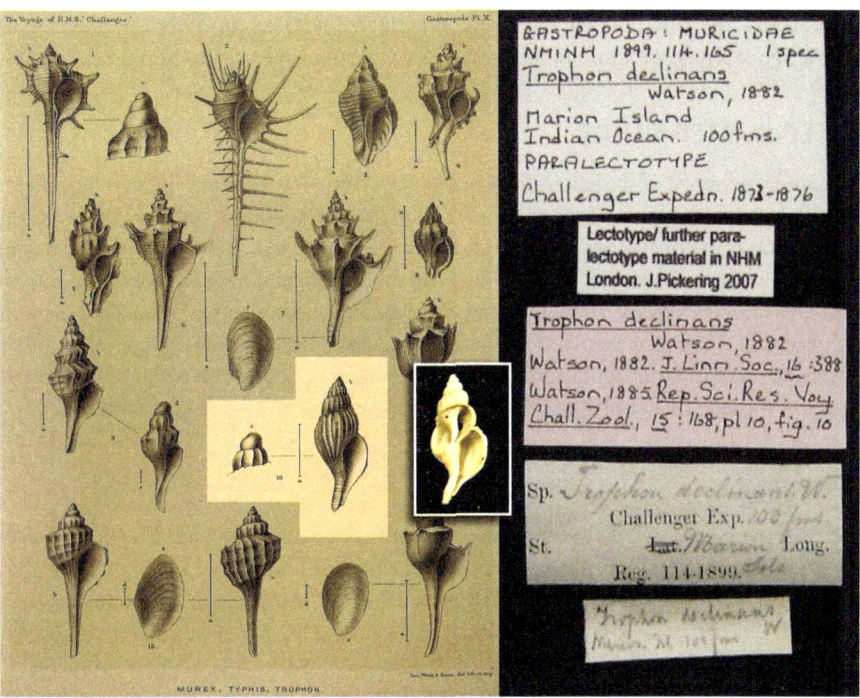

**FIGURE 8.1**   The paralectotype specimen (centre inset) of the gastropod *Enixotrophon declinans* (Watson, 1882), shown alongside the illustration from the original description (left) and the specimen labels (right) housed with the specimen in the National Museum of Ireland. The labels, from bottom to top, include the original pencil note '*Trophon declinans*/Marion Id, 100 fms', an original collection label from the H.M.S. *Challenger* Expedition collection, a note (on pink paper) with the publication details for the species, a printed note cross-referencing other type material, and an updated collection label showing classification, museum catalogue number, locality, type status, and collecting date. Subsequent taxonomic revision means that many of the species names used in the 1880s have been updated and altered to reflect our current understanding. *Trophon declinans* Watson, 1882 is now in a different genus, *Enixotrophon* (Barco et al., 2015), but the original specimens remain important reference points.

18th century, when it was a pursuit of those who could financially afford a pleasant interest in the natural world (Farber, 2000). This is essentially historical data being applied to modern problems in combination with modern scientific resources.

Taxonomy is a science with a long historical tradition. The 18th and 19th century Museum Movement was a product of the interaction of science and public enthusiasm for the aesthetic and knowledge-based aspects of scientific collecting. Specimens acquired in those early days of rigorous taxonomy are not historical relics but remain essential and active tools of modern science. This is particularly true in understudied taxa, where information about the groups has been acquired at a relatively slow pace through history. The limited number of scientists working on groups such as molluscs, and other far more neglected animals and plants, means that working scientists today are directly expanding the work of colleague-ancestors from a century ago, sometimes without any intermediary research having been undertaken. This community of knowledge extends across boundaries of space and time and reaches far into the future.

## A STEP-BY-STEP GUIDE

Like any experience of the scientific process, descriptive work is not linear but cyclical. This is an example of what Latour (1999) called the 'circulating reference', the series of incremental transformations in ideas and perception in the development of a scientific study; how the wonderful basic things we take entirely for granted (like maps) originally came from the meticulous work of others, and eventually lead to a new 'a ha!' moment. The steps involved in naming a new species are not separate events that have to happen in a particular order. The following sections are a rough reflection of the main threads that contribute to the overall workflow of describing a species.

### DISCOVERY

The 'discovery phase' of finding a new species is open to anyone; it usually starts when you attempt to identify a specimen, but find an individual or population that does not quite fit the available description. We use the relevant regional keys or guides to get the best possible identification. But it does not sound quite right, or a specimen is not quite like the illustration. There are various possible explanations: (1) I got it wrong, perhaps by misunderstanding the descriptive vocabulary used in the key; there may be strange anatomical terms and I guessed a bit about what they meant. (2) My specimen is missing key features (it might be damaged, or a juvenile, or not flowering right now), but, having never seen one before, I cannot know that. (3) The key or field guide was wrong or incomplete; every book contains errors. Or, (4) The organism I am looking at goes beyond the realm of knowledge presented in the key; because my specimen has not been found in this place before, my specimen is different from the expected range of variation for its species, or my specimen is from an entirely new species.

So, the very first step in finding or naming a new species, is to be an expert. Unfortunately, there is no realistic way to shortcut the process of this first step, because you have to establish a sufficient comparative context to be able to recognise what is different and potentially interesting. An expert is not necessarily a taxonomist or systematist; anyone who is attentive to natural history and spends many hours making observations in the field is likely to gain a holistic view of local species. Characteristics that are not used in traditional taxonomy, like behavioural ecology, may still be a giveaway for a new species. Becoming an expert involves observing the local organisms around you, learning about their relatives through other documentation, books, or specimens, and practising good noticing. Only by knowing the scope of available knowledge can you successfully identify something that really is novel. If your primary goal is to name a new species, it is also helpful to work on groups where there is a high proportion of undescribed species, such as nonvertebrate animals, and fungi. Although it may be harder to get started, the more obscure the group, the more likely you are to be welcomed with open arms by the small community of people who care passionately about them.

## PRESERVATION

New species are often found in museum collections. For marine invertebrate animals, recent calculations predicted 65,000 undescribed species already collected and housed in museum collections, but not yet published (Appeltans et al., 2012). The scientific collections housed in museums are usually physically arranged in systematic order, separated into major clades, ordered on shelves like books in a library, by families, genera, and species (Müller-Wille, 2006; Figure 8.2). At the end of each group is often a little shelf labelled 'sp.' or 'unidentified', containing the material that was donated without species identification. No museum or herbarium has sufficient staff to be expert in everything—there are far too many species—and unidentified material awaits the curiosity and generosity of visiting experts or a student project to examine them and connect specimens with identifications.

Expert knowledge about a particular group, understanding the technical details that enable you to describe a species, is arguably of more practical relevance to actually naming a species than any of the skills required to find a new species. Even outside museums, if you went to the lab of any invertebrate biologist who engages

**FIGURE 8.2**   A small section of one herbarium cabinet in the University Herbarium, UC Berkeley, California. The shelf holds specimens of the fern genus *Dryopteris*, each folio contains several herbarium sheets in the same species. An example herbarium sheet is shown to the right, *Notholaena ochracea* (Hooker) Yatskievych and Arbeláez. The specimen label records details of the collectors (CJ Rothfels, R Dyer, P Rothfels), the date (27 Aug 2009), and the collecting locality, including habitat notes and latitude and longitude.

with taxonomy or systematics, and said, 'I want to describe a species', they probably have one in a drawer that they have not yet got around to working on (though they would not necessarily be inclined to hand over responsibility). The life cycle of species descriptions can be long—a scale of several decades is a normal lag between discovery and description (Fontaine et al., 2012; Bebber et al., 2014). The end part of the process requires depositing designated name-bearing type material in a publicly accessible museum collection. Proper preservation of specimens is critical, and the method used has downstream consequences for what type of analysis or description can be accomplished. As death is irreversible, this requires some forethought.

You need to know in advance what features are important for species identification, in order to determine how to preserve a specimen of a living species. Identification of vertebrate species may depend on skeletal characters, which requires extensive preparation (Carter & Walker, 1999). Plants are traditionally pressed and dried on herbarium sheets, but the drying process has to protect them from decay or mould (Clark, 1986). Most soft-bodied animals, but also vertebrates and some plants, are preserved whole in a fluid preservative (Collins, 2014). Anatomical preservatives (e.g., glutaraldehyde or formalin, the latter being an aqueous solution of the gas formaldehyde) are 'cross-linking' fixatives that create chemical bonds between proteins and anchor cellular proteins to the cytoskeleton. This makes the tissue more rigid and maintains its original shape and structure, and also stabilises many pigments to prevent them degrading. However, the cross-linking also disrupts the structure of DNA molecules. Aldehyde solutions are acidic, and must be chemically buffered to prevent tissue damage. Specimens might be transferred to a buffer solution or to alcohol for longer term or permanent storage. (Seawater, contrary to popular belief, is *not* a sufficient buffer.) Alcohol is not a fixative, though it is a preservative. Unlike fixatives, alcohols (such as ethanol) dehydrate and denature proteins, which causes tissues to wither or shrink as the water in cells is replaced by alcohol. However, DNA is relatively stable in a high-concentration ethanol solution. The preservation of DNA is more effective when the alcohol is kept cold (Quicke et al., 1999), and cooler temperatures have added safety benefits for storing a flammable chemical. Alcohols are also less hazardous to human health than aldehydes. The choice of fixative is a trade-off: formalin is good for anatomy but bad for DNA; alcohol is good for DNA but bad for anatomy. Other chemical agents (RNA stabilisation solutions, or other buffers) are good for short term storage for specific assays, but not for long preservation. All choices have benefits and drawbacks (Nagy, 2010). Alcohol, for example, bleaches out almost all colour from specimens. Since we do want to be able to identify additional members of our new species, it is important to also be able to describe its morphology accurately. If an organism is large enough, you can keep a snip of tissue for sequencing and preserve the rest to describe the morphology.

Documentation is an essential aspect of the preservation process. Live photographs of organism and habitat are clearly useful, but even more important are the essential data that form the potential record of a specimen as a permanent museum specimen. Date and locality are paramount among these data. Typical museum specimen records include not only the collecting locality and collecting date but also the names of the people who collected the specimen, donated it to the museum, and identified the species (who may all be different people), and information about the organisms from when it was found, and other habitat information or sedimentary context for fossils.

## COMPARATIVE RESEARCH—LITERATURE

With an unusual specimen in hand, and suspicion about its identity in mind, the overarching question remains, is it really new? There are a lot of species buried in the primary literature, far more than in field guides or keys. Once you have your mystery organism roughly identified and know what it seems to be most similar to, the next step is to begin an exhaustive search to prove yourself wrong by finding the mystery species in the archives of knowledge. Even if it is not a new species, sometimes a strange organism can expand the range of variation in a known species, or it may prove to be a 'lost' species, which was previously named from poor or limited specimen material, or perhaps described minimally and therefore ignored or forgotten by subsequent workers.

Comparative research, using literature and other specimens (below), requires access to specialist resources. This is much easier to do if you live in a place where there are well-resourced universities and museums, and far more challenging if you do not (Agosti, 2006; Chapter 12). The taxonomic scientific community needs an expanding workforce (Bebber et al., 2014; Peterson et al., 2015); specialists will always help other researchers, and resources are becoming more dispersed through digital projects. But an experienced guide or mentor can be more informative than any digital tool. Bear in mind that asking someone simply to identify things is time consuming and unrewarding for them, but if you have good specimen photographs (positioning the specimen to see key features) and make a first attempt at identifying it yourself, and you have good reason to think it is novel in some way, then you are likely to get an answer relatively quickly.

In taxonomy, more than any other branch of science, it is important to go back to early sources. A lot of the taxonomic literature is old. Older literature is often hard to find and much of it still has not been digitised, so an exhaustive literature search almost always requires access to very good library resources. Mass digitisation projects are rapidly improving this situation and making rare or obscure publications available to a broad international audience through open access projects like the Biodiversity Heritage Library (http://www.biodiversitylibrary.org) and Gallica digital libraries (http://gallica. bnf.fr), and others (Wen et al., 2015). Increasingly, scientific publication is conducted in English in all nations, which is a particular boon for native English speakers but also provides a universal benefit of increasing communication across diverse countries. However, there is literature in many languages and alphabets, and it is entirely possible that someone writing in Chinese or Russian already described the species you think is new; the onus is on the researcher to eliminate the possibility of redundancy.

The phrase 'I can't get it online' causes some inter-generational friction to those who remember a pre-Internet era. Rapid and easy access to digital resources means that younger scientists may never have been trained in the habits that were required in ancient times. So, a few practical tips are warranted. First, ask a colleague (there are professional email lists for most groups of organisms, and many of us who work in institutions with big libraries feel an obligation to help source literature for colleagues

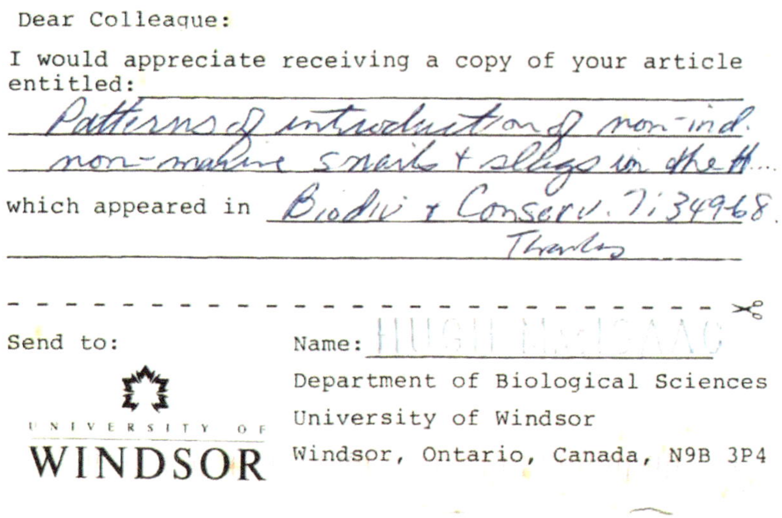

Dear Colleaque:

I would appreciate receiving a copy of your article
entitled:

*Patterns of introduction of non-ind.
non-marine snails + slugs in the H...*

which appeared in *Biodiv + Conserv. 7: 349 68.*

*Thanks*

- - - - - - - - - - - - - - - - - - - - - - - - ✂

Send to:                     Name: HUGH MCISAAC

                             Department of Biological Sciences
☩                            University of Windsor
UNIVERSITY OF                
**WINDSOR**                  Windsor, Ontario, Canada, N9B 3P4

**FIGURE 8.3**    An example of a pre-printed reprint request card, sent by Hugh McIsaac at the University of Windsor, to Robert Cowie requesting a copy of Cowie (1998). Today, sending the same sort of polite request by email to an author will usually gain a copy of the paper you need.

with overlapping research interests but who do not have direct access to the same infrastructure). People with the same interests may have previously needed to get a copy of the same obscure article and will have a photocopy or a digital scan they can share. For more recent literature, it is customary to email the corresponding author directly and ask for a copy of their paper (even if they are intimidatingly senior, they will usually be happy for the interest). Academic researchers are not generally paid for the articles they write, gain nothing from publisher-imposed paywalls, and are happy to send you their work for free if you ask. Some people still call these 'reprint requests' although articles are often not printed at all anymore; before email, many scientists had little postcards with 'please can you send me a copy of ...' and their return address preprinted, and blanks to write in the address of the author and the name of the article being requested (Figure 8.3). Before you give up hope, ask a librarian (a skilled librarian's sympathy is, surprisingly often, proportional to the obscurity of a request). You may find that a physical copy of the resource you need is right there on the shelf in your local university library building or through inter-library loan networks. University libraries are generally welcoming to members of the lay public who have a research need, though this may not extend to borrowing privileges, asking politely and explaining your interest will get you permission to read material within the library and access electronic resources in almost any academic library in the world.

## COMPARATIVE RESEARCH—SPECIMENS

The 'type' specimens mentioned above are the original individual specimens associated with a species description (Chapter 4; below). These are called 'name-bearing types', and they 'fix' the names, in the sense that they are a reference point that holds the meaning or interpretation of the species name. According to the codes of nomenclature every species has a holotype, a single specimen that becomes the reference point for the species (Chapter 4). The actual specimens are a permanent part of the history of science, they are held in perpetuity by museums, and they have a great and tangible power over the names we use for the species that fill our world. Type specimens are like relics, in the religious sense, and museums the reliquary. Museums hold many other specimens, and smaller museums can have important collections that have no type material at all. Specimens are used for many, many kinds of studies beyond taxonomy (McLean et al., 2016).

Most natural history museums function like international lending libraries. Some collections will not allow type specimens to leave the premises but will loan comparative material or may be willing to photograph type material on request. Comparative material can include specimens that may have been collected by the same scientist who described the species, which makes it a good bet it was correctly identified. But collections also contain material from earlier and later points in history, acquired and identified by other people, so not all identifications on museum specimens are guaranteed accurate. Museums always keep a complete history of identifications; in part, names change because of taxonomic revisions, but often a specimen comes to the museum without a species-level identification, then a curator identifies it, but later a specialist expert visits and perhaps realises it is something different, or more rare (or even a new species). All of these historical stages are retained as separate little slips of paper and packed away with the specimens (Figures 8.1 and 8.2). We always doubt, and hypotheses are always available for testing, so you will want to be able to reconstruct the logical process that led to the present identification.

Types are perhaps even more important in revisionary work than they are in the straightforward description of a new taxon (Chapter 4). To describe a new species, you mostly need to prove that it is not the same as anything that has already been described. If it is sufficiently different then this is not necessarily complicated, once you have mastered the comparative expertise required. Besides type specimens, museums hold comparative collections of many species from many places. There is great power in holding two objects side by side, in three dimensions, comparing them from all angles in shape, colour, and texture. Laying out multiple specimens reveals the variation and scope in morphology. You may see subtle details that were not documented in the written description; or features that the author of the original description thought were obvious and perhaps illustrated but forgot to spell out explicitly.

This is precisely why we have museum specimens, so that no one has to actually rely on their memory, and to maintain these records long after we lose our colleagues. Sometimes the type material is damaged or there is local morphological environmental plasticity, such that a feature in a newly discovered specimen is not as diagnostic as it might appear. Although we think of types as a total authority, that is a taxonomic legal standing; comparative material and the types should be considered together. The best source of information may still be the person who described a species, because no one ever looks at a species in so much detail as when they first describe it.

## Diagnosis and Description

A thorough literature search necessarily involves consulting the original descriptions of species and groups, and probably revisionary work that presents updated descriptions. Complete species descriptions contain several standard parts, which are different from other papers in life sciences (Notton et al., 2011). Descriptions start with the systematic details (higher classification, name, authority, and date), and references to figures, in a section header. This is followed by the synonymy (the accounting of previous use of names for the species; see Chapter 4), and then the sections of the description: type material (including designation of the holotype, see below), material examined, type locality (where the type specimen was found alive), distribution, etymology, diagnosis, description, and additional remarks. There are some variations on these sections, such as geological setting for fossil taxa, parasitic species would have a section about the host, phenology for organisms with seasonal patterns (such as flowering plants), or reference to molecular sequence data. There may also be an electronic name registration 'Life Science Identifier' (LSID) or Web link to the registration record (Figure 8.4). There is debate about species identification from DNA barcoding and the increasing importance of molecular data to resolving taxonomic problems. Species nomenclature applies to *all* species including fossils, and DNA sequences are not required by the codes of nomenclature for any new species name.

Nuculanidae H. Adams & A. Adams, 1858
*Ledella* Verrill & Bush, 1897
— Systematic placement (family, genus)

*Ledella knudseni* Taylor & Wiklund, sp. n.
http://zoobank.org/66E692B5-7C61-4ADC-9539-EFC085424147
— Taxonomic name and authority
— Electronic registration of name

**Material examined.** Paratype NHM_288A NHMUK 20170047.1-2, collected 2013-10-17, 13.75583 -116.48667, 4076 m. http://data.nhm.ac.uk/object/8aec47f4-dccc-4668-8398-9e4b0c28ecb8
    Holotype NHM_288C NHMUK 20170048, collected 2013-10-17. 13.75583 -116.48667, 4076 m. http://data.nhm.ac.uk/object/f1886d78-22bf-403e-bdb2-784b91c0eb12
— Specimen details, including museum catalogue numbers & type locality (latitude, longitude, depth)

**Description.** Shell relatively thick, robust. Ovoid with short rostrum, umbones broad, prominent; postero ventral margin sinuous; broad, shallow sulcus extending from umbones to posteroventral margin. Sculpture of low, relatively broad, closely
. . .
— Description, in telegraphic style

dissoconch large, ellipsoidal 0.3 mm long, with sharp rim, surface irregularly pitted. Holotype NHM_288C shell length 2.2 mm, width 1.5 mm; paratype NHM_288A shell length 2.1 mm, height 1.5 mm. (Figure 7).
    **Genetic data.** GenBank NHM_288A COI-MF157515; NHM_288C 18S-MF157491, COI-MF157516.
— GenBank registration for sequences from type specimens

    **Remarks.** Similar in form to *Ledella ultima* (Smith, 1885) widespread in the abyssal Atlantic (Allen 2008), but has a less massive hinge with more teeth, 8-9 compared with 6-8 in *L. ultima*. Also similar is the species identified by Knudsen (1970) as *L. ultima* from the Sunda Trench in Indian Ocean at 3810 m. The only species recorded from the deep eastern Pacific is *Ledella dicella* (Dall, 1908) from 734-1200 m off Ecuador but this lacks the short rostrum and has 12-13 hinge teeth on each side of the ligament (Coan and Valentich-Scott 2012 pl. 26). No genetic matches on GenBank. *Ledella knudseni* groups in a small subclade with but is distinct from the Atlantic species *L. ultima* and *Ledella jamesi* Allen & Hannah, 1989, as well as another *Ledella* species from this study in the Pacific, *Ledella* sp. (NHM_381) (Figure 12). The new species can be confused with juveniles of *B. calcar* (see above), but shell is less shiny and iridescent, and ribs are more pronounced. DNA may be required to confirm identification.
— Comparative remarks

    **Etymology.** Named for Jørgen Knudsen (1918-2009), deep-sea bivalve systematist and author of the Galathea Report on abyssal and hadal Bivalvia.
— Origin of the name

    **Ecology.** Found in polymetallic nodule province.
— Species habitat

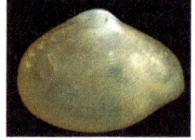

**FIGURE 8.4**    Anatomy of a species description. The left column shows the text of a recently published species description (reproduced from Wiklund et al., 2017). The right column contains annotations indicating the major parts of a typical zoological species description. This bivalve, *Ledella knudseni* Taylor & Wiklund, 2017, is shown in the lower right.

The *diagnosis* is a brief explanation of characteristics that identify the new species. (In older zoological literature this was formerly called the 'definition'.) The diagnosis is not necessarily of great biological or evolutionary relevance, but it is important because it usually presents practical features that facilitate a confident identification. Many authors see these two sections as redundant, but a good diagnosis is not just an abbreviated description (Winston, 1999). The diagnosis should contain some general characters that confirm the group and some specific characters that identify the new species. While the *description* section is as long as necessary, the diagnosis is concise. Until recently, but no longer, the Botanical Code stipulated that either the diagnosis or description, or both, had to be written in Latin (McNeill et al., 2012). (When that was still required, many botanical authors seemed to write the briefest possibly Latin diagnosis, and quickly move on to the rest of the paper in English.) Comparative notes explaining why this species is different from its relatives, are often placed in a different section, often titled 'remarks', at the end of the description. In some groups, comparisons are put in a special section called the 'differential diagnosis', separate to the ordinary 'diagnosis', that contrasts key features in similar taxa.

The *description* section of a species description is the longest part and serves to document all the observations about the new species. The description section must, as much as possible, refer specifically to the type material, and very specifically to the holotype. Among fossils, it is important to note whether the type is a fossil of the organism itself, or instead preserved as a mould. The type material fixes the name, so descriptions should not make any assumptions that any other specimens but the type are really the same species. Generalities, or observations about the range of variation in certain features, may be presented in the description or in other sections such as the diagnosis or remarks.

For future comparison, it is also important to present high quality figures of the type material and any relevant comparative material; every organism has a different set of key features that have to be identified. Drawings and photographs serve different purposes in scientific illustration; drawings may be necessary if features are not obvious from photographs (or scanning electron micrographs). Increasingly, aspects of morphology may be presented as digital 3D models (Sutton et al., 2014).

The description section, and the diagnosis, are usually written in 'telegraphic style'. This style removes unnecessary words (like verbs) in favour of the raw essential descriptive data, resulting in text that is fast and easy to read, even for those with limited fluency in English.

The challenge in writing a good description is to be thorough, but without wasting time and space on features that are redundant or irrelevant (Winston, 1999). Groups that have been the subjects of thorough, high quality monographic revisions often set a gold standard for style and even the presentation order of features to be included in future descriptions, but not all groups have that luxury. Later work will inevitably find new features that were not previously documented. For example, in chitons, we are very fortunate to have a detailed six-volume *Monograph of Living Chitons* that redescribed over 650 valid species known in the late 20th century (e.g., Kaas & Van Belle, 1985), and the style of that work provides a template for descriptions of new species. However, monographs are not permanently definitive, and it is the responsibility of the taxonomist to be aware of more recent ongoing developments in the group of interest. A new sense organ called the 'Schwabe organ' was described in 2014 that is actually found in more than 100 species in one clade of chitons (Sigwart et al., 2014), and now the presence and shape of that organ is included in descriptions of new species of that clade when suitable material is available (e.g., Güller et al., 2015). The features of interest in a species description are themselves a sort of hypothesis subject to refinement.

Notably absent from the standard description of a species is any phylogenetic context. Often, phylogenetic analyses of new species will appear in the same publication and inform the systematic placement of the new species in an appropriate genus, for instance. The analysis might be in the same paper, but it is not part of the description. Phylogenetic studies can be published separately, and often are (Chapter 6; Table 6.1). Because many species are discovered almost incidentally as part of the pursuit of phylogenetic or evolutionary studies, there is some concern about the lag between analytical results and descriptive results. An evolutionary tree could be published on the basis of molecular data, with terminals marked with just 'sp.' and numbers or incomplete identifications as a placeholder. If these are not clearly linked to museum voucher specimens, then it can lead to duplication of effort when the species is discovered again by someone else. Species need to be named. It may be a part of an evolutionary story, but without a name, it does not exist.

The Botanical and Zoological Code do not stipulate the total content of a species description, so presenting a diagnosis *or* a description are often considered alternative options (Renner, 2016), though many species descriptions provide both. Where there is sufficient evidence for a new species, an extended diagnosis, and appropriate designation of type material, is enough to fix the name (Butcher et al., 2012; Riedel et al., 2013). This approach is controversial, and traditionalists may think it prioritises speed over quality. But efficient basics can sometimes be the only way to manage, for example, in a case where the first proper revision of one species revealed 189 separate taxa (Lücking et al., 2017). Names are a first step to make sense of the chaos, and they need to be disseminated quickly (Maddison et al., 2012). Further description studies and analysis can be considered separate scientific endeavours. Putting a name on a taxon makes it available for the larger scientific community to continue working on it.

## TYPES

Because the type material lends an objective reference to the hypothesis of a new species, the type specimens, type locality, and the final description have to all line up, and connect to a single specimen from a known place. Centering the taxonomic framework on type specimens was not part of the practice of early taxonomists, and much of the apparent confusion over type designation comes from the detective work involved in deducing which individual specimens may have been directly examined by a scientist working over 200 years ago. The intervening centuries have revealed a constant stream of new features and new data that are important for species identification. The approach to preserving specimens has changed to capture those data, and to ensure the long-term conservation of individual specimens housed in museums. These historical variations have led to a confusing variety of different types of types.

The *holotype* is the single name-bearing individual specimen for a species. *Paratypes* are secondary type specimens, collected from the same type locality and at the same time as the holotype. Paratypes may be selected to represent additional life stages, or separate sexes (since the holotype can only be one individual), and paratype specimens may be deposited in different museums to increase accessibility of the specimen material. For historical names, *syntypes* were historically a set of multiple specimens that were all used as part of the original descriptive process (now that the rules require a single holotype, syntypes are no longer allowed). Revisionary taxonomy usually designates one member of such a type series as a *lectotype*, so a lectotype is like a *post hoc* holotype. Designation of a lectotype (and paralectotypes) fixes the name to a single specimen to prevent ambiguity. If the original type material was destroyed or permanently lost, and there is a need to designate a replacement specimen as a type, this new replacement is called a *neotype* (see also Chapter 6).

The Botanical Code has an additional provision for *epitypes*, supplementary type specimens for situations in which the original material does not preserve critical diagnostic characters. For example, if a common species is later discovered to be two cryptic species, but the holotype does not preserve whatever characteristic that separates the lineages (DNA usually, or a specific morphological feature), then an epitype can be designated that cites and effectively supplements the original type. This has been applied to fungi (Ariyawansa et al., 2014), and may eventually be extended to the Zoological Code (Schrödl & Haszprunar, 2016).

Early descriptions of soft-bodied animals were based on illustrations of live animals, because the preserved specimens did not retain any diagnostic features. What, then, is the 'type'? According to the Zoological Code of nomenclature, if a species name is based on an illustration, or a photograph, the type material is by definition the individual animal shown (not the illustration itself), even if that animal specimen was destroyed or never preserved (ICZN, 1999). The philosophy that there is a single individual name-bearing type organism, illustrated by images or other evidence, is an important extension of considering species names as hypotheses. A holotype must be designated for a new species (Notton et al., 2011); that names are based on individuals provides an objective basis for hypothesis testing. This does also allow for a name-bearing type individual to be a live animal, and other documentation can be presented to describe the species and crucially prove that the type individual existed, even if a specimen is never deposited in a museum. (It is generally not preferred, since not having access to the type organism limits its usefulness for future hypothesis testing.) This was refined in the 1999 Zoological Code with consideration for rare and endangered species, where sacrificing a live animal as a museum specimen would be considered unethical and unacceptably damaging to the species.

The ICZN makes an important philosophical distinction that a physical part of an organism can be type material, but illustrations of a type specimen are just that, illustrations, and not themselves types. This has been extended to DNA sequences; the actual basepair sequence of a DNA strand is an illustration of the DNA extracted from an organism, usually made by repeatedly copying the original through a process such as polymerase chain reaction (PCR). So, the sequence data, and the amplified DNA product, are important accessory data, but the type specimen is the individual organism that the DNA came from (ICZN Art. 72.5.6). Theoretically, a sample of pure DNA without amplification could be considered a type specimen (Jörger & Schrödl, 2013). Certainly, a DNA sequence can be an important part of the species diagnosis in living organisms.

It is relevant to reiterate that the type system is not an original feature of the Linnaean System of nomenclature (Chapter 4). Complications arise because of *post factum* application of names without reference to type material, or confusion over which specimen goes with a name (Bouchet & Strong, 2010). Although all species have types, and many early taxonomists (including Linnaeus) made reference to particular specimens, the idea that the species description is fixed by an individual specimen is a later innovation that allows taxonomy to be robustly evidence-based.

## PUBLISHING

Species descriptions are usually published in peer-reviewed papers, as is most of the primary literature in any other scientific discipline. In some cases, especially for important or charismatic fossils, these descriptions can appear in high profile scientific journals. However, the nomenclatural codes do not require peer review (books and monographs, for example, are not published through the same process as journal articles), though they do make explicit statements about what does or does not qualify as sufficient for publication (Notton et al., 2011). These rules are intended to make sure nomenclatural 'acts' are available to the whole scientific community, and also to prevent the potential chaos of people arguing that, for instance, a name written in a letter or on a Web site has some official standing.

In the early 21st century, there were long arguments about the electronic-only publication of scientific journals, especially in concern over the archiving standards of something that was published on the ephemeral Web (Nicolson et al., 2017). The transient compromise by some taxonomists was to take electronically-published journal articles, print them out, and donate them to academic libraries (Nature, 2012), but this was cumbersome and ridiculous. Another serious potential problem was establishing consensus about publication dates, since many scientific journals publish content online first, and assign it to a formal issue with page numbers later; so which publication event is the date of a new species name? The Botanical Code was updated in 2011 to include explicit provisions for electronic publication (McNeill et al., 2012). A short time later, a special amendment to the Zoological Code was published to the same effect to allow electronic publication of nomenclatural acts (ICZN, 2012). In both codes, the date of species description is the first online appearance of a publication.

Electronic registration of names is required for all new names of fungi (McNeill et al., 2012), and registration is mandatory for electronic-only works that contain new names of animals and strongly encouraged for all new names of animals (ICZN, 2012). Fungal names can be registered through MycoBank (Crous et al., 2004) or Index Fungorum, either of which can currently generate an official 'Life Science Identifier' (LSID) number which must be included in the publication of the taxonomic description. The LSID represents a persistent identifier which should provide an improved future way to trace the history of revisionary taxonomy (though it does not obviate the need for revision to names) (e.g., Page, 2008). For animals, the same process is managed through the online database ZooBank, which serves as the official registers of names for the ICZN. For animals and fungi, currently, electronic publication of new names is not valid without the LSID. These databases also serve as a central clearinghouse for data on described species, the associated publications, and type material.

## CRITICAL REFLECTION

After all the analyses are done, there is still scope to critically evaluate the evidence for a new species. The aims of describing a new species are to establish a case that the differences observed are sufficient evidence that a group of organisms represents an independent evolutionary trajectory. As discussed earlier in this book, this does not depend on 'species concepts' or single lines of evidence. It is more rational to see that all lines of evidence contribute to the common goal of understanding species (Chapter 7). Our understanding of what constitutes necessary and sufficient evidence is of course in a constant state of development (Latour, 1999).

The esoteric or technical details contained in a species description make the primary literature accessible only to a small group of trained specialists, but that is not to say that taxonomy is exclusive. No one can own or control a taxonomic group. Expanding on Blackwelder's (1967) dictum of humility, taxonomists work in service to our organisms, to our colleagues, and for the advancement of human knowledge of the natural world.

The forces that shape and define species are at the forefront of investigation in evolutionary biology. And the work of separating lineages and putting names on them is a separate but strongly interacting process with this scientific endeavour. The field of taxonomy is strongly connected to its heritage, and while all science proceeds by incremental improvements on the work of previous scholars, the connection to physical specimens makes this a real connection across scientific generations. The accumulated observations of experts, and accumulated specimen collections of museums, represent data storage in more interesting forms than in other disciplines. But a long history does not create stasis. Species names are hypotheses, yet taxonomy is permanent, and this generates a certain tension within the field of systematics. While you may never name a species, as a species user it is still beneficial to understand the labour and concern that feed into the process of how to name a species.

## REFERENCES

Agosti D. 2006. Biodiversity data are out of local taxonomists' reach. *Nature*. 439: 392–3.

Appeltans W, Ahyong ST, Anderson G, and 119 other authors. 2012. The magnitude of global marine species diversity. *Current Biology*. 22: 2189–202.

Ariyawansa HA, Hawksworth DL, Hyde KD, Jones EG, Maharachchikumbura SS, Manamgoda DS, Thambugala KM, Udayanga D, Camporesi E, Daranagama A, Jayawardena R. 2014. Epitypification and neotypification: Guidelines with appropriate and inappropriate examples. *Fungal Diversity*. 69: 57–91.

Barber L. 1980. *The Heyday of Natural History, 1820–1870*. Doubleday.

Barco A, Marshall B, Houart R, Oliverio M. 2015. Molecular phylogenetics of Haustrinae and Pagodulinae (Neogastropoda: Muricidae) with a focus on New Zealand species. *Journal of Molluscan Studies*. 81: 476–88.

Bebber DP, Wood JR, Barker C, Scotland RW. 2014. Author inflation masks global capacity for species discovery in flowering plants. *New Phytologist*. 201: 700–6.

Blackwelder RE. 1967. *Taxonomy: A Text and Reference Book*. Wiley.

Bouchet P, Bary S, Héros V, Marani G. 2016. How many species of molluscs are there in the world's oceans, and who is going to describe them? *Mémoires du Muséum national d'Histoire naturelle*. 208: 9–24.

Bouchet P, Strong EE. 2010. Historical name-bearing types in marine molluscs: An impediment to biodiversity studies? In: Polaszek A (editor). *Systema Naturae 250*. CRC Press.

Boxshall GA, Self D. 2010. *UK Taxonomy and Systematics Review*. Natural Environment Research Council, UK.

Butcher BA, Smith MA, Sharkey MJ, Quicke DL. 2012. A turbo-taxonomic study of Thai *Aleiodes* (*Aleiodes*) and *Aleiodes* (*Arcaleiodes*) (Hymenoptera: Braconidae: Rogadinae) based largely on COI barcoded specimens, with rapid descriptions of 179 new species. *Zootaxa*. 3457: 1–232.

Carter D, Walker AK. 1999. *Care and Conservation of Natural History Collections*. Butterworth Heinemann.

Clark SH. 1986. Preservation of herbarium specimens: An archive conservator's approach. *Taxon*. 35: 675–82.

Collins C. 2014. *Standards in the Care of Wet Collections. Cloth Makers Foundation Expert Workshop on Benchmark Standards for the Preservation on Wet Collections*. The Natural History Museum, London. http://conservation.myspecies.info accessed September 2017.

Costello MJ, May RM, Stork NE. 2013. Can we name Earth's species before they go extinct? *Science*. 339: 413–6.

Cowie RH. 1998. Patterns of introduction of non-indigenous non-marine snails and slugs in the Hawaiian Islands. *Biodiversity & Conservation*. 7: 349–68.

Crous PW, Gams W, Stalpers JA, Robert V, Stegehuis G. 2004. MycoBank: An online initiative to launch mycology into the 21st century. *Studies in Mycology*. 50: 19–22.

The Economist. 2016. All together now: Why research papers have so many authors. Nov 24 2016.

Farber PL. 2000. *Finding Order in Nature: The Naturalist Tradition from Linnaeus to EO Wilson*. JHU Press.

Fontaine B, Perrard A, Bouchet P. 2012. 21 years of shelf life between discovery and description of new species. *Current Biology*. 22: R943–4.

Guiry MD, Guiry GM. 2017. AlgaeBase. World-wide electronic publication, National University of Ireland, Galway. www.algaebase.org accessed October 2017.

Güller M, Liuzzi MG, Zelaya DG. 2015. A new species of *Leptochiton* (Polyplacophora: Leptochitonidae) from the Southwestern Atlantic. *Malacologia*. 58: 147–55.

Hill KD, Johnson LA. 1995. Systematic studies in the eucalypts 7. A revision of the bloodwoods, genus *Corymbia* (Myrtaceae). *Telopea*. 6: 185–504.

Hind KR, Saunders GW. 2013. A molecular phylogenetic study of the tribe Corallineae (Corallinales, Rhodophyta) with an assessment of genus-level taxonomic features and descriptions of novel genera. *Journal of Phycology* 49: 103–14.

Hutsell KC, Hutsell LL, Pisor DL. 1999. *Registry of World Record Size Shells*. Snail's Pace Productions.

[ICZN] International Commission on Zoological Nomenclature. 1999. *International Code of Zoological Nomenclature*, Fourth Edition. International Trust for Zoological Nomenclature, London.

[ICZN] International Commission on Zoological Nomenclature. 2012. Amendment of Articles 8, 9, 10, 21 and 78 of the International Code of Zoological Nomenclature to expand and refine methods of publication. *Zootaxa* 3450: 1–7.

Isley D. 1972. The disappearance. *Taxon*. 21: 3–12.

Jörger KM, Schrödl M. 2013. How to describe a cryptic species? Practical challenges of molecular taxonomy. *Frontiers in Zoology.* 10: 59.

Kaas P, Van Belle RA. 1985. *Monograph of Living Chitons (Mollusca: Polyplacophora) 2, Suborder Ischnochitonina, Ischnochitonidae: Schizoplacinae, Callochitoninae & Lepidochitoninae.* EJ Brill/W. Backhuys.

Knudsen KE. 2009. *Caloplaca obamae*, a new species from Santa Rosa Island, California. *Opuscula Philolichenum.* 6: 37–40.

Latour B. 1999. Circulating reference: Sampling the soil in the Amazon forest. In: Latour B (editor). *Pandora's Hope: Essays on the Reality of Science Studies*, pp. 24–79. Harvard University Press.

Leviton AE, Aldrich ML. 2000. India: A case study of natural history in a colonial setting. *California Academy of Sciences Memoirs.* 25: 51–80.

Lücking R, Dal Forno M, Moncada B, Coca LF, Vargas-Mendoza LY, Aptroot A, Arias LJ, Besal B, Bungartz F, Cabrera-Amaya DM, Cáceres ME. 2017. Turbo-taxonomy to assemble a megadiverse lichen genus: Seventy new species of *Cora* (Basidiomycota: Agaricales: Hygrophoraceae), honouring David Leslie Hawksworth's seventieth birthday. *Fungal Diversity.* 84: 139–207.

Maddison DR, Guralnick R, Hill A, Reysenbach AL, McDade LA. 2012. Ramping up biodiversity discovery via online quantum contributions. *Trends in Ecology & Evolution.* 27: 72–7.

McLean BS, Bell KC, Dunnum JL, Abrahamson B, Colella JP, Deardorff ER, Weber JA, Jones AK, Salazar-Miralles F, Cook JA. 2016. Natural history collections-based research: Progress, promise, and best practices. *Journal of Mammalogy.* 97: 287–97.

McNeill J, Barrie FR, Buck WR, Demoulin V, Greuter W, Hawksworth DL, Herendeen PS, Knapp S, Marhold K, Prado J, Prud'homme Van Reine WF (editors). 2012. *International Code of Nomenclature for algae, fungi and plants (Melbourne Code) adopted by the Eighteenth International Botanical Congress Melbourne, Australia, July 2011.* Regnum Vegetabile 154.

Minelli A, Kraus O, Tubbs PK. 2000. Names for cash. *Science.* 287: 1203.

Mora C, Rollo A, Tittensor DP. 2013. Comment on 'Can we name Earth's species before they go extinct?' *Science.* 340: 237.

Müller-Wille S. 2006. Linnaeus' herbarium cabinet: A piece of furniture and its function. *Endeavour.* 30: 60–4.

Nagy ZT. 2010. A hands-on overview of tissue preservation methods for molecular genetic analyses. *Organisms Diversity & Evolution.* 10: 91–105.

Nature. 2012. The name game. *Nature.* 489: 178.

Nelson JS, Stames WC, Warren ML. 2002. A capital case for common names of species of fishes—a white crappie or a White Crappie. *Fisheries.* 27: 31–3.

Nicolson N, Challis K, Tucker A, Knapp S. 2017. Impact of e-publication changes in the International Code of Nomenclature for algae, fungi and plants (Melbourne Code, 2012)-did we need to 'run for our lives'? *BMC Evolutionary Biology.* 17: 116.

Notton D, Michel E, Dale-Skey N, Nikolaeva S, Tracey S. 2011. Best practice in the use of the scientific names of animals: Support for editors of technical journals. *Bulletin of Zoological Nomenclature.* 68: 313–22.

O'Riordan CE. 1976. *The Natural History Museum, Dublin.* Stationery Office/Oifig an tSoláthair.

Page LM, Espinosa-Pérez H, Findley LT, Gilbert CR, Lea RN, Mandrak NE, Mayden RL, Nelson JS. 2013. *Common and Scientific Names of Fishes from the United States, Canada, and Mexico*, Seventh Edition. American Fisheries Society, Special Publication 34.

Page RDM. 2008. Biodiversity informatics: The challenge of linking data and the role of shared identifiers. *Briefings in Bioinformatics.* 9: 345–54.

Parks KC. 1978. A guide to forming and capitalizing compound names of birds in English. *The Auk.* 95: 324–6.

Peterson AT, Soberón J, Krishtalka L. 2015. A global perspective on decadal challenges and priorities in biodiversity informatics. *BMC Ecology.* 15: 15.

Poulin R. 2014. Parasite biodiversity revisited: Frontiers and constraints. *International Journal for Parasitology.* 44: 581–9.

Pyles T. 1939. Tempest in teapot: Reform in Latin pronunciation. *ELH.* 6: 138–64.

Quicke DL (editor). 2013. *Principles and Techniques of Contemporary Taxonomy.* Springer.

Quicke DL, Lopez-Vaamonde C, Belshaw R. 1999. Preservation of hymenopteran specimens for subsequent molecular and morphological study. *Zoologica Scripta.* 28: 261–7.

Renner SS. 2016. A return to Linnaeus's focus on diagnosis, not description: The use of DNA characters in the formal naming of species. *Systematic Biology.* 65: 1085–95.

Riedel A, Sagata K, Suhardjono YR, Tänzler R, Balke M. 2013. Integrative taxonomy on the fast track-towards more sustainability in biodiversity research. *Frontiers in Zoology.* 10: 0015.

Schrödl M, Haszprunar G. 2016. Do we need epitypes in Zoology? *Spixiana.* 39: 199–201.

Sheets-Pyenson S. 1987. Cathedrals of science: The development of colonial natural history museums during the late nineteenth century. *History of Science.* 25: 279–300.

Sigwart JD, Sumner-Rooney LH, Schwabe E, Heß M, Brennan GP, Schrödl M. 2014. A new sensory organ in 'primitive' molluscs (Polyplacophora: Lepidopleurida), and its context in the nervous system of chitons. *Frontiers in Zoology.* 11: 007.

Stearn WT. 1959. The background of Linnaeus's contributions to the nomenclature and methods of systematic biology. *Systematic Zoology.* 8: 4–22.

Steghaus-Kovac S. 2000. Researchers cash in on personalized species names. *Science.* 287: 421.

Stevens PF. 2002. Why do we name organisms? Some reminders from the past. *Taxon.* 51: 11–26.

Suarez AV, Tsutsui ND. 2004. The value of museum collections for research and society. *BioScience.* 54: 66–74.

Sumner-Rooney LH, Sigwart JD. 2017. Lazarus in the museum: Resurrecting historic specimens through new technology. *Invertebrate Zoology.* 14: 73–84.

Sutton MD, Rahman I, Garwood R. 2014. *Techniques for Virtual Palaeontology.* Wiley.

Sutton MD, Sigwart JD. 2012. A chiton without a foot. *Palaeontology.* 55: 401–11.

Tancoigne E, Dubois A. 2013. Taxonomy: No decline, but inertia. *Cladistics.* 29: 567–70.

Trivedi BP. 2005. What's in a species' name? More than $450,000. *Science.* 307: 1399.

Webster MS. 2017. The extended specimen. In: Webster MS (editor). *The Extended Specimen: Emerging Frontiers in Collections-Based Ornithological Research*, pp. 1–9. CRC Press.

Wen J, Ickert-Bond SM, Appelhans MS, Dorr LJ, Funk VA. 2015. Collections-based systematics: Opportunities and outlook for 2050. *Journal of Systematics and Evolution.* 53: 477–88.

Wiklund H, Taylor JD, Dahlgren TG, Todt C, Ikebe C, Rabone M, Glover AG. 2017. Abyssal fauna of the UK-1 polymetallic nodule exploration area, Clarion-Clipperton Zone, central Pacific Ocean: Mollusca. *ZooKeys.* 707: 1–46.

Wheeler QD, Knapp S, Stevenson DW, Stevenson J, Blum SD, Boom BM, Borisy GG, Buizer JL, De Carvalho MR, Cibrian A, Donoghue MJ. 2012. Mapping the biosphere: Exploring species to understand the origin, organization and sustainability of biodiversity. *Systematics and Biodiversity.* 10: 1–20.

Wheeler QU, Hamilton AN. 2014. The new systematics, the new taxonomy, and the future of biodiversity studies. In: Hamilton A (editor). *The Evolution of Phylogenetic Systematics*, pp. 287–301. University of California Press.

Winston JE. 1999. *Describing Species: Practical Taxonomic Procedure for Biologists.* Columbia.

# 9 Biodiversity and Extinction through Time

## EXTINCTION IS A NATURAL PART OF EVOLUTION

The urgency of modern biodiversity loss has created a crisis in modern taxonomy, and a compelling fear that we are losing Earth's biodiversity faster than it can be discovered and described (Maddison et al., 2012; Wheeler et al., 2012). This is the motivation for efforts to increase the pace of taxonomy. But this biodiversity crisis has a context. To better understand biodiversity loss, we can look to previous experience of extinction throughout Earth's history. All lineages have evolved, persisted, and most have perished, in environments that were quite different, at times much warmer or much colder than present average temperatures (Willis et al., 2010). Our ideas of what species mean, and how species work, cannot be inferred from the living biota alone: the species we interact with are just a horizontal slice through their larger, time-vertical distribution (Chapter 2). The process of comparing diversity through time intersects with understanding the nature of extinction, how the nature of extinction conceptually compares between fossil and living lineages, and how we measure both biodiversity and the loss of diversity.

Current threats to global diversity have been characterised as the 'sixth mass extinction' based on the current rate of species loss (Barnosky et al., 2011). This implicitly includes the notion that a 'mass' extinction is faster or more severe than the normal or background rate at which species are naturally going extinct. Extinction rates and speciation rates are variable, not a constant clock that marks out geological time. Concentrated extinction events interrupt long spans of lower species turnover, but those long intervals of general equilibrium still witness low levels of speciation and extinction, and elevated global extinction rates do not affect all taxa or environments equally (Raup & Sepkoski, 1982; Eldredge et al., 2005). Past extinction patterns in the fossil record, including five earlier mass extinction events but also regular, gradual loss of lineages over time, give us some context to understand what level of extinction is 'normal' and how current events fit in with global long-term trends. The geological record extends over incomprehensibly long stretches of time and many patterns that are described as single events are not equivalent to the fairly instantaneous experience of a human lifespan. The aims of this chapter are to introduce the differences between extinction in the modern realm and the fossil record.

In simplest terms, extinction refers to the end of a species lineage. Famous examples include the Dodo (*Raphus cucullatus* (Linnaeus, 1758)), a kind of giant flightless pigeon, or the Thylacine (*Thylacinus cynocephalus* (Harris, 1808)), a marsupial dog found in Tasmania until the early 20th century. It is relatively easy to envision the end of a large, visually obvious animal, especially one that lives on an island. There are other types of relevant local extirpation which are also significant concerns for conservation.

'Local extinction' refers to the removal of a geographically restricted population, while other members of the species persist in other parts of its range. A local extinction could be the demise of a metapopulation, or just that the species is extirpated within a human political boundary. 'Ecological' extinction or 'commercial extinction' describe cases where the species persists, and perhaps over its full natural range, but in much-reduced population densities. Climate change has already caused extensive range reductions and population contractions in many animal and plant species (Wiens, 2016). A decline in abundance can mean that a species cannot perform its former role, either in an ecological or commercial context. These functional extinctions have been observed in finfish stocks that have been depleted by overharvesting (Chapter 3). Large pelagic fish have been reduced by 74% compared to historic populations, such that many commercial species such as cod and sharks cannot be harvested sustainably (Jackson, 2008). The reduction of these predator populations creates a cascade that impacts populations of species in lower trophic levels (Shepherd & Myers, 2005). 'Extinct in the wild' is a designation officially recognised by the International Union for Conservation of Nature (IUCN), though the other subextinction categories are not. That commercial or local extinctions have no standard assessment criteria remains problematic for conservation legislation (NRC, 1995).

The IUCN Red List of endangered and threatened species, and its criteria, are a powerful tool to protect and restore wild species. Subextinction reductions also contract the scope for genetic and phenotypic variation within the species, with long-lasting effects. Populations of the cheetah (*Acinonyx jubatus* (Schreber, 1775)) have remarkably low genetic diversity and physiological abnormalities that are usually symptomatic of the type of severe inbreeding seen in domesticated species, which in this case resulted from significant reductions in the population ca. 10,000 years ago (Menotti-Raymond & O'Brien, 1993; Charruau et al., 2011). Species that are extinct in the wild represent declines with anthropogenic intervention, as they are reduced to captive individuals, so the species are *de facto* also ecologically or functionally extinct. A very small number of species that were extinct in the wild have been saved by the intervention of breeding programmes expanding on small captive populations, such as the Kihansi Spray Toad, *Nectophrynoides asperginis* Poynton et al., 1998, in Tanzania (Nahonyo et al., 2017), the California Condor, *Gymnogyps californianus* (Shaw, 1797), and others (Tilman et al., 2017).

There are around 900 recorded modern extinctions of animals and plants (IUCN, 2017), but thousands more are critically endangered (Table 9.1). The largest fraction of recorded recently-extinct species is represented by vertebrate animals, over 350 species including those that are extinct in the wild, but the larger assessment of extinct and critically endangered species takes in far more diversity. The large number of endangered molluscs reflects mainly efforts to understand freshwater mussels and island endemic landsnails, while there have been no assessments for most marine molluscs or marine animals in other invertebrate groups. Inaccessibility and low visibility of many species make assessment challenging and there is no systematic programme in place to complete assessments for total global diversity.

As an illustrative example, Hawaiian endemic landsnails are cryptic, inaccessible, and rare. Only historical records reveal the extent of their extinction—species that cannot be refound despite painstaking field work over many years of searching—and

**TABLE 9.1**

**Taxonomic Distribution of Extinct and Critically Endangered Global Species**

| | Extinct | Extinct in the Wild | Critically Endangered | Total High-Risk Species |
|---|---|---|---|---|
| **All Assessed Groups** | **844** | **68** | **5101** | **6013** |
| **Animalia** | **728** | **33** | **2613** | **3374** |
|   **Annelida** | **1** | | **1** | |
|     Clitellata | 1 | | | |
|     Polychaeta | | | 1 | |
|   **Arthropoda** | **81** | **2** | **1** | **84** |
|     Arachnida | 9 | | 47 | |
|     Branchiopoda | | | 6 | |
|     Chilopoda | | | 3 | |
|     Diplopoda | 3 | | 7 | |
|     Entognatha | | | 2 | |
|     Insecta | 58 | 1 | 223 | |
|     Malacostraca | 7 | 1 | 125 | |
|     Maxillopoda | 2 | | 7 | |
|     Ostracoda | 2 | | 2 | |
|   **Chordata** | **347** | **17** | **1600** | **1964** |
|     Actinopterygii | 63 | 6 | 437 | |
|     Amphibia | 33 | 2 | 533 | |
|     Aves | 147 | 5 | 184 | |
|     Cephalaspidiomorphi | | | 2 | |
|     Chondrichthyes | | | 20 | |
|     Mammalia | 78 | 2 | 190 | |
|     Reptilia | 25 | 2 | 232 | |
|     Sarcopterygii | | | 1 | |
|   **Cnidaria** | | | **7** | **7** |
|     Anthozoa | | | 6 | |
|     Hydrozoa | | | 1 | |
|   **Mollusca**[a] | **297** | **14** | **579** | **890** |
|     Bivalvia[a] | 32 | | 70 | |
|     Cephalopoda | | | 1 | |
|     Gastropoda[a] | 265 | 14 | 508 | |
|   **Nemertina** | **1** | | **1** | **1** |
|     Enopla | 1 | | 1 | |
|   **Onychophora** | | | **3** | **3** |
|   **Platyhelminthes** | **1** | | | **1** |
|     Turbellaria | 1 | | | |

[a] Assessed primarily on freshwater and terrestrial species, not marine species (see text).

*(Continued)*

**TABLE 9.1** (*Continued*)
**Taxonomic Distribution of Extinct and Critically Endangered Global Species**

| | Extinct | Extinct in the Wild | Critically Endangered | Total High-Risk Species |
|---|---|---|---|---|
| *Plantae* | *116* | *35* | *2480* | *2631* |
| **Bryophyta** | **2** | | **12** | **14** |
| Bryopsida | 2 | | 12 | |
| **Marchantiophyta** | **1** | | **11** | **12** |
| Jungermanniopsida | 1 | | 10 | |
| Marchantiopsida | | | 1 | |
| **Rhodophyta** | **1** | | **6** | **7** |
| Florideophyceae | 1 | | 6 | |
| **Tracheophyta** | **112** | **35** | **2451** | **2598** |
| Cycadopsida | | 4 | 53 | |
| Liliopsida | 4 | 4 | 416 | |
| Lycopodiopsida | | | 9 | |
| Magnoliopsida | 106 | 26 | 1892 | |
| Pinopsida | | | 27 | |
| Polypodiopsida | 2 | 1 | 54 | |
| **Chromista** | | | **4** | **4** |
| **Fungi** | | | **4** | **4** |

comparisons with museum collections provide only a minimum estimate, not capturing species that may have gone extinct without ever being collected (Lydeard et al., 2004). Other areas are even less explored than Hawai'i, and the scientific baseline data for Hawai'i and other Polynesian islands post-dates human impacts. One study estimated that the number of endemic bird species on Pacific islands was originally four times higher than the endemic species richness today, and that around 200 extinct species are missing, extirpated with no recognised remains (Curnutt & Pimm, 2001). The problem is not knowing how much was lost already when we started looking. These examples highlight the broader problem of 'shifting baselines', where damaged conditions in an environment can appear to be 'normal' if they are the only condition we have ever known (Dayton et al., 1998).

The absence of evidence is usually not sufficient evidence of absence. Rare species, that is, those with small populations or that live in a limited or restricted range, are fundamentally more vulnerable to extinction under normal background extinction rates (Jablonski, 2008). Rare species are not necessarily easier to assess in terms of extinction risk, however. If you are looking for a 1 mm long landsnail, even if it only lives on one island, when can you be sure that it is really gone and not hiding under the next leaf (Durkan et al., 2013)? The total evidence under consideration includes historical records, expert fieldwork to investigate the most likely and most pristine possible habitats, and accessory verification, such as widespread habitat destruction and the introduction of invasive snail-eating predators (Curry et al., 2016).

Extinction is not always a dramatic finale, the last polar bear that might someday float away on a lonely iceberg. Species extinction is often not a binary on/off switch.

The future facing the polar bears *Ursus maritimus* Phipps, 1774 in particular, is probably one where individuals in the dwindling population will find themselves sharing niche space with their closest related congeneric species, as the grizzly bears *Ursus arctos* Linnaeus, 1758 move northward into an increasingly ice-free Arctic. The populations of these two species, returned to contact after perhaps only 150,000 years of separation, will interbreed (Matsuhashi et al., 2001; Hailer et al., 2012). In this case, the horizontal (synchronic) separation of the species is clear, but their vertical (diachronic) separation is minimal, so the fuzzy edges of the two lineages can quite easily blend back together. At present, hybrid *Ursus maritimus* × *arctos* individuals have been traced to a small number of parents (Pongracz et al., 2017). In the future, genes and phenotypes that make polar bears distinctive will be drowned out by the majority grizzly bear genotypes, or certain features may be actively selected against as thick white coats become more hazardous than helpful. Over time, there will be fewer and fewer pale or web-footed bears among the Arctic *Ursus* populations. Although some distinct genetic markers may still be retained, *Ursus maritimus* would be extinct when it is no longer an identifiable population lineage with its own evolutionary trajectory. Extinction like this is not a blinking out, but a gradual fade and blend; the variance of the dominant lineage expands slightly to accommodate the new hybrids, and then probably subsides back to a morphological state closer to an earlier equilibrium.

In deep time, species successions are described in terms of the removal of a lineage creating opportunities for other species. This is an over-simplification of palaeoecology, and a misinterpretation that has been unfortunately transferred to modern environmental biology (e.g., Pyron, 2017). The idea also seems to be founded on an assumption that there is a finite number of niches, but this is not really true. The removal of one lineage does not automatically create a vacancy, and more to the point it confuses the time scales of modern extinction (fast), and speciation (slow). Moreover, extinction because of habitat destruction means that habitat is gone for all the species that might have potentially used it.

Extinctions occur when environmental perturbations destroy environment or niche spaces faster than a lineage can adapt, or to be more precise, the habitat changes beyond the scope of the physiological and morphological plasticity of the species (Chapter 5). All of the preceding examples for modern extinctions are directly attributable to anthropological influences. The impacts of humans are so ubiquitous it is almost impossible to identify any 'natural' extinction events in modern times. Yet extinction is a natural and inevitable part of the long-term trajectory of any species: This is the 'first law' of paleobiology (Marshall, 2017; Table 9.2).

## LIVING 'FOSSILS' ARE STILL EVOLVING

A few species and clades have persisted morphologically unchanged through long spans of geological time; they seem to have defied the odds of extinction or bring a piece of the past to life. Are these 'living fossils' more vulnerable to the rapid pace of the modern world? Or are they the great survivors that can weather any change?

Much attention is given to the extremes of evolutionary patterns: explosive radiations that result in large numbers of similar species on the one hand, and on

## TABLE 9.2

## Five Palaeobiological Laws that Explain the Evolution of Living Species

**First law:** Lineages become extinct.

**Second law:** Species are geologically short-lived. *Corollary*: The living biota represents only a tiny portion of all species that have ever lived.

**Third law:** Extinction and speciation rates are approximately balanced, over the long term.

**Fourth law:** Extinction and speciation rates are variable, over time, and within and among lineages.

**Fifth law:** The past cannot be understood from the present alone.

*Source:* Modified from Marshall, 2017.

the other, apparently isolated smaller lineages that persisted, apparently unchanged, for long spans of time. The term 'living fossil' is attributed to Darwin (1859), who wrote 'Species and groups of species which are called aberrant, and which may fancifully be called living fossils, will aid us in forming a picture of the ancient forms of life'. The idea was that certain mysteriously strange animals retain primitive characteristics, such as egg laying in the platypus *Ornithorhynchus*, and others that are representatives of clades with an extensive fossil record but relatively few living species such as the American lungfish *Lepidosiren* (Darwin, 1859).

The relationship between genotype and phenotype is nonlinear. There is evidence that changes in the genome (or parts of the genome) are clocklike and continuous (Martin & Palumbi, 1993), but changes in phenotype are saltatory. Rapid radiations represent clear morphological change in short timespans and with minimal genetic divergence, such as in chiclid fishes in Lake Victoria (Seehausen, 2006). 'Living fossils' are not static, but examples of lineages where there is proportionately low phenotypic change compared to genotypic change. Another way to consider these forms is in the model of 'adaptive landscapes' where peaks in an abstract landscape represent optimally adapted forms—changes in environment change this abstract landscape, meaning that previously advantaged phenotypes may find themselves suddenly in a valley instead of a mountaintop (*sensu* Simpson, 1944). 'Living fossils' are those that continued in a stable adaptive landscape, so there has evidentially been nothing to prompt a change in phenotype.

A more appropriate term for lineages with little long-term morphological change might be 'stabilomorph', which suggests a hypothesis that the lack of change is itself an adaptive strategy (Kin & Błażejowski, 2014). We must pause to reiterate a key point: everything alive today has been evolving for the same length of time. Evolutionary rates of change and speciation do vary among groups. However, all living species necessarily have a mix of 'primitive' or ancestral characteristics, and more recently derived features. While 'living fossils' are often very important to understand larger evolutionary questions, it is essential to remember that any species that is alive today is *de facto* not the ancestor of any other species also currently living.

Biology still carries various archaic ideas that hinder our deeper understanding of living systems: Two among these are, first, that evolution is 'progressive', moving deliberately toward increasing complexity (Gould, 1996; see Chapter 5; Figure 5.3), and second, that species are fixed entities (Chapter 7). These are deep biases that can be identified most clearly when we are confronted with counter-evidence. Species with fuzzy boundaries that sometimes hybridise are conceptually disconcerting if you were first taught to believe in interbreeding populations as a litmus test for separating species lineages (Chapter 7). Evolutionary 'reversals' or regressive traits are fascinating, especially if you were taught to expect evolution to be 'progressive'. Our enduring (and warranted) fascination with eye loss in cave animals must be inspired by the embedded idea that this is not the way things are supposed to happen; surely evolution should make things better, not worse? In fact, eye loss may be a relatively easy path to go down (Sumner-Rooney et al., 2016). Natural selection only guarantees that features are sufficient for the moment, not continuous objective improvement. But I have exploited the unconscious biases of my audience every one of the countless times I have presented a talk or a paper about 'primitive' animals and 'living fossils', as labels commonly applied to chitons.

In the original sense, the phrase 'living fossil' was used to signify taxa that retain dominant diagnostic features that represent ancestral states for the group. In the 20th century, several important discoveries added the idea of 'Lazarus taxa' (as in the myth of a man who was raised from the dead), animal and plant species that disappear from the fossil record and then unexpectedly manifest again much later (Jablonski, 1986). The coelacanth, Wollemi Pine, and monoplacophorans are famous examples of this phenomenon in the living biota. There are other less celebrated cases, such as a barnacle genus first described from fossils and later found alive (Newman & Ross, 1977), and far more examples of taxa that are known from fossils in one geological period, and then a much later period, but with nothing in between.

The Wollemi Pine (*Wollemia nobilis* Jones et al., 1995) has a modern natural range of about 1 km$^2$ in two small colonies in Australia discovered in the 1990s (Jones et al., 1995; Macphail et al., 1995). The most astounding fact about the discovery of this rare tree was its pollen, unlike any other living tree, but a near perfect match for the Cretaceous fossil pollen genus *Dillwynites*. The tiny native stands comprise less than 100 mature living trees, though the species is now widespread in cultivation and can be seen in almost any botanic garden. The wild population has a shockingly low level of natural genetic variability, leading conservation botanists to fear it may be highly vulnerable to disease or pests. The original genetic results from standard markers seemed to indicate the living trees were all effectively clones; more recent data from plastid genomes has shown that other genetic characters are more variable between individuals (Yap et al., 2015). Genetic comparison between *Wollemia* and its nearest living relative dates the split between those lineages to around 60 million years ago (Biffin et al., 2010), but the fossil record could push that date to 118 Mya (Chambers et al., 1998). The Wollemi Pine is a lineage that seems to have simply carried on, exactly as it was, without evidence of change in either genotype or phenotype (because of extreme reduction in population), now bounded by very unfuzzy boundaries of their lineage and their canyon refugium.

A parallel story is the discovery of the monoplacophoran molluscs (Tryblidia), known from ancient fossils in the Silurian and Devonian (ca. 400 Mya) and then recognised alive in the 1950s (Lindberg, 2009). The study of monoplacophorans is complicated by their inaccessible deep-sea habitats, though they are now known from about 30 living species worldwide (Sigwart et al., 2018). Where this example diverges from the 'Lazarus' story of the Wollemi Pine is the evidence for diversity among living monoplacophorans. Molecular evidence suggests that living monoplacophoran species share a common ancestor that lived around 80 million years ago in the Late Cretaceous (Kano et al., 2012), indicating that the living species comprise a relatively recent radiation.

Cycads, or tree ferns, are the iconic woodland plants at the time of the dinosaurs, with an impressive diversity of species especially from the Jurassic and Cretaceous periods. Cycads still survive, but the group has around 300 living species. These were considered another group of dogged survivors, but genetic evidence shows that each of the 11 living genera is a radiation in its own right, which each experienced a burst of speciation 12 Mya or even more recently (Nagalingum et al., 2011). Just like the monoplacophorans, these are not refugees from ancient times, they are modern radiations of recently derived species that also have a prestigious pedigree.

In the broad scope of species called living fossils, some have been called a 'missing link' or interpreted as a transitional form between different forms. Darwin (1859) mentioned the lungfish *Lepidosiren*, a member of the lobefin fish clade Sarcopterygii, which are the closest living relatives to early tetrapods. Another member of Sarcopterygii, the Coelacanth (*Latimeria chalumnae* Smith, 1939), was known from fossils but then famously discovered alive by Marjorie Courtenay-Latimer (b. 1907–d. 2004) in 1938. Coelacanths seem to have undergone far less evolutionary change than lungfish (Marshall & Schultze, 1992). The seemingly limb-like fins of the Coelacanth earned it the nickname 'old fourlegs', but it is emphatically not the ancestral tetrapod. Indeed, it is still an open question whether coelacanths or lungfish are more closely related to tetrapods (Takezaki et al., 2004). A second species of *Latimeria* in Indonesia also punctured the myth of total stasis (Inoue et al., 2005). Coelacanths have their own adaptations to their particular deep-water lifestyle: they are viviparous and birth live young, and like many living and fossil fish species they have a well-developed electric sense to sense prey in soft sediment (Casane & Laurenti, 2013). The fact that coelacanths are living things, still subject to evolutionary pressure, obscures their ancestors' relationship with early tetrapods in the Carboniferous.

The historical view of these lineages that preserve plesiomorphic or ancestral characters was that they were somehow evolutionary static relics. In that case, it would seem that they could provide a window to directly observe the past. In some ways, this is true. Analogue taxa, organisms that have similar morphology to fossil counterparts, may give us a direct insight to the functional morphology (e.g., Orr et al., 2007) and plasticity of traits in their ancestors (e.g., Bacon et al., 2016), and some idea of how adaptive landscapes have changed or not over time. Variability in morphologically constrained living forms helps pick apart microevolutionary (within-lineage) changes from morphological changes that signal substantial macroevolutionary separations among species. This is not equivalent to assuming that one plesiomorphic trait in a living species means that all other facets of its anatomy and physiology have ceased to evolve. A comparative approach including living and extinct taxa enhances the insights gleaned from the fossil record.

# EXTINCT SPECIES OUTNUMBER LIVING SPECIES

The history of life has seen the natural rise and ultimate fall of previously diverse and dominate groups like trilobites and ammonoids, as well as monoplacophorans and tree ferns that are reduced from the former species richness of their clades, but not extirpated. Each geological period, including the present day, is characterised by a distinct environmental range and associated assemblages of flora and fauna. Very broadly, the marine fossil record is separated into three 'evolutionary faunas'—a Cambrian fauna dominated by trilobites, a late Paleozoic fauna dominated by brachiopods, and a Modern fauna from the Mesozoic onward dominated by molluscs (Sepkoski, 1981)—but there are no sharp breaks where one fauna ends and the next begins (Figure 9.1).

As lineages go extinct, they are superseded ecologically or phylogenetically by later forms, but the total number of species that ever existed in the past dramatically

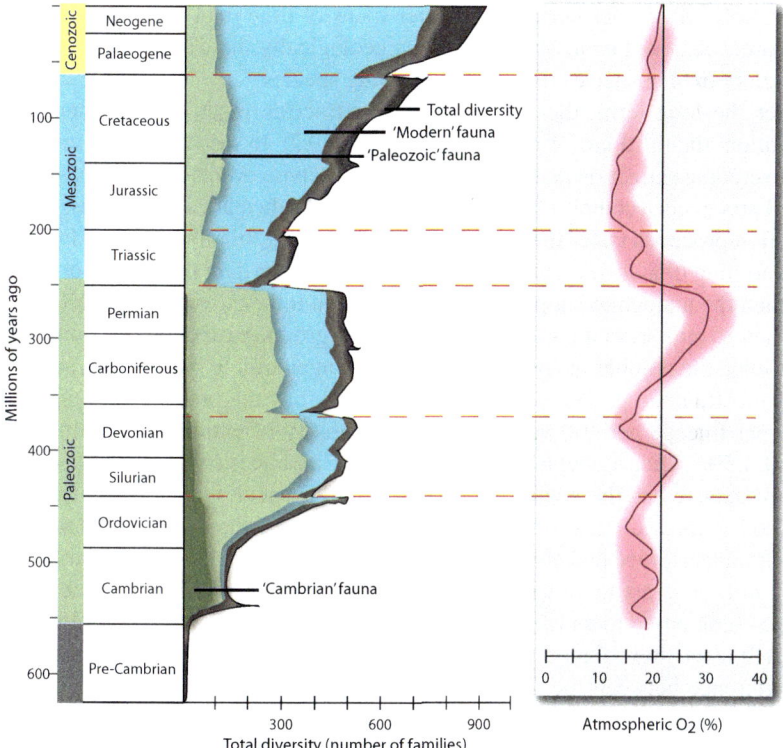

**FIGURE 9.1** Increasing marine animal diversity over time (left, redrawn from Sepkoski 1984) and fluctuating levels of atmospheric oxygen in the Phanerozoic (right, data from Berner et al., 2007). The approximate total diversity is indicated by the encompassing dark grey envelope, subdivided into groups of families with similar temporal persistence: the 'Cambrian' types (dark green), 'Paleozoic' types (green), and the Mesozoic/Cenozoic or 'Modern' types (blue). The five major mass extinction events are highlighted with dashed lines. Fluctuations in diversity (both marine and terrestrial) have been associated with fluctuations in climate and atmospheric composition in the past as well as the present.

outnumbers the species richness of the present: this is the second law of palaeobiology (Table 9.2). The total *known* fossil species do not outnumber the *known* living species; both are incomplete reckonings, but the fossil species are far more under-represented because of multiple confounding factors including preservation potential (which organisms may become fossilised), exposure (whether rocks of the relevant age are accessible at the surface), and recognition (if species lineages can be recognised from the parts that are preserved). A few well-studied living groups are nearly all named, but because the fossil record is incomplete it will never even theoretically be possible to discover direct evidence of all species that ever lived.

So-called 'living fossils' are members of groups that have persisted for a long time without obvious morphological change, yet the living species may have originated relatively recently (for example, in cycads, monoplacophorans, and *Nautilus*; Combosch et al., 2017). Even these species, which superficially seem to have defied evolutionary pressure to change, are short-lived in comparison to geological time. Very simply, all species persist for shorter spans of time than the overall history of their parent clade or taxon. Most species are geologically short-lived. There are many, many more extinct species than there are living species.

Over the long term, the rate of species extinction must balance their rate of origination (the third law of palaeobiology, Table 9.2). In the short term, origination rates overcome extinction rates by a small margin. These events are stochastic, a noisy natural process in a complex system. But by simple mathematical properties, if overall extinction proceeds faster than speciation for very long, biodiversity rapidly crashes out to nothing (that is, if extinction rate is overall greater than the species origination rate, then the mathematical probability of eventual total extinction is 100%). On the other hand, if the speciation rate exceeds the background extinction rate for too much or too long, the number of species increases exponentially to biologically unrealistic numbers[*] (Raup et al., 1973; Sigwart et al., 2018). Say the probability of speciation is 5% per lineage per 100 years, but the probability of extinction is a little lower, around 4.5%. After a couple hundred centuries, an ancestor lineage will have had time to speciate and there will be 2 or 3 species (2.7 species on average per ancestor at 20,000 years). Run that forward to 100,000 years, and there are ~150 descendants for every ancestor we had at the start. But the numbers rapidly get very large, and at only 1 million years there would be $10^{21}$ living species from that first one ancestor lineage—compared to an estimated total of $10^8$ large eukaryotic species living now on Earth after more than 500 million years of animal evolution (Chapter 10). If the difference between the speciation rate and extinction rate is any greater, then this explosion happens much faster (change our experiment from a 0.5% difference between the speciation and extinction rates to a 1% difference, after 1000 intervals you get 22,000 living descendants per ancestor instead of 150).

Speciation and extinction rates have to be balanced, but they are not steady. Variation within the rates of evolution, and rates of extinction, are fundamental mechanisms that explain the patterns of diversity on Earth (the fourth law of paleobiology, Table 9.2; Marshall, 2017). We can observe variety in rates of evolution among groups of living

---

[*] The number of descendants in a 'birth-death' branching model at time $t$ with speciation rate $\lambda$ and extinction rate $\mu$ is calculated as $e^{t \cdot (\lambda - \mu)}$.

species, and chaotic responses to environmental change, such as the differences discussed above between stabilomorphs and rapid radiations. So, it is parsimonious to assume that such variability has been a consistent feature of the evolution of organisms. Species longevity and background extinction rates are linked, and these rates vary in different clades. Generation time, body size, and other life history factors contribute to how fast a lineage evolves, and these processes are synergistically linked to the environment.

Species origination is as much part of the overall dynamics of biodiversity as extinction. Origination rates can be calculated from stratigraphic evidence in the fossil record (Figure 9.2) or from phylogenetic evidence. Lineage-through-time

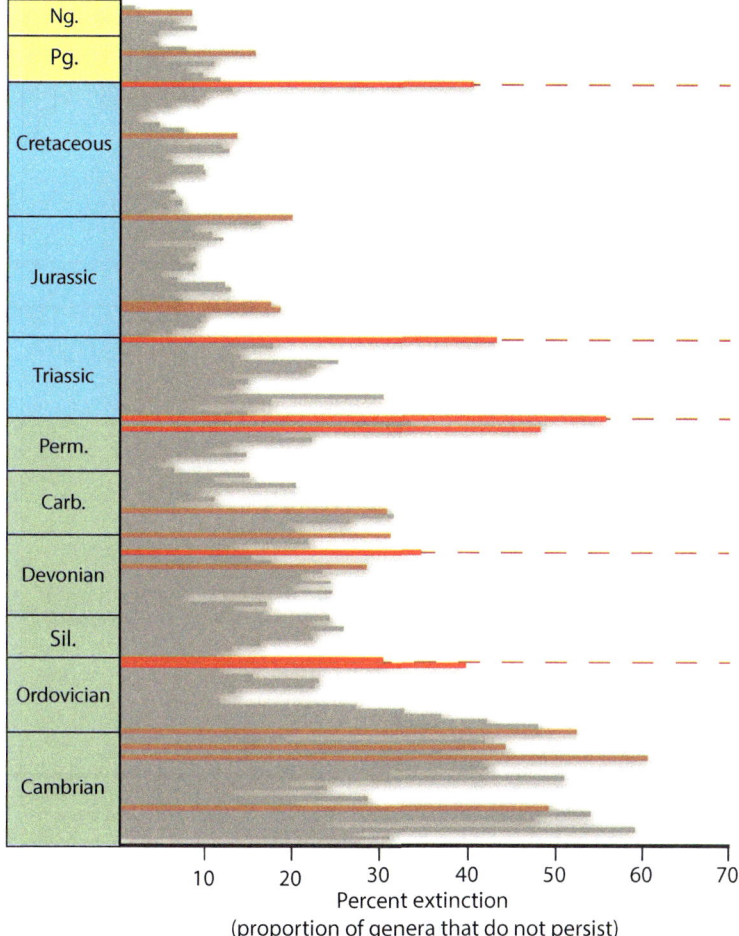

**FIGURE 9.2** Extinction over time in the Phanerozoic. (Data from Sepkoski, 1984, 1986, and Bambach, 2006.) The approximate number of genera lost in each of 165 substages is indicated by the grey bars, with intervals that putatively qualify as major extinctions in red. In addition to the commonly accepted 'Big 5' mass extinctions (highlighted with dashed lines), there are a total of 18 intervals that show aberrantly high extinction using multiple analytical approaches.

(LTT) plots visualise these dynamics usually based on reconstructions of the inter-relationships of living species (Figure 9.3). In this approach, time-calibrated molecular phylogenies are used to look for correlations between inferred speciation rates and ecomophological diversity in present and past species, to test whether rates slow down as more niches become occupied (e.g., Mahler et al., 2010). The figure here is based on a simulation model, where all of the branches are fully known because they were generated by a computer. The comparison of the same tree with and without extinct lineages illustrates how much data may be missing if only living species are available

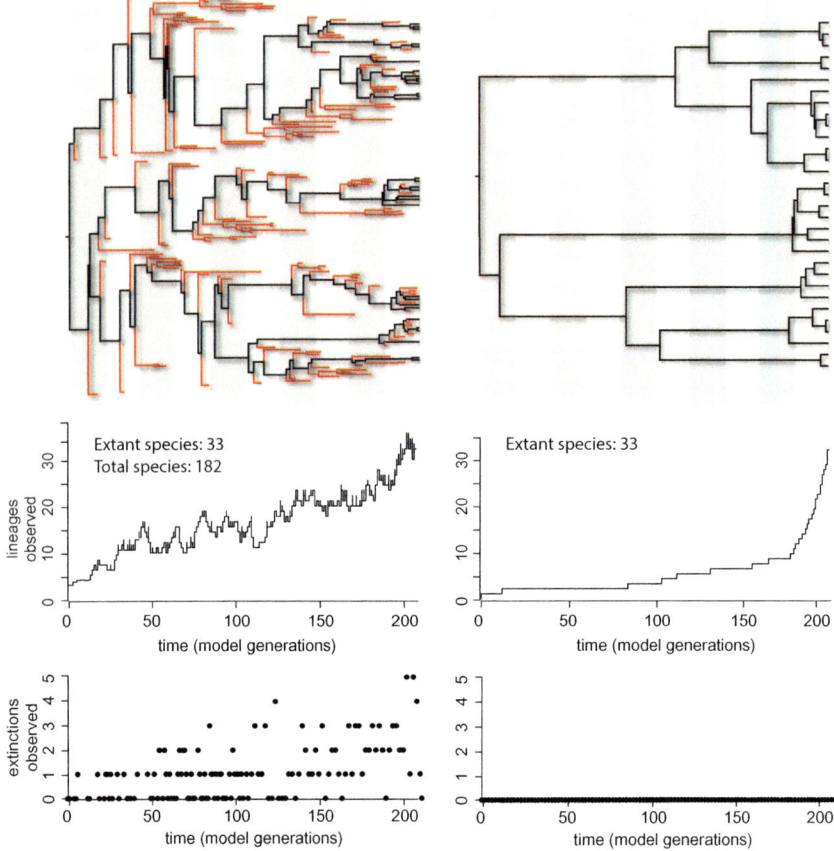

**FIGURE 9.3**   A simulated phylogenetic tree, generated from a birth-death model using fixed rates of speciation and extinction. The tree in its complete form is shown top left (lineages that go extinct before the model is sampled are shown in red), and at top right, the same tree is shown with only the 'living' species retained. The middle panels show 'lineage through time' (LTT) plots for each version of the tree, the number of branches present at each time slice in the original tree, and the tree reconstructed from the living species. In many studies these would be illustrated on log-transformed plots and the curve would appear as a straight line. The lower panels reflect the number of extinction events occurring (left) or reconstructed to have occurred (right) based on the two trees. This illustrates the diversity that is lost, and cannot be fully reconstructed, from data from a single time slice.

for analysis. The nodes that can be reconstructed through the inter-relationships of living species is only a subsample of total diversity.

Over the evolutionary history of a clade, rates of speciation wax and wane. 'Early burst' dynamics are the evolutionary signature of some adaptive radiations. An early burst is a process that is conceptually easy to understand; colonising a new niche space provides an opportunity to diversify, refine, and specialise, which leads to splitting new specialist lineages. Subsequent slowdowns are more complex and confounded by the loss of extinct lineages (Moen & Morlon, 2014). There are a number of cautions to be considered in linking process to pattern. As shown in a simple model tree (Figure 9.3), removing data from extinct lineages can give a false impression of how rapidly the living species radiated. Phylogenetic models that estimate rates of diversification increasingly incorporate sophisticated statistical inference to account for missing data (Cusimano & Renner, 2010). There are limits on what can be inferred about extinct lineages from data where extinct lineages are not represented, because it is impossible to know how much is missing (Quental & Marshall, 2010).

## FOSSILS FILL THE GAPS

Data from the past and present are complimentary; living species inform our understanding of form and function in fossils, but we cannot infer the past history of biodiversity from living species alone (the fifth law of palaebiology, Table 9.2). The phenotypic, genetic, and geographical distances that separate living species are explained by their history; 'horizontal' species in a single time slice are connected by time-vertical lineages. This is the core idea in biogeography, understanding historical distributions and the geographic movement of species through time. There is no way to truly understand the current distribution of species in their genetic diversity, their geographic ranges, and morphological diversity, without connecting them to the time-vertical dimensions that explain their origination.

There are natural gaps in the total tree of life—not every species blends one into the next, genetically or morphologically—these gaps are a major part of the evidence for species as biologically 'real' products of evolution rather than artificial boundaries imposed by scientists. We previously considered a thought experiment about a phylogeny of all individuals of all living species (Chapter 7). Now, what if we could compare *all* the species that ever lived? Is there a complete set of transitional forms that fills in all the gaps? Even if all extinct taxa could be found, this would not create a complete blur between all of the species. First, just as in the present, extinct species were not all alive at the same time. Second, fossil species are under the same dynamic constraint as living species.

Macroevolution is by definition the study of speciation and extinction, but also the persistence of lineages (Pennell et al., 2014; Lieberman & Eldredge, 2014). Many of us are preoccupied with the origins and termination of species, but in fact, most species are persisting in an equilibrium state, most of the time (Eldredge & Gould, 1972). This is similar to the paradigm in animal behaviour studies on activities related to mating, feeding, and competition. These are very important to questions of fitness, but organisms demonstrably spend a dominant proportion of their time resting and 'just hanging out'. The model of equilibria explains observed genetic and phenotypic variation within lineages that are variable but still a species, as microevolutionary fluctuations around the vertically connected population-lineage segment as it progresses through time (Chapter 7).

The model of punctuated equilibrium (Eldredge & Gould, 1972) helps to explain the morphological gappiness of the fossil record, in contrast with alternative models of 'gradualism'. Proponents of evolutionary gradualism accept that the rate of change can vary over time but suggest that this results in continuous accumulating change. Given the intrinsic limitations of the fossil record, it is difficult to test the difference between fuzzy boundaries of species shifting over time, versus gradually accumulating change. In the geological record, there are 'dead end' extinctions where a lineage or clade simply ends, that fossil form just stops at a certain point in time. There are also cases where an earlier species seems to be replaced by later slightly modified versions, akin to anagenesis, sometimes called 'chronological extinctions'. Some series of fossils capture morphological variability over time, where morphology is demonstrably different at later time points than earlier ones and a long series captures all the intermediate forms, for example, in Ordovician trilobites (Sheldon, 1987). But these series are almost by definition limited in time and in geographic scope, and the variation can be understood as short term changes that would be reversible over genuinely macroevolutionary timescales (Gingerich, 1983).

All species persist for a limited duration (Table 9.2). Lineage duration probably also follows a skew distribution and only a small fraction would be expected to be long-lived in geological time. Some fossil species may be found in a huge range, like the abundant and well-studied trilobite *Eldredgeops rana* (Green, 1832), which ranges from 419 to 372 Mya (e.g., Glazier et al., 2012). But most palaeospecies are known from far fewer fossils, or only one or a handful of specimens.

Fossils are more data-limited than living specimens, and so evidence from fossils is sometimes treated as an accessory to support conclusions from molecular phylogeny or other analysis of the living biota. But the living fauna, and their genomes, only capture one single present time slice. The known and described living species dramatically outnumber described fossil species, because not all fossil individuals, and not all fossil species are preserved. There are even fine-scale differences between closely related species that may control whether their remains enter the fossil record (Sigwart et al., 2014).

Both fossil and living species vary substantially in the quality of available data; some living species are only known from a single specimen, but some fossils are known from exquisite material with 'exceptional' preservation that retains soft parts or evidence of behaviour. Mineralised hard parts, such as calcium carbonate in the skeletal elements of molluscs, echinoderms, and vertebrates, have a high potential to be preserved as fossils even though the soft tissues of those organisms typically decay and skeletal elements could disarticulate. Exceptionally preserved specimens in specific depositional contexts called *Lagerstätten* (singular *Lagerstätte*) retain soft part features even through the process of remineralisation (Allison & Bottjer, 2011). These exceptionally preserved faunas occur throughout the geological record. The Solnhofen Limestone in Bavaria, Germany, preserves freshwater and terrestrial fauna from the Jurassic including the fossil bird *Archaeopteryx* (Seilacher et al., 1985). The Herefordshire Lagerstätte is older, Silurian in age, with many three-dimensionally preserved marine invertebrates that have revealed important information about the early evolution of multiple phyla (Sutton et al., 2001). The Burgess Shale in British Columbia, Canada, and similar Cambrian-era deposits such as those at Chengjiang, China, and Sirius Passet, Greenland, are Lagerstätten that record some of the most important evidence of the early evolution of complex animal life, and many of those early species were broadly distributed (Hendricks, 2013). Exceptionally preserved fossils are less data limited than conventional fossilised remains of hard parts and may be the only fossil record for organisms without any hard parts; conventional fossils make up most evidence for palaeospecies.

Local environments with low disturbance (like the bottom of a lake) are more conducive to preservation. By contrast, organisms living on the intertidal shore, pounded by waves, or in tropical forests with rapid rates of decay, have quite low preservation potential. Further, rocks of all ages are not exposed in all parts of the world, and some rock formations are lost through subsidence or other processes of plate tectonics. This is not necessarily dissimilar to the differential in information quality for living environments depending on accessibility: we know much more about the living terrestrial flora and fauna than the marine, and much more about shallow coastal habitats than the deep sea, and more about temperate latitudes than polar ecosystems. In the fossil and in living records, data from rare and inaccessible places or species can offer substantial new insights.

Fossils give us new insights to the limits of evolution—there are numerous body forms and lifestyles that were viable in the past, but not present in the living biota. Morphological turnover can remove adaptive phenotypes that seem like they would have been useful to the organism and were lost without apparent selective cause (Strathmann, 1978). Traits can be lost through founder effects, where it is simply missing in the genomes of the individual organisms that become the direct ancestors of a new lineage. Or, traits can be lost through adaptation associated with changes in environmental conditions; if the environment reverts to its previous configuration, it may be impossible to regain or re-evolve that trait (Dollo, 1893). Adaptation is not necessary to explain diversification (Gould & Vrba, 1982). Just because a trait is missing from the fossil record, or from the present biota, does not imply that a 'missing link' is waiting to be found. Many hypothetical niches remain empty at any given time, because of the turnover from background extinction and speciation rates (Walker & Valentine, 1984).

## IS DIVERSITY ON EARTH DECREASING?

Summarising the evolution of life in broad strokes, life started at a single origin point and then later diversified, so there are more species alive now than there were in the primordial ooze. This is too broad a timescale to be particularly informative, more an example of how any two points define a line (the upward slope from species on Earth at time zero, and species on Earth at present). The modern biota is better studied than any past period in the fossil record, which provides a potentially biased reference frame for identifying past forms as distinct species, and the geological record provides fewer fossils of usually poorer resolution as the rock record becomes more ancient (Benton & Emerson, 2007). These underlying biases may skew diversity estimates higher in younger parts of the rock record, which drives a trend called the 'Pull of the Recent' (Raup, 1979). Despite a potential skew in the total diversity preserved in the fossil record, and a further bias in what subsample of species is discovered and described, there were nonetheless rises and falls in overall taxon richness and not a continuous climb to the present biota (Bennett, 2013; Figure 9.1). In the last few hundreds or thousands of years, the expanding impact of humans has clearly depleted the living biota in excess of the usual short-term stochastic rise and fall of lineage diversity.

The last 540 million years have been punctuated by five previous mass extinction events that affected global biodiversity; the durations, magnitudes, characteristics, and causes of each of these events are subjects of extensive study and debate (Jablonski, 1986). Some experts contend that the effects of various of these events were more or less limited, and there may be other points in the fossil record with diversity perturbations of similar magnitude (Bambach, 2006; Figure 9.2). The termination of a single taxonomic group is not a sufficient criterion for a mass extinction, instead, these events are categorised based on geographically and taxonomically widespread decreases in diversity.

The 'big five' were periods when over 75% of contemporary species were lost within a relatively short interval in geological time. These big five mass extinctions were first statistically categorised based on the global fossil record of hard-shelled marine invertebrates (Raup & Sepkoski, 1982; Sepkoski, 1984). The primary description of these events was based on marine faunas, and the experience of the terrestrial biota was in many cases quite different. Other events had greater impact on plants, or other terrestrial organisms (Figure 9.2). The biotic shifts marked by these intervals were known long before, and had been used to recognise breaks that define geological eras and periods. Studies in the 1980s formalised the palaeobiological approach to mass extinctions and transitioned the field toward more quantitative approaches (Bambach, 2006). The metrics used to assess past extinction events include: extinction magnitude and intensity, duration, forcing mechanisms, geographic and taxonomic selectivity, and ecological turnover. The nuance in how these are measured is relevant for establishing comparisons with modern biodiversity loss.

Extinction intensity in the fossil record is usually reported as a percentage of taxa lost over a specified time period. Fossil species are data-limited and very frequently fossils cannot necessarily be identified confidently as a particular species but can be put in a species group (like genera, or families; Chapter 6). The issue of differing 'species concepts' has been raised as a potential confounding factor in comparing past extinctions with modern biodiversity loss (Quental & Marshall, 2010); there are differentially available evidence to identify a precise number of species lineages in a given time slice, and that is true for comparing different parts of the fossil record to each other, as well as making comparisons between fossil and recent taxa. The marine fossil record remains the best and most detailed data on global biotic change; however, this stands in contrast to a strong terrestrial bias in the knowledge base about modern fauna and flora. It is not possible to trace individual lineages with precision through the fossil record, while extinction of modern species is typically considered at the level of tracing individual organisms. Given that limitation, diversity dynamics in the fossil record are usually summarised as the number of genera or families that occur within given time intervals (Figures

9.1 and 9.2). So, there are two important things to note about the resolution of these data about loss over time; the taxonomic resolution represents groups of species, not species, and the resolution of time in geological scales is much coarser than in the present. Although these look like very short intervals in the scope of geological time, many extinction 'events' span more than 100,000 years and combine both decreases in species origination and increases in extinction rates. Interestingly, these lengths of time mean that such biodiversity crises were sustained for periods longer than the cycles of major orbital oscillations. The pace of extinction and recovery as recorded in the rock record is probably slower than the events as experienced by species in reality (Lu et al., 2006). The magnitude and intensity of geological extinctions represent the frequency of species groups being lost over time spans of thousands or even millions of years; this is very different to the loss of individual living species over spans of at most hundreds of years.

Species origination is as much a contributing factor to diversity as extinction. Over long spans of time, there is ample scope for natural background extinction rates to end many lineages; if extinction is matched or outpaced by speciation then there will be no net decline in overall species richness even if many lineages are lost. Based on extinction magnitude alone, the percent of taxa lost in the marine record, there are only three events in Earth's history that can unequivocally be separated from normal stochastic fluctuations in extinction rates (Bambach et al., 2004): at the end-Ordovician (455–430 Mya), end-Permian (251 Mya), and end-Cretaceous (65 Mya). The other two of the 'big 5' events, the late Devonian and end-Triassic extinction events, are also characterised by dramatic declines in global diversity but these are linked to declines in speciation rates as well as increases in extinction rates.

Abrupt changes in the global fauna are also correlated with other evidence for major changes in the environment. The rock record preserves direct and indirect evidence of past climate from isotopic signatures, mineral composition, signatures of volcanic activity, landscape features, and microfossils (Bambach, 2006).

The end-Cretaceous extinction event was triggered by the impact of a meteorite near the Yucatan Peninsula in Mexico, where the 26,000 $km^2$ crater is centred near Chicxulub (Hildebrand et al., 1991). This extinction event is the best-characterised for a number of independent reasons: it is the most recent of the largest mass extinctions, a layer enriched with extraterrestrial iridium preserved in the rock record represents an unequivocal signature of the impact and helps to correlate the timing of worldwide geological sections (Smit & Hertogen, 1980), and there is a great popular interest in its effect on non-avian dinosaurs. The rate of destruction by the impact can be estimated at the scale of hours, days, or decades, as the initial impact caused dramatic initial destruction, followed by massive dust clouds, and longer-term disruption to atmospheric chemistry (Pope et al., 1994). But this is not necessarily the sole cause of the extinction; some analyses suggest that speciation rates in dinosaurs may have been in decline prior to the impact and other factors combined to create the full effect of a genuine mass extinction (Archibald et al., 2010).

Earlier, the 'Great Dying' at the end Permian was by far the most devastating past mass extinction event, when up to 95% of species went extinct (Benton & Twitchett, 2003). The primary cause of the end-Permian event has been linked to massive volcanic activity in Siberia, eruptions that covered eastern Russia in a layer of lava up to 3 km thick at the deepest part and spanning an area of 1.6 million km$^2$ (Benton & Twitchett, 2003). Radiometric dating from the top and bottom of this lava indicates the volcanic period lasted around 600,000 years, though the extinctions were more concentrated in time (Mundil et al., 2001). This duration is orders of magnitude longer than the end-Cretaceous event, but in geological timescales it is nonetheless still very rapid. The subsequent release of light carbon in the form of volcanic gases caused a catastrophic long-term feedback loop of atmospheric greenhouse warming and widespread ocean anoxia, and acidification on land (Sephton et al., 2015). The event also severely impacted terrestrial biota but the effects in flora and fauna are not tightly correlated (Neveling et al., 2016). Greenhouse effects, excess carbon in the atmosphere, and oceanic 'dead zones' are reminiscent of familiar signatures of modern biomes.

Global climate change is a common feature in all mass extinction events. Substantial and rapid perturbations in the global environment create a tipping point, where the majority of lineages cannot manage in a changed world. Climate change is a natural part of the Earth system. There are climate oscillations caused by long-term orbital variations in the relationship of Earth to our Sun, known as Milankovich cycles (Bennett, 1990). Some dramatic geological events are apparently intrinsic, like the mass eruption of the Siberian Traps at the end-Permian or the forthcoming explosion of Yellowstone (Lowenstern et al., 2006), while other major volcanic events may be triggered extrinsically, as with the meteor impact that caused the end-Cretaceous extinction. The end-Ordovician mass extinction was also a period of mass glaciation and climate change. However, not all climate change causes mass extinctions. These are permanent, astronomical features of the solar system, and phenomena of geological processes, not only part of the current era. These give us context to measure the excess change from anthropogenic activity. The natural oscillations in the expansion and contraction of glaciated ice caps in the Cenozoic has been correlated with significant shifts in global temperature regimes in a geologically rapid sequence (Hewitt, 2000), but prior to the recent emergence of humans, these Cenozoic patterns were not associated with mass extinction. The rates of magnitude of past climate perturbations give us a potential baseline to compare future biodiversity change (Willis et al., 2010).

In past extinctions as in the living flora and fauna, 'climate change' is a complex set of long-term parameters, which is not a direct mechanistic explanation of the extinction individual species. For example, in the end-Cretaceous mass extinction, animal species that directly consumed photosynthetic primary producers perhaps suffered much higher losses than those that consumed detritus (Sheehan et al., 1996). This is not a complete explanation, however, as terrestrial detritus comes from a food web based on photosynthesis. The atmospheric disruption that followed the impact of the meteorite blocked sunlight, causing a cascade in the food chains dependent on photosynthesis. During this extinction event, 81% of venerid bivalve groups (at the subgenus level) went extinct (Lockwood, 2004). So, while the extinction event severely impacted the bivalves, the bivalves are inferred to have suffered from a global food shortage rather than the actual meteorite impact. In fact, it is not clear whether the short-term single- or multi-generational experience of those bivalve species (or most other lineages incorporated into analyses of extinction events) may have been more stressed by food availability, or changes in sea surface temperature, ocean chemistry, turbidity, or interactions with other species.

In the interval after a mass extinction event, new species diversify to fill ecological gaps left by extinction (Edie et al., 2018). Post-extinction diversity of emergent clades can rapidly expand to match or exceed the diversity recorded before a mass extinction horizon. This is a pattern that characterises adaptive radiations in general (Chapter 5), but some of the best-studied rapid bursts of morphological evolution, with evidence from the fossil record, are diversification events that follow mass extinction events. However, some important subtleties cloud this generalisation. One example of such a burst is the 'Cretaceous terrestrial revolution', a period with notable diversification in angiosperms and terrestrial vertebrates. This was before the end-Cretaceous extinction, so that a burst of speciation follows a mass extinction only inasmuch as any period when a mass extinction is not happening is *de facto* following after a mass extinction (Hull, 2015). One example of diversification that is robustly correlated to post-extinction recovery is the radiation of ammonoids after the end-Permian, but that diversification also tracked a period of around 1–2 million years (McGowan, 2004). 'Rapid' bursts in the fossil record are not happening at modern 'ecological' timescales (Chapter 5).

In terms of extinction magnitude, or the percent of lineages lost, species extinctions in the present do not yet match the intensity of the 'big 5' mass extinction events (Barnosky et al., 2011). Earlier we noted there are around 900 species that represent the known modern extinctions over the last several hundred years. If there are around 7,000,000 species on Earth, present extinctions represent a loss of 0.01% of species diversity, compared to mass extinction losses of 75% of *genera*. Because modern extinctions are assessed at the species level, many lineages have other related species in their genus or family that are still present. However, these losses have happened far more rapidly than the pace of past mass extinctions. The larger number of species already assessed as endangered, and their potential extinction even in the next few hundred years, represents a significant threshold that could tip the global system into a state of fully-fledged mass extinction (Barnosky et al., 2011, 2012).

It is difficult to document the proximate cause of extinction in living species, and the problem is redoubled for fossil lineages. Pattern and process are separate issues, and in macroevolution the patterns of diversity can be attributed to multiple possible processes. A decrease in speciation rates can lead to the same rates of biodiversity loss as an increase in extinction rate. In the modern 'sixth' mass extinction event, we have much better knowledge, and much stronger control, over the environmental parameters causing extinction. Species originate and go extinct on geological timescales, and past mass extinction events have played out over spans of thousands to hundreds of thousands of years. What is lost will not be recovered in any timeframe that has any meaning to us. But the study of extinction in Earth's history provides room for optimism. First, extinction and environmental change are natural parts of the Earth system. 'Living fossils' in particular show that lineages can persist, and some that survive an extinction horizon by a whisker can carry on to diversify in later environments. Emergent patterns, and particularly the gaps in knowledge left by extinct taxa in the past, are important to understanding patterns of biodiversity and extinction in the present.

# REFERENCES

Allison PA, Bottjer DJ. 2011. *Taphonomy: Process and Bias Through Time.* Springer.

Archibald JD, Clemens WA, Padian K, Rowe T, Macleod N, Barrett PM, Gale A, Holroyd P, Sues HD, Arens NC, Horner JR. 2010. Cretaceous extinctions: Multiple causes. *Science.* 328: 973–6.

Bacon KL, Haworth M, Conroy E, McElwain JC. 2016. Can atmospheric composition influence plant fossil preservation potential via changes in leaf mass per area? A new hypothesis based on simulated palaeoatmosphere experiments. *Palaeogeography, Palaeoclimatology, Palaeoecology.* 464: 51 64.

Bambach RK. 2006. Phanerozoic biodiversity mass extinctions. *Annual Review of Earth & Planetary Sciences.* 34: 127–55.

Bambach RK, Knoll AH, Wang SC. 2004. Origination, extinction, and mass depletions of marine diversity. *Paleobiology.* 30: 522–42.

Barnosky AD, Matzke N, Tomiya S, Wogan GO, Swartz B, Quental TB, Marshall C, McGuire JL, Lindsey EL, Maguire KC, Mersey B. 2012. Approaching a state shift in Earth's biosphere. *Nature.* 486: 52–8.

Barnosky AD, Matzke N, Tomiya S, Wogan GO, Swartz B, Quental TB, Marshall C, McGuire JL, Lindsey EL, Maguire KC, Mersey B. 2011. Has the Earth's sixth mass extinction already arrived? *Nature.* 471: 51–7.

Bennett KD. 1990. Milankovitch cycles and their effects on species in ecological and evolutionary time. *Paleobiology.* 16: 11–21.

Bennett KD. 2013. Is the number of species on Earth increasing or decreasing? Time, chaos and the origin of species. *Palaeontology.* 56: 1305–25.

Benton MJ, Emerson BC. 2007. How did life become so diverse? The dynamics of diversification according to the fossil record and molecular phylogenetics. *Palaeontology.* 50: 23–40.

Benton MJ, Twitchett RJ. 2003. How to kill (almost) all life: The end-Permian extinction event. *Trends in Ecology & Evolution.* 18: 358–65.

Berner RA, VandenBrooks JM, Ward PD. 2007. Oxygen and evolution. *Science.* 316: 557–8.

Biffin E, Hill RS, Lowe AJ. 2010. Did Kauri (*Agathis*: Araucariaceae) really survive the Oligocene drowning of New Zealand? *Systematic Biology.* 59: 594–602.

Casane D, Laurenti P. 2013. Why coelacanths are not 'living fossils'. *BioEssays.* 35: 332–8.

Chambers TC, Drinnan AN, McLoughlin S. 1998. Some morphological features of Wollemi pine (*Wollemia nobilis*: Araucariaceae) and their comparison to Cretaceous plant fossils. *International Journal of Plant Sciences.* 159: 160–71.

Charruau P, Fernandes C, Orozco-Terwengel P, Peters J, Hunter L, Ziaie H, Jourabchian A, Jowkar H, Schaller G, Ostrowski S, Vercammen P. 2011. Phylogeography, genetic structure and population divergence time of cheetahs in Africa and Asia: Evidence for long-term geographic isolates. *Molecular Ecology.* 20: 706–24.

Combosch DJ, Lemer S, Ward PD, Landman NH, Giribet G. 2017. Genomic signatures of evolution in *Nautilus*—An endangered living fossil. *Molecular Ecology.* 26: 5923–38.

Curnutt JO, Pimm ST. 2001. How many bird species in Hawaii and the Central Pacific before first contact? *Studies in Avian Biology.* 22: 15–30.

Curry PA, Yeung NW, Hayes KA, Meyer III WM, Taylor AD, Cowie RH. 2016. Rapid range expansion of an invasive predatory snail, *Oxychilus alliarius* (Miller, 1822), and its impact on endemic Hawaiian land snails. *Biological Invasions.* 18: 1769–80.

Cusimano N, Renner SS. 2010. Slowdowns in diversification rates from real phylogenies may not be real. *Systematic Biology.* 59: 458–64.

Darwin CR. 1859. *On the Origin of Species by Means of Natural Selection, or the Preservation of Favoured Races in the Struggle for Life.* John Murray, publishers.

Dayton PK, Tegner MJ, Edwards PB, Riser KL. 1998. Sliding baselines, ghosts, and reduced expectations in kelp forest communities. *Ecological Applications.* 8: 309–22.

Dollo L. 1893. The laws of evolution. *Bulletin de la Société Belge de Géologie de Paléontologie et D'hydrologie.* 7: 164–66.

Durkan TH, Yeung NW, Meyer WM, Hayes KA, Cowie RH. 2013. Evaluating the efficacy of land snail survey techniques in Hawaii: Implications for conservation throughout the Pacific. *Biodiversity & Conservation.* 22: 3223–32.

Edie SM, Jablonski D, Valentine JW. 2018. Contrasting responses of functional diversity to major losses in taxonomic diversity. *Proceedings of the National Academy of Sciences.* 2018: 201717636.

Eldredge N, Gould SJ. 1972. Punctuated equilibrium: An alternative to phyletic gradualism. In: Schopf TJM (editor). *Models in Paleobiology*, pp. 82–115. Freeman, Cooper.

Eldredge N, Thompson JN, Brakefield PM, Gavrilets S, Jablonski D, Jackson JB, Lenski RE, Lieberman BS, McPeek MA, Miller W. 2005. The dynamics of evolutionary stasis. *Paleobiology.* 31: 133–45.

Gingerich PD. 1983. Rates of evolution: Effects of time and temporal scaling. *Science.* 222: 159–61.

Glazier DS, Powell MG, Deptola TJ. 2012. Body-size scaling of metabolic rate in the trilobite *Eldredgeops rana. Paleobiology.* 39: 109–22.

Gould SJ. 1996. *Full House: The Spread of Excellence from Plato to Darwin.* Belknap Press.

Gould SJ, Vrba ES. 1982. Exaptation—a missing term in the science of form. *Paleobiology.* 8: 4–15.

Hailer F, Kutschera VE, Hallström BM, Klassert D, Fain SR, Leonard JA, Arnason U, Janke A. 2012. Nuclear genomic sequences reveal that polar bears are an old and distinct bear lineage. *Science.* 336: 344–7.

Hendricks JR. 2013. Global distributional dynamics of Cambrian clades as revealed by Burgess Shale-type deposits. *Geological Society, London, Memoirs.* 38: 35–43.

Hewitt G. 2000. The genetic legacy of the Quaternary ice ages. *Nature.* 405: 907–13.

Hildebrand AR, Penfield GT, Kring DA, Pilkington M, Camargo A, Jacobsen SB, Boynton WV. 1991. Chicxulub crater: A possible Cretaceous/Tertiary boundary impact crater on the Yucatan Peninsula, Mexico. *Geology.* 19: 867–71.

Hull P. 2015. Life in the aftermath of mass extinctions. *Current Biology.* 25: R941–52.

Inoue JG, Miya M, Venkatesh B, Nishida M. 2005. The mitochondrial genome of Indonesian coelacanth *Latimeria menadoensis* (Sarcopterygii: Coelacanthiformes) and divergence time estimation between the two coelacanths. *Gene.* 349: 227–35.

[IUCN] International Union for Conservation of Nature. 2017. IUCN Red List of Threatened Species. Version 2017.2. www.iucnredlist.org accessed September 2017.

Jablonski D. 1986. Causes and consequences of mass extinctions: A comparative approach. In: Elliott DK (editor). *Dynamics of Extinction*, pp. 183–229. Wiley.

Jablonski D. 2008. Species selection: Theory and data. *Annual Review of Ecology, Evolution, & Systematics.* 39: 501–24.

Jackson JB. 2008. Ecological extinction and evolution in the brave new ocean. *Proceedings of the National Academy of Sciences.* 105: 11458–65.

Jones WG, Hill KD, Allen JM. 1995. *Wollemia nobilis*, a new living Australian genus and species in the Araucariaceae. *Telopea.* 6: 173–6.

Kano Y, Kimura S, Kimura T, Waren A. 2012. Living Monoplacophora: Morphological conservatism or recent diversification? *Zoologica Scripta.* 41: 471–88.

Kin A, Błażejowski B. 2014. The horseshoe crab of the genus *Limulus*: Living fossil or stabilomorph? *PLoS ONE.* 9: e108036.

Lieberman BS, Eldredge N. 2014. What is punctuated equilibrium? What is macroevolution? A response to Pennell et al. *Trends in Ecology & Evolution.* 29: 185–6.

Lindberg DR. 2009. Monoplacophorans and the origin and relationships of mollusks. *Evolution Education & Outreach.* 2: 191–203.

Lockwood R. 2004. The K/T event and infaunality: Morphological and ecological patterns of extinction and recovery in veneroid bivalves. *Paleobiology.* 30: 507–21.

Lowenstern JB, Smith RB, Hill DP. 2006. Monitoring super-volcanoes: Geophysical and geochemical signals at Yellowstone and other large caldera systems. *Philosophical Transactions of the Royal Society A.* 364: 2055–72.

Lu PJ, Yogo M, Marshall CR. 2006. Phanerozoic marine biodiversity dynamics in light of the incompleteness of the fossil record. *Proceedings of the National Academy of Sciences of the United States of America.* 103: 2736–9.

Lydeard C, Cowie RH, Ponder WF, Bogan AE, Bouchet P, Clark SA, Cummings KS, Frest TJ, Gargominy O, Herbert DG, Hershler R. 2004. The global decline of nonmarine mollusks. *BioScience.* 54: 321–30.

Macphail M, Hill K, Partridge A, Truswell E, Foster C. 1995. Wollemi Pine—old pollen records for a newly discovered genus of gymnosperm. *Geology Today.* 11: 48–50.

Maddison DR, Guralnick R, Hill A, Reysenbach AL, McDade LA. 2012. Ramping up biodiversity discovery via online quantum contributions. *Trends in Ecology & Evolution.* 27: 72–7.

Mahler DL, Revell LJ, Glor RE, Losos JB. 2010. Ecological opportunity and the rate of morphological evolution in the diversification of Greater Antillean anoles. *Evolution.* 64: 2731–45.

Marshall CR. 2017. Five palaeobiological laws needed to understand the evolution of the living biota. *Nature Ecology & Evolution.* 1: 00165.

Marshall C, Schultze HP. 1992. Relative importance of molecular, neontological, and paleontological data in understanding the biology of the vertebrate invasion of land. *Journal of Molecular Evolution.* 35: 93–101.

Martin AP, Palumbi SR. 1993. Body size, metabolic rate, generation time, and the molecular clock. *Proceedings of the National Academy of Sciences.* 90: 4087–91.

Matsuhashi T, Masuda R, Mano T, Murata K, Aiurzaniin A. 2001. Phylogenetic relationships among worldwide populations of the brown bear *Ursus arctos. Zoological Science.* 18: 1137–43.

McGowan AJ. 2004. Ammonoid taxonomic and morphologic recovery patterns after the Permian–Triassic. *Geology.* 32: 665–8.

Menotti-Raymond M, O'Brien SJ. 1993. Dating the genetic bottleneck of the African cheetah. *Proceedings of the National Academy of Sciences.* 90: 3172–6.

Moen D, Morlon H. 2014. Why does diversification slow down? *Trends in Ecology & Evolution.* 29: 190–7.

Mundil R, Metcalfe I, Ludwig KR, Renne PR, Oberli F, Nicoll RS. 2001. Timing of the Permian–Triassic biotic crisis: Implications from new zircon U/Pb age data (and their limitations). *Earth & Planetary Science Letters.* 187: 131–45.

Nagalingum NS, Marshall CR, Quental TB, Rai HS, Little DP, Mathews S. 2011. Recent synchronous radiation of a living fossil. *Science.* 334: 796–9.

Nahonyo C, Goboro E, Ngalason W, Mutagwaba S, Ugomba R, Nassoro M, Nkombe E. 2017. Conservation efforts of Kihansi spray toad *Nectophrynoides asperginis*: Its discovery, captive breeding, extinction in the wild and re-introduction. *Tanzania Journal of Science.* 43: 23–35.

[NRC] National Research Council (U.S.) Committee on Scientific Issues in the Endangered Species Act. 1995. *Science and the Endangered Species Act.* National Academy Press.

Neveling J, Gastaldo RA, Kamo SL, Geissman JW, Looy CV, Bamford MK. 2016. A review of stratigraphic, geochemical, and paleontologic data of the terrestrial end-Permian record in the Karoo Basin, South Africa. In: Linol B, de Wit M (editors). *Origin and Evolution of the Cape Mountains and Karoo Basin*, pp. 151–7. Springer.

Newman WA, Ross A. 1977. A living *Tesseropora* (Cirripeda: Balanomorpha) from Bermuda and the Azores: First records from the Atlantic since the Oligocene. *Transactions of the San Diego Society of Natural History.* 18: 207–16.

Orr CM, Delezene LK, Scott JE, Tocheri MW, Schwartz GT. 2007. The comparative method and the inference of venom-delivery systems in fossil mammals. *Journal of Vertebrate Paleontology.* 27: 541–6.

Pennell MW, Harmon LJ, Uyeda JC. 2014. Is there room for punctuated equilibrium in macroevolution? *Trends in Ecology & Evolution.* 29: 23–32.

Pongracz JD, Paetkau D, Branigan M, Richardson E. 2017. Recent hybridization between a polar bear and grizzly bears in the Canadian Arctic. *Arctic.* 70: 151–60.

Pope KO, Baines KH, Ocampo AC, Ivanov BA. 1994. Impact winter and the Cretaceous/ Tertiary extinctions: Results of a Chicxulub asteroid impact model. *Earth & Planetary Science Letters.* 128: 719–25.

Poynton JC, Howell KM, Clarke BT, Lovett JC. 1998. A critically endangered new species of *Nectophrynoides* (Anura: Bufonidae) from the Kihansi Gorge, Udzungwa Mountains, Tanzania. *African Journal of Herpetology.* 47: 59–67.

Pyron RA. 2017. 'We don't need to save endangered species. Extinction is part of evolution'. The Washington Post. 22 November 2017.

Quental TB, Marshall CR. 2010. Diversity dynamics: Molecular phylogenies need the fossil record. *Trends in Ecology & Evolution.* 25: 43–441.

Raup DM. 1979. Biases in the fossil record of species and genera. *Bulletin of the Carnegie Museum of Natural History.* 13: 85–91.

Raup DM, Gould SJ, Schopf TJM, Simberloff DS. 1973. Stochastic models of phylogeny and the evolution of diversity. *Journal of Geology.* 81: 525–42.

Raup DM, Sepkoski JJ. 1982. Mass extinctions in the marine fossil record. *Science.* 215: 1501–3.

Seehausen O. 2006. African cichlid fish: A model system in adaptive radiation research. *Proceedings of the Royal Society B.* 273: 1987–98.

Seilacher A, Reif WE, Westphal F. 1985. Sedimentological, ecological and temporal patterns of fossil Lagerstätten. *Philosphical Transactions of the Royal Society.* 311: 5–24.

Sephton MA, Jiao D, Engel MH, Looy CV, Visscher H. 2015. Terrestrial acidification during the end-Permian biosphere crisis? *Geology.* 43: 159–62.

Sepkoski JJ. 1981. A factor analytic description of the Phanerozoic marine fossil record. *Paleobiology.* 7: 36–53.

Sepkoski JJ. 1984. A kinetic model of Phanerozoic taxonomic diversity III. Post-Paleozoic families and mass extinctions. *Paleobiology.* 10: 246–67.

Sepkoski JJ. 1986. Phanerozoic overview of mass extinction. In: Raup DM, Jablonski J (editors). *Patterns and Processes in the History of Life*, pp. 277–95. Springer-Verlag.

Sheehan PM, Coorough PJ, Fastovsky DE. 1996. Biotic selectivity during the K/T and Late Ordovician extinction events. *Geological Society of America Special Paper.* 307: 477–89.

Sheldon PR. 1987. Parallel gradualistic evolution of Ordovician trilobites. *Nature.* 330: 561–3.

Shepherd TD, Myers RA. 2005. Direct and indirect fishery effects on small coastal elasmobranchs in the northern Gulf of Mexico. *Ecology Letters.* 8: 1095–104.

Sigwart JD, Carey N, Orr PJ. 2014. How subtle are the biases that shape the fidelity of the fossil record? A test using marine molluscs. *Palaeogeography, Palaeoclimatology, Palaeoecology.* 403: 119–27.

Sigwart JD, Sutton MD, Bennett KD. 2018. How big is a genus? Towards a nomothetic systematics. *Zoological Journal of the Linnean Society.* 183: 237–52.

Sigwart JD, Wicksten MK, Jackson MG, Herrera S. 2018. Deep-sea video technology tracks a monoplacophoran to the end of its trail (Mollusca, Tryblidia). *Marine Biodiversity.* doi.org/10.1007/s12526-018-0860-2.

Simpson GG. 1944. *Tempo and Mode in Evolution.* Columbia University Press.

Smit J, Hertogen J. 1980. An extraterrestrial event at the Cretaceous–Tertiary boundary. *Nature.* 285: 198–200.

Strathmann RR. 1978. Progressive vacating of adaptive types during the Phanerozoic. *Evolution.* 32: 907–14.

Sumner-Rooney L, Sigwart JD, McAfee J, Smith L, Williams ST. 2016. Repeated eye reduction events reveal multiple pathways to degeneration in a family of marine snails. *Evolution.* 70: 2268 95.

Sutton MD, Briggs DE, Siveter DJ, Siveter DJ. 2001. Methodologies for the visualization and reconstruction of three-dimensional fossils from the Silurian Herefordshire Lagerstätte. *Palaeontologia Electronica.* 4: 1–7.

Takezaki N, Figueroa F, Zaleska-Rutczynska Z, Takahata N, Klein J. 2004. The phylogenetic relationship of tetrapod, coelacanth, and lungfish revealed by the sequences of forty-four nuclear genes. *Molecular Biology & Evolution.* 21: 1512–24.

Tilman D, Clark M, Williams DR, Kimmel K, Polasky S, Packer C. 2017. Future threats to biodiversity and pathways to their prevention. *Nature.* 546: 73–81.

Walker TD, Valentine JW. 1984. Equilibrium models of evolutionary species diversity and the number of empty niches. *American Naturalist.* 124: 887–99.

Wheeler QD, Knapp S, Stevenson DW, Stevenson J, Blum SD, Boom BM, Borisy GG, Buizer JL, De Carvalho MR, Cibrian A, Donoghue MJ. 2012. Mapping the biosphere: Exploring species to understand the origin, organization and sustainability of biodiversity. *Systematics & Biodiversity.* 10: 1–20.

Wiens JJ. 2016. Climate-related local extinctions are already widespread among plant and animal species. *PLoS Biology.* 14: e2001104.

Willis KJ, Bennett KD, Bhagwat SA, Birks HJ. 2010. 4° C and beyond: What did this mean for biodiversity in the past? *Systematics & Biodiversity.* 8: 3–9.

Yap JY, Rohner T, Greenfield A, Van Der Merwe M, McPherson H, Glenn W, Kornfeld G, Marendy E, Pan AY, Wilton A, Wilkins MR. 2015. Complete chloroplast genome of the Wollemi pine (*Wollemia nobilis*): Structure and evolution. *PLoS ONE.* 10: e0128126.

# 10 How Many Species are There?

## THE GREAT GLOBAL CENSUS

The question of how many species inhabit Earth has been the focus of scientific curiosity for centuries. Recently this has been presented as a conservation issue, an important unknown factor in planning actions to protect global biodiversity (Bickford et al., 2007). There are other accessory benefits to pursuing the project to catalogue and understand life on Earth, but at the root it is primarily an intellectual grand question, the number and variety of species that exist is the type of information that we would generally like to know about how the world works. At the turn of the 21st century, quantitative estimates for global eukaryotic diversity ranged over several orders of magnitude, from $10^6$ to $10^8$ living species (May, 1988, 2010). Several recent analyses have produced apparently congruent estimates at a number between 5 and 8 million species (Mora et al., 2011; Costello et al., 2013), but there is no good evidence that large-scale species richness estimates are genuinely converging at a precise or robust answer (Caley et al., 2014). The true number of living species is meanwhile slowly shifting downward as species are driven to extinction by anthropogenic activities.

This chapter considers the problem of a census of global species, and what has made it so difficult to resolve. As in the rest of this book, we are only considering the influence of macroscopic eukaryotes—animals, plants, and fungi—which is notably limited compared to *all* living species on Earth. Even within our subgroup of mainly macroscopic, multicellular organisms, the bounds of global total species richness estimates are not much more precise than for all species.

Most approaches to answering how many species there are on Earth have used subsampling, taking the number of taxa in a particular group or place, or the relationships of relative diversity among different groups of organisms, and multiplying species-per-subgroup to extrapolate to a global total. Any process like this that uses logical induction, taking a specific case and expanding it to a general conclusion, is prone to error and controversy. Taxonomy, the field at the fundament of recognising species, may seem especially prone to controversy, and resistant to generalities. In the last decade, new evidence has emerged to support universal patterns of higher-ranked taxa (phyla, classes, orders, etc.) that could be exploited to predict the total number of species they contain (Williams & Gaston, 1994; Benton, 1997; Mora et al., 2011; Chapter 2). The number of species per genus follows a skew or hollow curve distribution that is very consistent across all animals and plants (Maruvka et al., 2013; Sigwart et al., 2018; Chapters 2 and 6). This is the most promising approach to robust predictions of global species richness, and it should inform research priorities for taxonomy. Working toward a common goal of finding out how many species there are, we should probably focus as much energy on revising higher classification as in actually naming species.

Understanding species richness is only one facet of the knowledge deficits about biodiversity (Hortal et al., 2015), but naming species cross cuts other aspects of biology. Knowing how many species remain unnamed is an important baseline to establish a strategy to discover and document undescribed diversity.

## UNDER-STUDIED TAXA

Taxonomic effort is not proportionate to taxonomic need, or to total species richness in different clades: there are far more scientists working on vertebrate animals than all other animal phyla combined. This is not anecdotal, or carping, it is a well-documented fact (May, 1988; Figure 10.1). There are many reasons that scientists work on particular groups of organisms, and taxonomy is only one minor part of this. The fact that parasites and other potentially repulsive animals have professional devotees at all is testament to our collective fascination with the natural world and its workings; only by studying these animals in close detail can essential scientific questions be addressed (such as understanding human health and disease, as well as global biodiversity). Parasites are just as prone to extinction as other taxa (Carlson et al., 2017). Beautiful and striking body forms, such as molluscs (seashells),

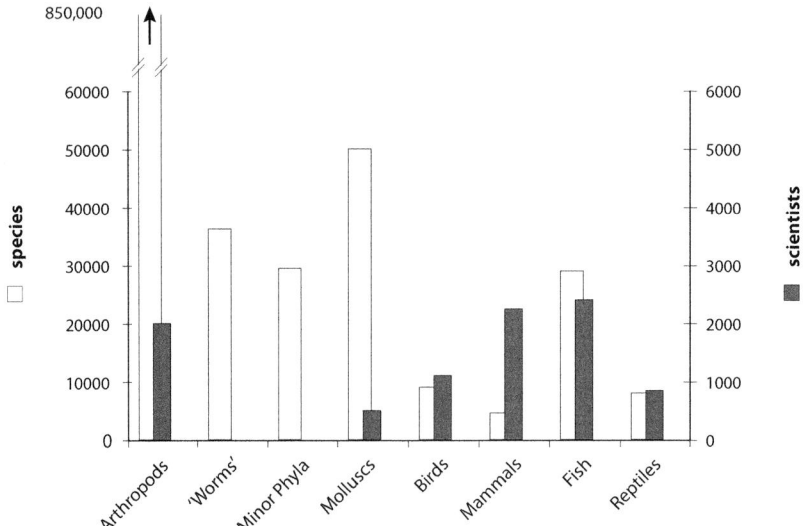

**FIGURE 10.1**    The numbers of described species in some animal groups (white bars, scaled to left vertical axis) compared to the number of scientists studying those groups (grey bars, scaled to right vertical axis). Human numbers here are based on the combined membership in major American and UK-based professional scientific societies. Taxonomists, naturally, make up only a fraction of the scientific community; for many groups of organisms, such as many invertebrate phyla, there are no scholarly societies or their memberships are so small as to not be shown on this simple chart.

insects (butterflies, beetles), flowering plants, and of course the vertebrates (birds and mammals) have a larger fan-base, both professional and recreational. So it is natural that better-studied and better-described groups attract a bigger community of scientists producing an ever-larger literature. Nonprofessional scientists who are nonetheless experts in a particular taxon or a local fauna or flora are hugely important to documenting biodiversity (e.g., Gardiner & Bachman, 2016) and also to the practice of naming new species (e.g., Bouchet et al., 2016). And those many eyes will spot a new species and recognise its importance. Less-studied groups of organisms are not neglected out of spite, they are simply on a different point in the path of scientific development.

Larger human scientific communities also facilitate historical continuity. Among minor groups, the remit of an 'expert' may be very broad, covering many species or higher taxonomic diversity; there are probably about 30 living people who are experts for all five 'minor' classes of molluscs apart from gastropods, bivlaves and cephalopods (class being the next rank below phylum). Over time and persisting to the present day, the community of scientists contributing to taxonomic efforts is growing (Poulin, 2014). However, many people counted in that community by co-authorship on a taxonomic paper have other primary expertise (Chapter 8). At any one time in history, there are not living experts on all groups of animals or plants. And each new generation of scientists adds experts on newly-discovered groups.

There is a strong differential between expertise and need across biological groups, but these examples have focussed on animal taxa. In fact, the most under-represented group in any global diversity census, and the wildcard for estimating global species richness, is the fungi. The present estimate for global fungal species richness is around 2.27 million species, although less than 100,000 have been described (Hawksworth, 2001; Hawksworth & Lücking, 2017). Fungi contain a respectably large number of species and evidentially an astonishing number of unrecognised species. The number of described species is lower than for either animals or plants, but still a substantial contribution that may be dramatically under-represented in global species richness estimates (Bass & Richards, 2011). Molecular methods have revolutionised species recognition in all groups, but perhaps most dramatically in fungi. Fungal organisms are almost impossible to find when the sporocarp (fruiting body) is absent; DNA surveys of soil organisms even in well-described, well studied areas have turned up unexpectedly high numbers of unrecognised fungi (Blackwell, 2011; Aime & Brearley, 2012). One study found evidence for around 800 species of fungi from DNA barcode fragments isolated from soil; a small volume of soil at one point in time found the same species richness as an analogous temperate forest plot that had been meticulously surveyed for fruiting bodies for over 20 years (O'Brien et al., 2005). As it is very unlikely that every species present was captured in that small molecular survey, it is more likely that the total diversity of forest fungi is significantly higher. The variation in the estimated number of fungi also has a huge effect on estimating global species richness (Figure 10.2).

Many more studies have tackled the problems of how many species are in particular groups of interest—fungi, molluscs, insects, and subsets of those in particular habitats—than have expanded to a global estimate for all species. Methods, approaches, and the data gaps in particular groups or regions vary so much that

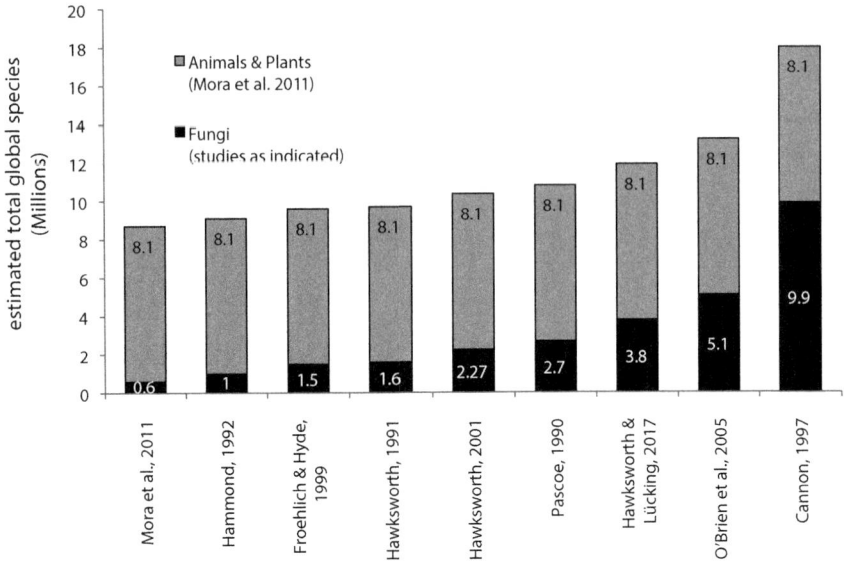

**FIGURE 10.2** Uncertainty over fungal species diversity impacts global total species richness estimates. Even accepting a given estimate for global animal and plant diversity (e.g., Mora et al., 2011), a very broad range of global fungal species numbers has been suggested in the literature (black bars, data from Bass & Richards, 2011; Hawksworth & Luecking, 2017).

they are largely not comparable (Caley et al., 2014). Considering the availability of taxonomic effort, and the different states of development of various organismal groups, large scale analyses of species description effort should be treated with caution. Even if the literature covers the taxonomy of global species, different groups of taxa are drawn from different scholarly ecosystems.

## APPROACHES TO ESTIMATING GLOBAL SPECIES RICHNESS

Analyses of global species richness sometimes are unclear about what sort of species are being considered—is it only animals, or the eukaryotes, or all organisms? In total, there are around 1.6 million eukaryotic species that have already been described and included in the Catalogue of Life Annual Checklist (Roskov et al., 2016). That list is not a complete reckoning of the available published species descriptions, so the current tally for eukaryotes is around 1.9 million species named (Bouchet et al., 2016). The Catalogue of Life project and some other global species estimates nominally include microbes (e.g., Mora et al., 2011), but the importance and diversity of microbial groups has generally been dramatically under-estimated (Hug et al., 2016). Nonetheless, different analyses have proposed wildly different counts of global species—Linnaeus honestly thought he would name all of the living species on Earth and he did document more than 10,000 (Stearn, 1959). Modern estimates range from around 2–3 million (Raven, 1983; Costello et al., 2013) up to 100 million (Erwin, 1982; May, 1992).

Many studies of species patterns come from arthropods, because arthropods and especially insects represent a very large fraction of animal diversity (Gaston, 1991). This was explicitly articulated in a seminal study by Erwin (1982) which produced one of the highest ever serious global species estimate, at 100 million ($10^8$). The logic is that arthropods are overwhelmingly the largest animal phylum, and beetles, particularly in the tropics, are the largest subset, so patterns of tropical beetle species richness could be taken as a proxy for the total count of animals. Finding a very large number of species endemic to a small area, Erwin multiplied this number by the total available habitat in tropical trees and arrived at a very large number indeed, two orders of magnitude higher than most current estimates. Subsequent additional work surveying canopy insects has continued to encounter a nearly overwhelming diversity of species, but not the same levels of fine scale endemism that was originally suspected.

The importance of endemism has a parallel with host-specificity in parasitic taxa. Molecular identification of parasite animal species has found an accelerating number of new taxa. Several parasite species that were thought to be more wide-spread turn out to be species complexes where each individual lineage is tied to the host organism (Poulin, 2014), or higher than expected host-endemism.

The extents of species ranges are clearly a major factor in calculating large-scale estimates of species richness. In deep-sea ecosystems, many species are collected only once because of the technological difficulty of accessing those habitats; thus, ranges and endemism are difficult to reconcile. Multiplying the continuously increasing rates of discovery in deep-sea animals has also suggested very large numbers of undiscovered species (Grassle & Maciolek, 1992). In fact, one estimate based on rarefaction and nonsaturation for the discovery of marine animals arrived at an estimate of 100 million marine species alone (Lambshead, 1993). Given that there is a problem with differential effort, and discovery, in various taxonomic groups and habitats, other researchers turned to ecological relationships to find patterns in how many species occur in a geographical area, or a food web, or in host-parasite relationships. These may be robust and relevant to particular lineages or ecosystems. Unfortunately, the stochastic variation in these systems, not unlike studies of canopy beetles, hampers expansion to truly global estimates.

## WHY IS THIS PROBLEM SO HARD?

It may seem like there should be a very specific answer, to the extent that the naive question of global species richness concerns just counting a finite number of things, in a given instantaneous time slice. Species are continuously evolving, which adds uncertainty and intra-lineage variation (Chapter 5). The fuzzy boundaries of species, and intrinsic data limitations in studying any species lineage, mean that there will be indefinite arguments about certain species designations. The project to describe life on Earth is, in part, an effort to refine our understanding of what species are. Beyond these additional layers of embedded problems in terms of how species are defined and recognised, there are still many species that remain undiscovered and undescribed. In terms of assessing the current state of progress toward estimating global species richness, it is therefore worth considering both the sources of data used to attempt to answer the question, and some fundamental limitations on how a large total number can be estimated when many elements of the population have not been observed.

Global patterns in biodiversity have been quantified in a number of different ways, by combining data compiled for different taxonomic groups, ecosystems, or regions, or extrapolating patterns of novelties. These represent processes of inductive logic, where each step of reasoning moves toward increasing generality. But each step also introduces potential error, and an accurate conclusion is entirely dependent on the accuracy of each contributing element.

Scientific investigations generally all proceed via inductive reasoning: specific observations are used to form generalisations, which then present testable hypotheses. The number of trees in a given area, for example, can be multiplied over the available habitat space to estimate the total global population of trees. Inductive approaches are prone to potential error—what if some area has been clear-cut? What if our sample area is not representative of forest density for other countries? These errors can be overcome, as in one recent study that used highly replicated sampling and satellite images, to guard against over-generalisations concluding with a robust estimate of 3.04 trillion trees on Earth (Crowther et al., 2015). This is the logical process that achieves most new understanding of really broad scientific phenomena; however, the larger the generalisations at any stage in an inductive process, the more the potential for error increases. Deductive arguments by contrast are considered to be truth-preserving. Beginning with generalities that are true (trees grow in most parts of the world, forests are composed of trees), we can deduce specific instances of the pattern without risk to accuracy (this hectare of forest in India is characterised by trees). However, valid deduction is nonampliative—it does not propose new knowledge—this limits its ability to solve outstanding general problems.

Experimental approaches are sometimes explained as a deductive process, as in a general hypothesis tested with a specific experiment, but this is somewhat misleading. Specific deductive tests are designed as attempts to falsify a working hypothesis. In fact, scientific knowledge grows as experimental results provide more individual observations that contribute to the *inductive* formation of general principles. Reliable inductive reasoning, moving from small observations to global patterns in biodiversity, depends on the accuracy and correct interpretation of all those individual observations.

In under-studied clades, taxonomy and phylogenetics remain at the stage of first pass in documenting basic observations. The rate of discovery of new species of gastropods is much higher than new species of birds, since effectively all birds have been described (May, 1988; Barrowclough et al., 2016; Figure 10.3). The rate of discovery in fungi is so high that the number of sequences produced by environmental sampling had started to outpace specimen-based voucher sequences by 2008 (Blackwell, 2011). Guessing how many species remain undocumented or unobserved in a big group, such as insects, may be quite arbitrary (May, 1988). The error in estimating species richness increases nonlinearly with clade size, that is, our collection of fundamental observations has more embedded uncertainty, so this error has a multiplying effect on any global species richness estimate.

Many estimates have used the proportion of new species in a quantitative sample of biodiversity as a microcosm for global patterns, using samples of particular taxa (heminopteran insects: Hodkinson & Casson, 1991), or environments (the deep sea:

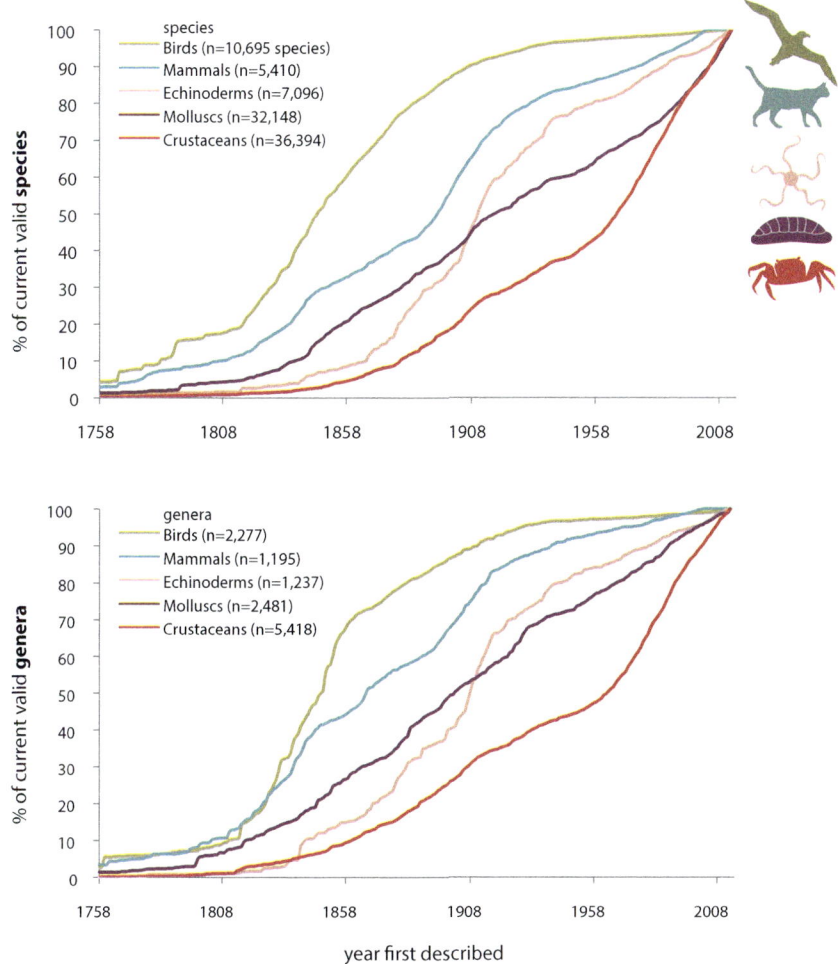

**FIGURE 10.3**  The rate of published descriptions of new species (top) and new genera (bottom) in several major groups of animals, shown as the percent of current valid species described cumulatively to the data sampling point in 2014. For birds, the curve smooths out to near horizontal as it approaches the present day, whereas invertebrates are ever increasing. (Data from Sigwart et al., 2018.)

Grassle & Maciolek, 1992). Apparently consistent ratios of the number of taxa per host-plant have been extrapolated to larger-scale estimates for arthropod species in the tropics (Erwin, 1982), or global fungi (Hawksworth, 1991). The major criticism of this approach is that there may be no evidence that such a ratio (the number of fungal species per plant, for example) would stand up when extrapolated to different environments (Storks, 1993; Lambshead & Boucher, 2003) or even to other, closely related taxonomic groups.

Another set of basic observations used  for estimating global species richness is the history of the rate of scientific description of new taxa, which can be extrapolated to make forward predictions of when new discoveries could become saturated (Joppa et al., 2011; Costello et al., 2012). This may be effective in a few groups that are approaching an asymptote—that is, where effectively all living species have been described (Bebber et al., 2007). This is the case for birds and possibly mammals at species and genus level (Figure 10.3). The method is not transferable to groups where the rate of new discoveries continues to increase (e.g., for molluscs or crustaceans, Figure 10.3). It is also problematic when scientific effort is temporally patchy, or the sample size of discoveries is too small, where the impact of one monograph or dissertation can suddenly make a step change in the diversity estimate. The quality of data available for different taxonomic groups is highly variable; birds for example may be well studied and reasonably well understood, whereas the estimates of how many feather lice species remains an open question. Yet all species must be counted equally in accounting for the species richness of the whole planet.

In all cases, attempts to estimate large-scale patterns of species richness have been criticised for embedding some sort of erroneous assumptions in their process. There is too much missing data, and too much uncertainty, and too much difference between groups (in terms of sampling and in terms of expected dynamics). Such errors are relevant, but more or less inevitable for any inductive process. Put another way, criticism of these studies generally reflects concern over sample size, and whether the number of individual observations warrants the next inductive step to a generality. If you accept that the bottom up approach is not a good path forward—there may not be a viable way to get to a total number either by cataloguing everything or by averaging across known groups—then the next step is to tackle the problem from a different angle.

Mora and colleagues (2011) proposed an alternative method, taking observed patterns in the species richness of higher clades and extrapolating the diversity they contain. The idea is that there are far fewer genera than species, fewer families than genera, etc., and that while most species may still be undiscovered, the class- or order-level groups have long since been identified. While hundreds of new species are named every year, new higher ranked groups are named infrequently enough that it becomes a bit of an event. The number of mammal species is continuing to increase, for example, but the number of mammal genera is more asymptotic (Figure 10.3). Furthermore, there seem to be consistent quantitative relationships between nested groups (Figure 10.4). The same approach had been tried earlier in combination with species-area effects (Williams & Gaston, 1994). These authors harnessed skew

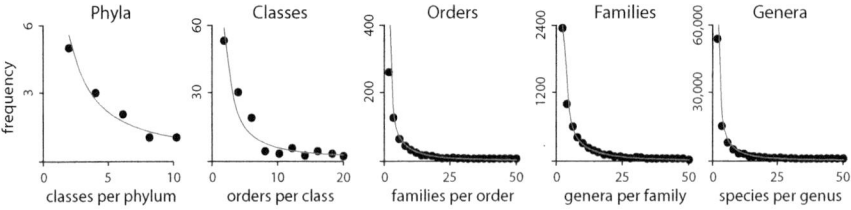

**FIGURE 10.4**   The skew distributions that relate to nested sets in Linnaean hierarchical ranks (data for animal taxonomy redrawn from Mora et al., 2011). Nested ranks are self-similar in animal and plant taxonomy; small units are very common and larger units, whether genera, or phyla, are relatively uncommon.

distribution patterns to project total species richness based on the established discovery of higher groups. This 'top-down' approach is a bit more like a deductive process, and as such it could produce more robust estimates. This may or may not depend on higher taxonomic ranks having some comparability, or meaning that is transferable between lineages with different taxonomic traditions. If a group of rank 'order' has an entirely different meaning for different organisms, this might limit its predictive power, but nested groups may yet follow deeper evolutionary patterns. Simulation approaches to taxonomy have now shown that genus level data, at least, are concordant with phylogeny (Sigwart et al., 2018; Chapter 6), lending new strength to the key data for this top down approach to estimating global species richness.

## THE IMPORTANCE OF HIGHER RANKS

Many studies have observed apparently very conserved, skewed frequency distributions of species richness distributions at least for genera across animals and plants. The shape of the frequency distributions seems subtly different at different ranks, and of course it is much less strong for phyla because there are fewer data points (Figure 10.4). Using these patterns for species richness calculations rests on the idea that there are consistent mathematical relationships between the number and size of groups, and these patterns have some global consistency that would apply to all living organisms.

Harnessing universal patterns about higher ranked groups may be the most powerful approach proposed to date to solve the question of global diversity, but the data being fed into the model still may be inadequate. Higher-ranked groups find themselves on shaky footing in any group where phylogenetic framework is poorly resolved (which is most of them). There is also a potential cultural bias among scientific readers; most taxonomists and organismal biologists tend to communicate most closely with other people who work on similar organisms. Knowing about a higher-ranked group does not guarantee strong detailed knowledge of the contained species, or understanding group-specific taxonomic uncertainty. The identification and establishment of higher ranked groups in some organisms is still contentious and subject to major shifts with phylogenetic revision.

Sometimes, the discovery of a new species can prompt the erection of a new higher taxon, or the revision of higher-ranked groups based on phylogenetic insights provided by studying the new organism (Chapter 4, Chapter 6). Returning to the idea of high levels of uncertainty in under-studied groups, new higher ranked groups are not uncommon, especially in fungi (where a new phylum was described only a few years ago: Bauer et al., 2015). We might be intrinsically more able to recognise relevant differences among animal taxa. (Although anthropomorphism is probably as much hindrance as help.) Still, recent analyses have made phylum level changes within Animalia and questions of whether some taxa should be phyla remain unresolved. Two groups of worms formerly considered distinct phyla were subsumed into Annelida—Pogonophora, the hydrothermal-vent endemic tube worms, and Echiura, marine 'spoon worms' with an extensible eponymous proboscis—are actually derived groups with modified body plans different from typical segmented worms (Struck et al., 2011; Weigert et al., 2014; Figure 10.5). This does not mean that these groups are not valid taxa; Echiura is unequivocally a monophyletic clade.

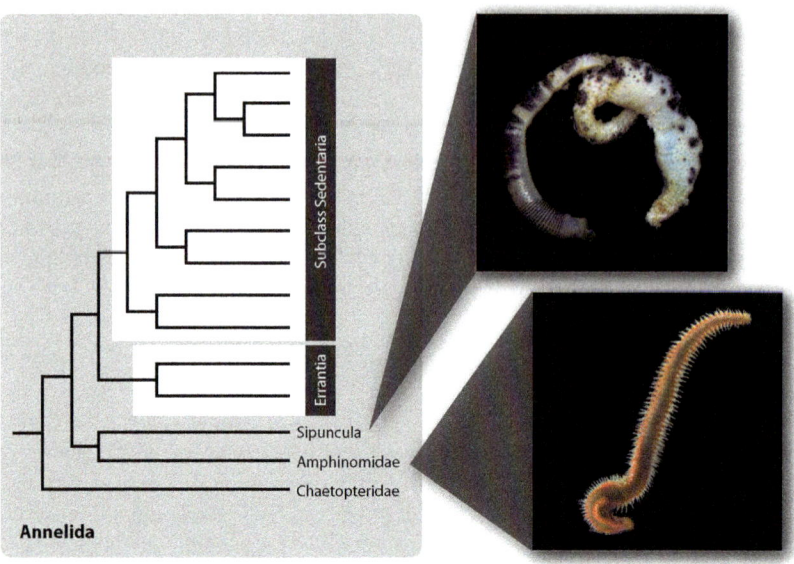

**FIGURE 10.5** Diversity of the phylum Annelida *sensu lato*: An example of the unsegmented Sipuncula *Phascolosoma scolops* (Selenka & de Man, 1883), equivocally classified in a separate phylum (top right), and a more 'typical' segmented marine polychaete *Eurythoe mexicana* Berkeley and Berkeley, 1960 (lower right). Molecular phylogenetic results indicate that Sipuncula is nested within the phylum Annelida (left, redrawn from Weigert et al., 2014).

It means that the level of variability in annelid body plans is much greater than was first understood, and segmentation, the original annelid synapomorphy, has been lost probably four separate times within the evolution of the phylum. This in fact is why it is relevant to assign hierarchical ranks to these clades, to clearly indicate that these features represent diversity within the phylum, not alongside it. Another related clade, Sipuncula, is also a derived branch within Annelida. But these 'peanut worms' have a deep fossil record and many other strange features that have led some experts to maintain them as a separate phylum (Lemer et al., 2015). These are all relatively large animals, dominating many benthic marine environments, which makes them easy to observe and collect, yet their status as potential phyla is in flux.

Taxonomic groups that are well described in terms of species richness often progress to a later scientific phase of revisionary work as the phylogeny becomes better understood, prompting the rearrangement and establishment of new higher-ranked groupings. Updated global estimates based on refined taxonomic models for higher taxa across all groups could help to converge estimates of global species richness.

## IF IT DOESN'T HAVE A NAME, IT DOESN'T EXIST

Identifying Earth's diversity is intimately tied to preventing loss and extinction: what is unnamed, goes unrecognised, and may be lost without our ever knowing it existed (Raven, 1985). But identifying species is also linked to understanding the evolutionary

**TABLE 10.1**

**Terms for Similar, Closely-Related Species**

| | |
|---|---|
| Cryptic species | Indistinguishable |
| Pseudocryptic species | Similar enough to be thought the same, but once you look again, you can see the differences |
| Species complex | A group of closely related species, where delimitation is difficult or contentious, and species-level taxonomy has not been fully resolved |
| Sibling species | Phylogenetic sister lineages |

processes that control lineage diversification, whether or not it is articulated as such. The identification of species is a process of consensus, and assessing total evidence of evolutionarily relevant differences among related lineages (Chapter 8). Our ability to recognise relevant features may become weaker as we examine things that are more distant to ourselves and our own experience. It is inevitable that there are many species that are not recognised, and this seriously handicaps the basic idea of counting all the species on Earth.

Taxonomic uncertainty is immensely frustrating the researchers who want a name to reliably refer to a particular species. The process of completing an analysis about what features 'count' as relevant and which do not is time-consuming and sometimes controversial (Chapter 8). A placeholder solution for this scenario is to refer to a 'species complex' or 'cryptic species complex', to acknowledge that the organisms represent multiple lineages, but that there is not (yet) an easy way to articulate the features that identify them. In the case of genuinely cryptic species, which cannot be distinguished except by DNA evidence, field identification may be impossible (Table 10.1). Cryptic species and cryptic species complexes have been called the 'wildcard' in biodiversity assessment.

Molecular techniques inarguably revolutionised the data and the process of systematics of living species. The ability to directly compare DNA sequence data has revealed additional genetic diversity that separates species which look superficially similar. Species that are apparently indistinguishable from morphology alone are called 'cryptic species'. In many cases, new evidence prompts a second look, and obvious differences reveal themselves to be relevant and the newly distinguishable species are called 'pseudocryptic' (Table 10.1). Different colours or superficial features are often plastic, but when colour is paired with a substantial difference in gene sequences (or distribution or reproductive habits), this is evidence for lineages on different evolutionary paths (Chapter 5; Figure 5.1). This has been a constant part of the process of taxonomy since the 18th century (Winker, 2005), but is increasingly recognised as a critical gap in estimating diversity (Knowlton, 1993).

Cryptic species may be associated with known subspecies, or names that had been subsumed into synonymy (Chapter 4). Re-instating a prior synonym is not 'taxonomic inflation' as has been suggested by some authors (Isaac et al., 2004), it is incremental revision that reveals greater diversity than was previously understood. Indeed, the volume of available names that are considered junior synonyms affects estimates of total diversity and stymies taxonomic progress (Bouchet & Strong, 2010).

Within the realm of marine invertebrates in particular, the importance of molecular differentiation has anecdotally led some workers to delay publishing species descriptions, for example where an organism is only known from a single specimen. Those specimens may languish in a lab, known to only one expert, and not available for comparison by anyone else who may stumble upon the second specimen. Where there is sufficient evidence to propose a well-supported hypothesis that a specimen represents a novel lineage, it should be given a name so that others can refer to it (Chapter 8). In groups that are critically under-studied such as marine invertebrates, the need for documentation of primary evidence for biodiversity—species— is probably more beneficial than the potential cost of introducing a junior synonym.

This proposes a double-ended approach, both bottom-up and top down. The (bottom-up) quest to name all the species on Earth continues. If we want an answer to the question of how many species are on Earth, before that task is complete, then we need multiple lines of evidence to estimate the answer, including both naming more undescribed species (bottom-up) and especially understanding relationships and patterns in global nested ranked groups (top-down). Quantifying the uncertainty and the frequency in cryptic species is important to making primary estimates of species richness in any region or clade. But the systematic arrangement and classification of those species is equally important. Authors of phylogenetic analyses often hedge their results, calling different groups as 'unranked clades' without committing to a Linnaean rank. Linnaean ranks help these modern hypotheses to be incorporated into other fields outside of phylogenetics. Some new clades cannot and should not be shoehorned into old names. It may lead to the abandonment of historic names like 'Pisces' when it becomes clear that fish are not a clade, but new names can be assigned to appropriate ranks with due consideration (Chapter 4). Others may see the taxon sampling or statistical support from a specific analysis as too sparse to allow inductive interpretation of the classification of the whole group.

The plea for names that are explicitly tied to appropriate ranks is a reminder that these names are of interest to a broader group of users. There is a popular idea that we need phylogeny more than we need ranked nomenclature, but that is comparing apples and orangutans. We need ranked groups to be useful to a broad audience, and having higher ranks named and tested will facilitate meta-analyses that may refine global species richness estimates.

## HOW MANY SPECIES ARE STILL UNDISCOVERED?

The task of discovering, describing, and naming all living species is an ambitious goal. Understanding how many species are probably sharing our planet is the first step toward deciding whether this task can be achieved in any kind of reasonably timeframe. The more interesting question may be, where are the knowledge gaps that significantly hamper global estimations, using any of the methods that have been attempted to date (Hortal et al., 2015)? If there are universal laws that govern the size-frequency distribution of nested groups of evolutionary radiations or clades, then higher ranked groups can be used to predict the species richness they contain. Thus this problem becomes in part a question of the discovery, description, and assignment of novel higher ranked groups. The most urgent gaps to fill are the recogition of  species that will resolve the higher classification in understudied groups, such as fungi, and species in remote environments that may represent unique lineages, such as in the deep sea.

Using the correct scientific species names contributes to the foundation of other important scientific questions (Chapter 3). As taxonomy becomes more diffuse, with a broader variety of people engaged and fewer scientists working as specialists in taxonomy and systematics (Chapter 7), it is critical to develop better communication about how to manage nomenclature. That management and planning may be the most important next challenge in our lifetime, before the ultimate project of describing life on Earth that may take hundreds more years to complete.

Technology will continue to increase the rate of discovery and description in a nonlinear trajectory. Species have a very long shelf life; Fontaine and colleagues (2012) showed the average time from discovery of the first novel specimen to publication of a new species is over 20 years. This lag means that the ongoing increases in rates of description actually may not yet fully capture discoveries being driven by molecular results (Costello et al., 2013; Bouchet et al., 2016). Technology to support in genome sequencing, better (more complete) reference databases for comparative identification, and technology to extract poor quality DNA from museum specimens (to ease comparison with types and historical material), will all continue to get better. It may be another 20 years before we see the real impact these technologies will have on taxonomy.

If something does not have a name, it does not exist, so all new taxa need to be described. Useful progress to answer the global species richness question is less a matter of actual primary description of species, and more a question of intensive, data driven systematic revision. It is not good enough just to find species and say they are new. But more to the point, those new species need to be put into a systematic context. Ongoing work in this vein will continue to attempt to falsify our working hypothesis, around 8 million living eukaryotic species, and refine that number to a higher degree of precision.

The catalogue of life is an important, grand intellectual goal, but we are far from achieving it, and we cannot confidently predict how long it will take to complete. That catalogue will eventually enable grand experiments on unimaginable scales: whole planet spatial ecology. It may be hubris to think we can catalogue and atomise even a single ecosystem, including microbes, parasites, and hidden life, let alone the whole of Earth. Aspects of the project are tantalisingly close to completion, such as birds (Figure 10.3, Chapter 6). Large projects are only achieved when every participant keeps moving forward. This should provide optimism for the larger global effort in other taxa but puts birds in a unique position as role models for other taxonomic groups in how to redirect their taxonomic efforts as the accumulation of new species reaches its final asymptote. That is the main purpose of having a solid number for how many species share our planet: not for the satisfaction of knowing, but for grasping the scope of the problem so that we can develop an adequate strategy to tackle it.

## REFERENCES

Aime MC, Brearley FQ. 2012. Tropical fungal diversity: Closing the gap between species estimates and species discovery. *Biodiversity & Conservation.* 21: 2177–80.

Barrowclough GF, Cracraft J, Klicka J, Zink RM. 2016. How many kinds of birds are there and why does it matter? *PLoS ONE.* 11: e0166307.

Bass D, Richards TA. 2011. Three reasons to re-evaluate fungal diversity 'on Earth and in the ocean'. *Fungal Biology Reviews*. 25: 159–64.

Bauer R, Garnica S, Oberwinkler F, Riess K, Weiß M, Begerow D. 2015. Entorrhizomycota: A new fungal phylum reveals new perspectives on the evolution of fungi. *PLoS ONE*. 10: e0128183.

Bebber DP, Marriott FHC, Gaston KJ, Harris SA, Scotland RW. 2007. Predicting unknown species numbers using discovery curves. *Proceedings of the Royal Society B*. 274: 1651–8.

Benton MJ. 1997. Models for the diversification of life. *Trends in Ecology & Evolution*. 12: 490–5.

Berkeley E, Berkeley C. 1960. Notes on some Polychaeta from the west coast of Mexico, Panama, and California. *Canadian Journal of Zoology*. 38: 357–62.

Bickford D, Lohman DJ, Sodhi NS, Ng PK, Meier R, Winker K, Ingram KK, Das I. 2007. Cryptic species as a window on diversity and conservation. *Trends in Ecology & Evolution*. 22: 148–55.

Blackwell M. 2011. The Fungi: 1, 2, 3 … 5.1 million species? *American Journal of Botany*. 98: 426–38.

Bouchet P. 2006. The magnitude of marine biodiversity. In: Duarte CM (editor). *The Exploration of Marine Biodiversity: Scientific and Technological Challenges*, pp. 31–62. Fundación BBVA.

Bouchet P, Bary S, Héros V, Marani G. 2016. How many species of molluscs are there in the world's oceans, and who is going to describe them? *Tropical Deep-sea Benthos*. 29: 9–24.

Bouchet P, Strong EE. 2010. Historical name-bearing types in marine molluscs: An impediment to biodiversity studies? In: Polaszek A (editor). *Systema Naturae 250*. CRC Press.

Caley MJ, Fisher R, Mengersen K. 2014. Global species richness estimates have not converged. *Trends in Ecology & Evolution*. 29: 187–8.

Carlson CJ, Burgio KR, Dougherty ER, Phillips AJ, Bueno VM, Clements CF, Castaldo G, Dallas TA, Cizauskas CA, Cumming GS, Doña J. 2017. Parasite biodiversity faces extinction and redistribution in a changing climate. *Science Advances*. 3: e1602422.

Costello MJ, May RM, Stork NE. 2013. Can we name Earth's species before they go extinct? *Science*. 339: 413–6.

Costello MJ, Wilson S, Houlding B. 2012. Predicting total global species richness using rates of species description and estimates of taxonomic effort. *Systematic Biology*. 61: 871–83.

Crowther TW, Glick HB, Covey KR, Bettigole C, Maynard DS, Thomas SM, Smith JR, Hintler G, Duguid MC, Amatulli G, Tuanmu MN. 2015. Mapping tree density at a global scale. *Nature*. 525: 201–5.

Erwin TL 1982. Tropical forests: Their richness in Coleoptera and other arthropod species. *Coleopterists Bulletin*. 36: 74–5.

Fontaine B, Perrard A, Bouchet P. 2012. 21 years of shelf life between discovery and description of new species. *Current Biology*. 22: R943–4.

Gardiner LM, Bachman SP. 2016. The role of citizen science in a global assessment of extinction risk in palms (Arecaceae). *Botanical Journal of the Linnean Society*. 182: 543–50.

Gaston KJ. 1991. The magnitude of global insect species richness. *Conservation Biology*. 5: 183–96.

Grassle JF, Maciolek NL. 1992. Deep-sea species richness: Regional and local diversity estimates from quantitative bottom samples. *American Naturalist*. 139: 313–41.

Hawksworth DL. 1991. The fungal dimension of biodiversity: Magnitude, significance and conservation. *Mycological Research*. 95: 641–55.

Hawksworth DL. 2001. The magnitude of fungal diversity: The 1.5 million species estimate revisited. *Mycological Research*. 105: 1422032.

Hawksworth DL, Lücking R. 2017. Fungal diversity revisited: 2.2 to 3.8 million species. *Microbiology Spectrum*. 5: FUNK-0052-2016.

Hodkinson ID, Casson D. 1991. A lesser predilection for bugs: Hemiptera (Insecta) diversity in tropical rain forests. *Biological Journal of the Linnean Society*. 43: 101–9.

Hortal J, de Bello F, Diniz-Filho JA, Lewinsohn TM, Lobo JM, Ladle RJ. 2015. Seven shortfalls that beset large-scale knowledge of biodiversity. *Annual Review of Ecology, Evolution, & Systematics*. 46: 523–49.

Hug LA, Baker BJ, Anantharaman K, Brown CT, Probst AJ, Castelle CJ, Butterfield CN, Hernsdorf AW, Amano Y, Ise K, Suzuki Y. 2016. A new view of the tree of life. *Nature Microbiology*. 1: 16048.

Isaac NJ, Mallet J, Mace GM. 2004. Taxonomic inflation: Its influence on macroecology and conservation. *Trends in Ecology & Evolution*. 19: 464–9.

Joppa L, Roberts DL, Pimm SL 2011. How many species of flowering plants are there? *Proceedings of the Royal Soceity B*. 278: 554–9.

Knowlton N. 1993. Sibling species in the sea. *Annual Review of Ecology & Systematics*. 24: 189–216.

Lambshead PJD. 1993. Recent developments in marine benthic biodiversity research. *Oceanis*. 19: 5–24.

Lambshead PJD, Boucher G. 2003. Marine nematode deep-sea biodiversity: Hyperdiverse or hype? *Journal of Biogeography*. 30: 475–85.

Lemer S, Kawauchi GY, Andrade SC, González VL, Boyle MJ, Giribet G. 2015. Re-evaluating the phylogeny of Sipuncula through transcriptomics. *Molecular Phylogenetics & Evolution*. 83: 174–83.

Maruvka YE, Shnerb NM, Kessler DA, Ricklefs RE. 2013. Model for macroevolutionary dynamics. *Proceedings of the National Academy of Sciences*. 110: E2460–9.

May RM. 1988. How many species are there on Earth? *Science*. 241: 1441.

May RM. 1992. How many species inhabit the Earth? *Scientific American*. 267: 18–24.

May RM. 2010. Tropical arthropod species, more or less? *Science*. 329: 41–2.

Mora C, Tittensor DP, Adl S, Simpson AG, Worm B. 2011. How many species are there on Earth and in the ocean? *PLoS Biology*. 9: e1001127.

O'Brien HE, Parrent JL, Jackson JA, Moncalvo JM, Vilgalys R. 2005. Fungal community analysis by large-scale sequencing of environmental samples. *Applied & Environmental Microbiology*. 71: 5544–50.

Poulin R. 2014. Parasite biodiversity revisited: Frontiers and constraints. *International Journal for Parasitology*. 44: 581–9.

Raven PH. 1983. The challenge of tropical biology. *Bulletin of the Entomological Society of America*. 29: 4–13.

Raven PH. 1985. Disappearing species: A global tragedy. *Futurist*. 19: 8–14.

Roskov Y, Abucay L, Orrell T, Nicolson D, Flann C, Bailly N, Kirk P, Bourgoin T, DeWalt RE, Decock W, De Wever A (editors). 2016. *Species 2000 & ITIS Catalogue of Life, 2016 Annual Checklist*. Species 2000. Digital resource at www.catalogueoflife.org/annual-checklist/2016 accessed September 2017.

Sigwart JD, Sutton MD, Bennett KD. 2018. How big is a genus? Towards a nomothetic systematics. *Zoological Journal of the Linnean Society*. 183: 237–52.

Stearn WT. 1959. The background of Linnaeus's contributions to the nomenclature and methods of systematic biology. *Systematic Zoology*. 8: 4–22.

Storks N. 1993. How many species are there? *Biodiversity & Conservation*. 2: 215–32.

Struck TH, Paul C, Hill N, Hartmann S, Hösel C, Kube M, Lieb B, Meyer A, Tiedemann R, Purschke G, Bleidorn C. 2011. Phylogenomic analyses unravel annelid evolution. *Nature*. 471: 95–8

Weigert A, Helm C, Meyer M, Nickel B, Arendt D, Hausdorf B, Santos SR, Halanych KM, Purschke G, Bleidorn C, Struck TH. 2014. Illuminating the base of the annelid tree using transcriptomics. *Molecular Biology and Evolution*. 31: 1391–401.

Williams PH, Gaston KJ. 1994. Measuring more of biodiversity: Can higher-taxon richness predict wholesale species richness? *Biological Conservation*. 67: 211–7.

Winker K. 2005. Sibling species were first recognized by William Derham (1718). *The Auk*. 122: 706–7.

# 11 Dynamic Patterns in Biodiversity

## WHAT DETERMINES BIODIVERSITY PATTERNS?

Different environments clearly support varying diversity and densities of both species and individuals. Regardless of how many species remain undiscovered and cryptic, it is still trivial to conclude that there are more species and more biomass in a tropical coral reef than the same size region on a polar ice sheet. Even within one region, you will encounter patches of relatively higher or lower species richness. What are the mechanisms that control the magnitude of regional biodiversity? This is a question that sits at the meeting point of ecology and evolutionary biology.

Regional species richness is not just a census of species or lineages, it relates to the type and quality of local ecosystems. Species richness alone may not account for trophic complexity; on the other hand, linkages and ecological interactions play a role in the evolution of species diversity. Unfortunately, the simplifications required to analyse food webs take on a strong flavour of 'A place for everything and everything in its place', which is antithetical to the real variability and flexibility of species (Chapter 2). Conservation biology aims to protect diversity in a self-sustaining system. This requires identifying both a sufficient set of linkages to provide a balanced ecosystem, and a strong understanding of what is appropriate to the region.

Global biogeographic patterns, the way diversity gradients form through evolutionary history, and how they are maintained, are intimately linked with macroevolution (Chapter 5). The evolution of a species lineage is not an isolated process; species evolve in concert with the other members of their ecosystem. Why are there a certain number of species in a place, and not more, or less? This connects to other scientific questions, such as understanding how many species have been lost to extinction over recent and more distant time scales (Chapter 9) and indeed how many species remain to be discovered (Edie et al., 2017; Chapter 10).

The relative diversity in a region correlates with energy availability and stability: rich topical ecosystems are associated with high solar energy, low seasonality, and regular precipitation in terrestrial habitats. This resource dependency suggests that there are ecological limits—a ceiling to how many species can be accommodated in an ecosystem (Rabosky & Hurlbert, 2015). However, proposing that limits exist, and that limits vary with ecogeographic factors, does not presuppose that ecosystems are saturated and functioning at full capacity. Equilibrium models suggest that more than 12% of niches are unfilled, now and in the fossil record of marine animals (Walker & Valentine, 1984). Ecosystems operate far below their potential limits for species diversity (Harmon & Harrison, 2015). But that does not mean limits are not in place, or that limits do not differ among habitats: the tropics contain more species now, and this has been true for effectively the whole history of life. To understand how the present 'horizontal' time slice responds to life in tropical environments, we have to look to the 'vertical' component of lineage history.

Global biodiversity as we see it today is shaped by major patterns in the Earth system and by human influences. Environmental destruction is accelerating with human population growth, but the impacts of human movements from only a few hundred years ago have echoes in present biodiversity, which we have already forgotten. A previous chapter mentioned the phenomenon of 'shifting baselines', where our expectations become normalised to current conditions, rather than to a more objective ideal (Chapter 9). But nostalgia, or mythologised versions of history can also warp perception and even undermine agreement over common goals (Stuckey, 2017). This chapter proposes a two-part approach to understand contemporary biodiversity. First, it may be useful to look at our recent history to understand modern ecosystems and the assumptions we make about biodiversity. Second, understanding the mechanisms that underpin observed global biogeographic patterns from evidence in modern and fossil faunas can give us an idea about the deeper origins of living biodiversity and its distribution on Earth.

## IMPACTS OF THE VERY RECENT PAST

Our post-Darwinian, long-term view of species as evolving, dynamic units is fundamentally intertwined with seeing species as flexible, and subject to change with altered environmental conditions. Mutability occurs at different scales—species can be highly variable without leading to speciation. What a species can and will put up with in terms of environmental change is important in considering our ethical responsibility to limit anthropogenic damage to biodiversity. Conservation biologists have long recognised that it does not matter how much planning, environmental modeling, restoration, labour, and protection enforcement get invested in the design of a reserve area if the wildlife decides to get up and move away.

The foundation of the modern conservation movement came from the late 19th and early 20th century investments in the protection of undeveloped lands in the USA (Child, 2009). The ideal of this movement is laudable, to protect wild places in perpetuity without the influence of encroaching human presence and urbanisation. The deep flaw in this philosophy is that humans are everywhere, and certainly were present in North America in vast numbers prior to the impact of European colonisation. Waves of infectious disease preceded European settlers, killing millions of people long before any face to face contact with foreigners (Mann, 2005). This hidden, unrecorded pandemic led to a long-held misconception that the Americas were effectively devoid of human habitation prior to European occupation, because, by the time Europeans started extensively exploring the 'New World', the majority of the people who had lived there were already dead and buried. The growing European colonies in 18th century North America were not the first 'civilisation' on the continent. In fact, European new arrivals made for substantially less impact and less human presence than there had been already from indigenous cities established before 1492 (Denevan, 1992). The scale of population—and population loss—in the history of the Americas was dramatically under-estimated by those who survived to write the history books, and a new wave of research findings is uncovering ever more evidence of forgotten cities that supported millions of people before European contact (e.g., Clynes, 2018; Davis, 2018). Landscape and wildlife management by indigenous peoples in the Americas changed the environment permanently. Forests in the Northeast of the USA and prairies in the Midwest were planted, created, and managed by people—not Europeans, but by the other indigenous nations that lived in the Americas thousands of years earlier—as part of wide scale land management and an early agrarian economy (Abrams & Nowacki, 2008). The idea that species assemblages should now be reconstructed and protected exactly as they were in the 18th or 19th century has many advantages over unfettered destruction and urbanisation, but it also arbitrary, and rooted in the flawed ideal of a pre-European American Eden.

The Passenger Pigeon (*Ectopistes migratorius* (Linnaeus, 1766)) is usually invoked as a tragic story of anthropogenic extinction, documented by 18th century colonists in North America for their vast sky-blackening flocks (Fuller, 2014). They were hunted indiscriminately and sold as cheap food, with whole flocks roosting in a tree taken down at once. This abundance, however, is in stark contrast to evidence of modest populations prior to the arrival of Europeans. There is no fossil evidence of abundance for Passenger Pigeons. In fact, early pigeon populations were probably small, perhaps capped by competition for food with humans, or by more deliberate exclusion or management by humans (Neumann, 1985). When North American human populations were decimated, the competition was removed, but people had planted and cultivated orchards of nut trees that were left behind, so the pigeon populations had limitless food and exploded into the vast flocks seen by Europeans arriving slightly later. These were outbreak populations, unnaturally high numbers, not in equilibrium with environmental resources; a hypothesis first proposed from midden remains and historical records (Neumann, 1982), and later corroborated by population genetic evidence from museum specimens (Hung et al., 2014). Other more recent research on the passenger pigeon, curiously, has been interpreted without context of the pigeons' interaction with human populations other than European colonists who hunted them (Murray et al., 2017). The same pattern of outbreak populations misinterpreted as natural is true for other iconic aspects of North American wildlife, such as the herds of American Bison or buffalo (*Bison bison* (Linnaeus, 1758)), were hunted to near extinction in the 19th century, but their rise and fall is also more nuanced than the myth. Overcrowding that could have led to disease and food shortages, combined with rampant hunting, led unstable faunal populations to collapse. Animal and plant species that were managed and domesticated by people in North and South America for millennia prior to European contact were turned back to the 'wild' when their human masters were struck down by disease epidemics. And some, like the bison, expanded out of control (Davidson & Lyttle, 2004). The American explorers Lewis and Clark admired vast herds of buffalo in 1805 but complained about prickly ground cover that could pierce their leather shoes; highly defended plants that characteristically thrive in areas of severe over-grazing (Kirby, 2010). This abundance of megafauna was not the 'natural' state of North America.

Forests in the Americas, but also worldwide, carry hallmarks of human cultivation. Archaeological evidence of settlements in the Upper Xingu region of Brazil shows former large, densely-populated settlements that were abandoned during the catastrophic depopulation in the 1600s (Heckenberger et al., 2003). These areas are in what is now commonly considered 'untouched' Amazonian rainforest (Willis et al., 2004). Many plant species in North and South America were historically cultivated as crops, inducing all the hallmarks of domestication on other continents, such as increased productivity, larger fruit, hybrids, and specialty varieties (Dillehay et al., 2007). Varieties of some plants (such as tomatoes and maize and more) are now treasured as 'heirloom' strains, but connection to the first plant breeders are largely obscure. Domestication has introduced evolutionary novelties to many species (Inderjit et al., 2017); the development of domestication was ongoing in many places and long before it was ever recorded. Meanwhile, an uncounted number of selectively bred plants, developed by indigineous nations, may have descendants in the 'wild' centuries later.

If species are ever-changing and human interference shaped what we naively assumed were untouched landscapes, how do we draw any distinction between natural and unnatural, or between urban landscapes and wildlife preserves? This struggle is felt more acutely in Europe, where the high population density of humans and a long and well-recorded history of human activity belies any idea of genuinely untouched places. European conservation efforts also show that diversity can be found, and protected, even in completely developed landscapes. When I was an undergraduate student in rainy western Canada, our professor explained that in Europe there was an important movement to conserve hedgerows, the woody and herbaceous floral borders around small farm fields, which encourage diverse insect pollinators and create wildlife corridors for small mammals. Our whole class laughed out loud. Most of us were actively engaged in campaigning against logging activity encroaching on the last remaining patches of old growth temperate coastal rainforest. The contrasting idea of deliberately preserving a human-fabricated habitat seemed totally absurd. (My comeuppance came later, in Ireland, teaching first year undergraduate students to key out the flowers living in beautiful 150-year-old hedgerows.) But the human alteration of landscapes is not absent from the pre-European history of the west coast of North America, it is not a phenomenon restricted to Europe, and not only modern history.

The last 10,000 years of land management in North America teaches us that landscapes can be managed sustainably—unlike most other global civilisations there is no evidence of repeated cycles of over exploitation and famine in North America prior to European contact (Mann, 2005). This carefully nurtured abundance is not equivalent to the myth of Native Americans living a life of easy luxury in their bountiful landscape; the landscape was bountiful because it was meticulously crafted to be that way, and astutely managed.

There is no strict dichotomy between 'wild' and 'anthropogenic' landscapes. Certainly now, in the post-Industrial era, the impacts of humans are apparent in every remote place on the planet. Plastic particles litter the open ocean (Carpenter & Smith, 1972) and the deep sea (Van Cauwenberghe et al., 2013). There is scope now for modern societies to both preserve wild lands, following the laudable ideals of the conservation movement, and to embrace landscape management that enhances the participation of humans and other species. Urban wildlife is still wildlife, hedgerows support diversity including not only attractive wildflowers but their pollinators that support the surrounding crop plants, and ancient giant trees inspire wonderment.

## IMPACTS OF INVASIVE SPECIES

So, what is a 'natural' ecosystem, and a 'natural' distribution range? Technologies of various kinds have facilitated travel and dispersal of people throughout the whole history of our species (Jablonski, 2012). Interconnectivity has accelerated dramatically in the post-Industrial era, for us and therefore for other species. In 2011, an earthquake and tsunami in eastern Japan sent out a flotilla of debris including lost ships, pieces of docks, and countless smaller objects, carrying nearly 300 species of marine coastal organisms from Japan across the Pacific Ocean to North America (Carlton et al., 2017). Tsunamis are natural events. Dispersal by rafting on floating material is a relevant hypothesis that has explained the colonisation of isolated oceanic volcanic islands (Briggs, 1974). For most species, the odds of a hitchhiker attaching to and then surviving an oceanic crossing on a natural raft such as a fallen tree would be near zero, though one in a million is sufficient on evolutionary timescales. Technological intervention, in the form of fabricated marine debris that starts pre-loaded with a well-developed colony of encrusting organisms, shifts that hitchhiker's odds to near-guaranteed survival.

Not all introduced or non-native species become destructive invaders. The native biogeographic provenance of an introduced species does not itself predict how that species will behave in a novel environment (Buckley & Catford, 2016). Some who work in environmental management have suggested that origin is unimportant, and that introduced species should be assessed and mitigated based only on the apparent symptoms of their interaction with native flora and fauna. Their argument is, basically, things are moving around all the time anyway, so if we cannot see it doing any harm, then just leave it alone and focus mitigation and improvement efforts to symptoms of environmental degradation. However, it is well established that non-native taxa are more likely than native species to cause negative environmental impacts (Rejmanek & Simberloff, 2017). This is not a tautology: native species are by definition a natural part of the environment, but they can have destructive impacts, such as outbreak populations. On the other hand, not every individual arrival of a non-native organism to a new environment will spark an invasion. 'Naturalised' species, non-native species with a long established history in an area, and considered to have no measurable economic or environmental negative impacts, are often left alone and their anthropogenic introduction alone may be considered insufficient reason to warrant a major removal effort, such as small mammals and amphibians in Ireland (Kane, 1893). But the compounding effects of tolerating multiple introduced species, and an increasingly stressed environment under climate change, can escalate to real environmental damage (Montgomery et al., 2012). After a species becomes established and the destructive consequences become apparent, it is then perhaps too late, or too expensive, to control or eradicate it.

Many invasive species cause massive economic damage through direct impact and the cost of control or removal. Animals, plants, and fungi that are a normal part of their home ecosystem, seem to become far more rapacious when released through human intervention into an environment free from their natural predators or competitors. Zebra Mussels (*Dreissena polymorpha* (Pallas, 1771)), European rabbits (*Oryctolagus cuniculus* (Linnaeus, 1758)), Kudzu (*Pueraria* spp.), or Japanese

Knotweed (*Sargassum* spp.), can choke and totally dominate environments outside their natural range. It is not clear to what extent this domination represents a release from competition that throttles a species in its natural range. Bollache and colleagues (2008) showed that an aquatic amphipod had a consistent functional response in feeding rates when they compared a native population given native prey or an invasive population given local prey. That specific case implies the invaders' ecological impacts are more about species traits, rather than plasticity, and an introduced species becomes invasive when it happens upon a niche without effective competition.

The key factors to controlling the effects of an invasive species are: reaction time, dispersion, and knowledge of the species' natural history (Pluess et al., 2012). This latter part is particularly challenging. When a species is newly discovered in an environment, it may be invasive, or it may be an undescribed species, or, it may be both. A new species of terrestrial planarian was discovered in Europe in 2003; however, the species clearly belongs to a genus that was otherwise only known from New Zealand (Jones & Sluys, 2016). The new species extends the range of the genus in a pattern that is untenable in terms of natural biogeography, whereas soil animals including planarians have been frequently introduced around the world with horticultural material since the 19th century (Winsor et al., 2004). The native habitat of the new species, and its potential impact on the environment in its known range, remain unknown.

The evolutionary and biogeographic history of a lineage is clearly important to ecology; whether a species is native or not is a foundational idea to environmental management for good reason. The time-vertical axis of species has an impact on the time-horizontal, present ecology of species interactions. The fact that there is scientific debate about whether origins and native ranges are important, or whether this can be worked around by addressing symptoms rather than cause, comes from an assumption that species have a certain fixed assigned role in ecosystems. This has conceptual origins in effectively creationist ideas of a species as a fixed essence, or at best an end point product of evolution, rather than active, dynamic agents in ongoing evolutionary processes.

Co-occurring species compete for resources, even in the context of a natural species assemblage without any introduced organisms. The ability to successfully complete for resources does not equate to ecological dominance or eradicating the less competitive species in a natural assemblage (Liow et al., 2016). Inequality is intrinsic to ecological interactions; the party with greater advantage (predator over prey, for example) tends to be relatively larger, and have a greater metabolic demand (Vermeij, 1999). The easy counter-examples to this broad generality do not undermine the assymetry of interactions. In natural systems, the interactions of 'winner' and 'loser' are also secondarily asymmetrical: if a predator wins, the prey dies, but if the predator loses, there is normally no significant harm to either party (Dawkins & Krebs, 1979). The skew to these inequal interactions in the long term drives the evolutionary escalation of competitive arms races, such as the increasing elaboration of anti-predatory defences, or adaptations to more efficiently secure limited resources. The impacts of invasive species are difficult to predict, because non-native species *de facto* have a different evolutionary history than native counterparts. Two species may have an apparent ecological or functional equivalency in their separate native ecosystems, but those different contexts of their lineage histories could tip the balance when they confront each other in a competitive interaction between native and non-native.

Biogeographic origins and history—both at short term and evolutionary timescales—are reflected in legacy effects in species functional traits. This heritage may constrain the scope for species to respond to further environmental change (Cavender-Bares et al., 2016). Species change all the time, but part of that for many lineages is their movements and relationship with humans over spans of the last decades to thousands of years. Current movements are increasing the encroaching homogenisation of ecosystems with urbanisation and invasive species (Vitousek et al., 1997; Qian & Ricklefs, 2006). In the face of all this change, it is perhaps remarkable there are pervasive and persistent large-scale biogeographic patterns.

## THE LATITUDINAL DIVERSITY GRADIENT

The differences among biomes in Earth's regions have fascinated scientists for centuries. Broadly speaking there are more species in the tropics than in temperate latitudes; this is the 'first order' diversity gradient. There are a number of generalities and rules that have been proposed to explain global patterns in the relative distribution of species richness (Table 11.1). The strongest, most consistent pattern is the latitudinal diversity gradient (LDG; Figure 11.1). Probably far more than half of the world's species live on 40% of the globe. Further, the real species richness of the tropics may still be dramatically underestimated because of disparity in scientific effort and resources (Chapter 12), thus future data may skew the LDG even further toward lower, more tropical latitudes.

There are numerous exceptions to the 'diverse tropics', influenced by precipitation and altitude (in the terrestrial realm), or depth (in marine areas), human disturbance (everywhere), and small-scale variation in environmental quality. In spite of exceptions and some complexity, the LDG is a clear global trend. Yet the mechanistic explanation for the LDG is more elusive. The LDG is not a modern phenomenon; the tropical latitudes were also more diverse than higher latitudes in past geological eras, a pattern that has been true since the Cambrian (Powell, 2009). The skew of the LDG—how much more diverse tropics are compared to higher latitudes—has fluctuated throughout Earth's history, and the latitude of peak diversity has also shifted at different points in time (Huang et al., 2014). The LDG is not a constant fixed planetary feature of latitude, there are other compounding factors in the Earth system. Thus, a good mechanistic understanding of the LDG stands to illuminate what exactly limits the capacity of an ecosystem to support species diversity.

The first question to be addressed is whether the LDG is somehow artefactual (if it is not a real phenomenon but an accident of measurement, that would make it much less useful for understanding the evolution of ecosystems). One idea is that there are more species in the tropics because there is more area in the tropics (e.g., Terborgh, 1973). Measuring distributions on a slightly oblate spherical body, centred around the equator, there is more surface area in the equatorial band than at higher latitudes, and furthermore the higher latitude regions of the northern and southern hemispheres are not in contact. The division of surface area on a sphere is about 25% northern temperate, above the Tropic of Cancer, and 25% southern temperate (beyond the Tropic of Capricorn). Polar areas each account for 5% of the total surface (actually closer to 4% and shrinking), although these are broad approximations that

## TABLE 11.1
## Summaries of Key Biogeographic Rules or Trends

| | **Biogeographic Truisms** |
|---|---|
| Latitudinal Diversity Gradient | Species richness is inversely correlated to latitude: There are more species in the tropics. |
| Land and Sea | Species richness of the terrestrial realm (in the modern era) is greater than the marine realm. |
| Biodiversity begets biodiversity | Species richness and ecological complexity are compounding, as more diverse assemblages create new specialised niches. |
| Sarawak Law | 'Every species has come into existence coincident both in time and space with a pre-existing closely allied species' (Wallace, 1855: p. 196). |
| | **Biogeographic 'Rules'** |
| Rapoport's rule (Rapoport, 1982) | Species distribution range is inversely correlated to latitude: Lineages at lower latitudes have smaller latitudinal ranges, and those at higher latitudes are more broadly distributed. |
| Thorson's rule (Mileikovsky, 1971) | Maternal investment is correlated with latitude (in benthic marine invertebrates): Species at lower latitudes have planktotrophic, widely dispersing larvae ($r$-selected life history), while species at higher latitudes tend to have lecithotrophic, direct developing, or brooded larvae ($K$-selected life history; see also MacArthur and Wilson, 1967). |
| Bergmann's rule (Bergmann, 1847) | Body size is inversely correlated to latitude: Within a clade, species and populations have larger body sizes at higher latitude (or colder environments) and relatively smaller body sizes at lower latitude (warmer regions). |
| Jordan's rule (Jordan, 1892) | Iterative morphological characters are inversely correlated to latitude (in fish): Among species or populations of fish, the number of vertebrae, fin rays, or scales, are higher at lower latitude or in warmer environments. |
| Allen's rule (Allen, 1877) | Allometry of mophological features is inversely correlated to latitude: In species and populations, the sizes of limbs and appendages are relatively longer at lower latitude than those at higher latitude. |
| Gloger's rule (Gloger, 1833) | Pigmentation is inversely correlated to latitude: Species and populations tend to be darker (more densely pigmented) in warmer and more humid environments. |
| Fosters's rule (Foster, 1964) | Isolation disrupts larger trends: lineages endemic on islands exhibit different evolutionary trends than would be predicted from their larger parent clade. |
| Jordan's Law/Wagner's Law (Jordan, 1908) | Sister lineages do not co-occur. |
| Van Valen's Law of Constant Extinction (Van Valen, 1973) | Extinction risk for a species lineage is not correlated to its age, the probability is constant (applies equally) for all lineages in a given time. |

Genus Richness

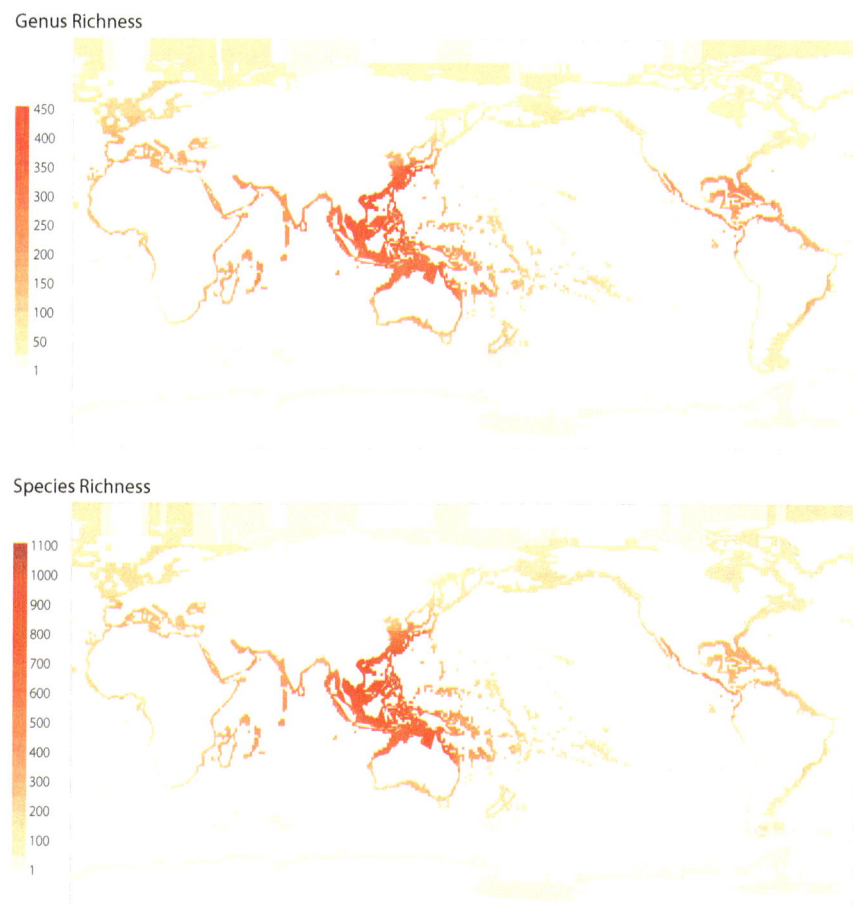

FIGURE 11.1    The geographic distribution of the diversity of living bivalves. These heatmaps illustrate the increase in diversity at lower (equatorial) latitudes, measured at both genus-level (top) and species-level taxonomic resolution. (Genus data modified from Edie et al., 2018; species data provided by Stewart Edie and colleagues.)

also ignore surface complexity and ocean basins. The combined tropical surface area is 40% of the Earth including land and oceans. Still, the species richness in most tropical habitats is far greater than the combined diversity in the rest of the world, and the magnitude of the difference could not be explained away by biases in available habitat area. Considering terrestrial assemblages, there is far less area suitable for tropical forests than for the vast boreal forests of Canada and Siberia, yet the species richness of northern coniferous forests is conspicuously less than the forests of the Amazon (Adams, 2010). Data that seemed to indicate that most species of insects (the global species-richness champions) were found in the northern hemisphere was used convincingly to show that tropical insects are under-sampled and under-described—faced with data for lots of northern temperate species, the conclusion is that the northern hemisphere insects are not dominant; it is rather that the tropical insects are mostly missing from available samples (May, 1992). The LDG is not an artefact. Availability of space over long periods of time is, however, clearly important to establish diverse and stable ecosystems (Fine, 2015).

There is no single mechanistic explanation for tropical biodiversity, but a number of interacting patterns that reinforce the LDG (Valentine & Jablonski, 2015; Jablonski et al., 2017). There have been over 30 independent hypotheses proposed to explain the LDG (summarised in Fine, 2015); a number of them are redundant or overly specific so they are not repeated here, but we will consider some general issues. These emergent patterns in biodiversity, increasing density and number of species in the tropics, are largely a matter of scale, both in space and in time. Much of the literature on the LDG has set up a dichotomy between 'biotic' or ecological explanations, or abiotic or 'historical' mechanisms. This is an argument over whether biotic interactions are more influential than environmental effects on the evolution of species and ecosystems (Barnosky 2001). Which aspect has a stronger emphasis in a particular study depends largely on the interests of the authors; palaeontologists may not have access to detailed data about biotic interactions in previous time slices, whereas ecologists may be working only on present day distributions without consideration of the time scale since the pattern was established (Benton, 2009; Liow et al., 2011). Any study or claim of a single explanation is misleadingly simplistic (Jablonski et al., 2017).

Broadly distributed clades, with wide ranges both in space *and* time, provide ideal tools to test hypotheses about the underlying mechanisms that generate the LDG pattern. Most clades of animals and plants are more speciose in the tropics than at temperate latitudes, in both terrestrial and marine realms (Hillebrand, 2004). There are exceptions in clades that are characteristic of polar regions, like penguins, but these do not inform global patterns. Some studies have used such patterns as exceptions that prove the rule (Krug et al., 2007). Several studies of widely distributed clades allow for empirical tests of what abiotic factors are most strongly associated with the relative density of species richness. In shallow marine gastropods, solar energy availability, higher in the tropics, is the single factor that best explains their increased tropical diversity (Roy et al., 1998). Differentiating solar energy from primary productivity is important, because gastropods include both primary and secondary consumers, and primary productivity can be highly variable on local and regional scales, as well as seasonally. Furthermore, direct measures of primary productivity are not correlated to species richness in shallow marine systems; chlorophyll in coastal and shelf systems is actually inversely related to diversity along the LDG (Dickson & Wheeler, 1993). But solar energy and sea surface temperatures are strongly correlated to species and genus-level diversity in global shallow marine communities (Belanger et al., 2012). High latitude environments get many more hours of sunlight during the summer months, but the radiation is diffused by the atmospheric angle, thus limiting atmospheric and sea surface temperatures. Solar energy, however, has further implications beyond converting energy via photosynthesis. In some plants and fish, higher species diversity in warmer ecosystems is correlated to comparatively higher rates of genetic evolution (Wright et al., 2011). However, in other clades, such as tropical frogs, phylogenetic diversification rates are related to biogeographic history without obvious correlation to latitude or climate (Wiens et al., 2011). Higher metabolic rates, little seasonal senescence, shorter generation times, and higher mutation rates all feed into increasing rates of speciation and diversification among lineages living at tropical latitudes (Brown, 2014).

Tropical diversity is correlated to temperature. The kinetic explanation for the LDG suggests that this is not a direct effect accelerating rates of evolution, but rather a compounding effect of increasing 'Red Queen' co-evolution (Chapter 5). Higher temperatures and higher metabolic rates lead to an increase in the rate and number of biotic interactions that drive diversification (Brown, 2014). However, it is not clear from theory or empirical data why increasing interactions would lead to a universal increase in diversity, and not elevate extinction rates, especially in the long term (Valentine & Jablonski, 2015). Metabolism is not inextricably linked to growth, development, or body size; these processes can be disassociated by physiological or environmental changes, and longer-term processes from aging to molecular rates of evolution may not be related to metabolic rate (Glazier, 2015; Fine, 2015). The pace of life is not driven by a simple metabolic clock, yet kinetics and thermal regime clearly underpin some aspects of the LDG.

Another essential property of the latitudinal gradient in solar radiation is its seasonality. Seasonal fluctuations in temperature and nutrient availability increase with latitude. Life history traits, and community structure, are related not just to climate, but to phenology (Williams et al., 2017). A periodically hostile environment favours species that are more generalists, adjusting their activity to suit available resources. Populations with larger ranges are buffered from stochastic local extinction events, which may be more likely in areas with seasonal fluctuation in nutrient availability (Valentine & Jablonski, 2015).

Adverse conditions tend to favour generalists. Simpson (1944) articulated this as 'rule of the survival of the relatively unspecialised'. Consequently we would predict the converse for tropical communities—with a warm and stable climate—that they are comprised of more species, that are each more specialised, and more restricted in range and niche, and this is broadly true (Valentine & Jablonski, 2015). More species fit into an ecosystem comprised of narrowly-defined, specialised niches; or, higher evolutionary rates and competitive effects may push species into specialisation.

Diversity begets diversity. Quite apart from the LDG, this is a universal principle of biodiversity, which sees its apex in tropical ecosystems. For example, the main predictor for parasite diversity in a region is the diversity of free-living animals (Poulin, 2014). Compounding diversity represents a synergy of various abiotic and biotic effects. It would also follow that tropical communities include more functional groups, however those are defined. Species create niche spaces for other species.

All of these factors can be folded in to an 'out of the tropics' hypothesis (Jablonski et al., 2006; Figure 11.2). Tropical areas have comparatively high speciation rates, lower extinction rates, and adaptive landscapes that accommodate large numbers of specialised niches, and crucially this whole scenario is sustained over long spans of time via climate stability without seasonal fluctuations (Jablonski et al., 2013; Valentine & Jablonski, 2015). The phylogeographic centre of origin of many temperate clades is in lower latitudes, even if the modern diversity is distributed elsewhere. Many clades of tropical origin remain at low latitudes, but others, especially those with high speciation rates, expanded poleward. The fossil record of marine bivalves indicates

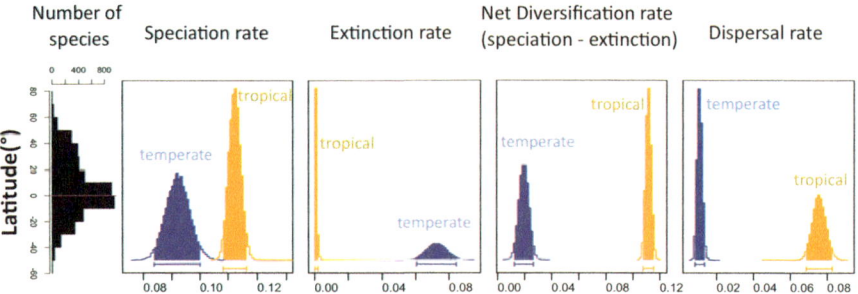

**FIGURE 11.2** Phylogenetic data for mammals support the 'out of the tropics' hypothesis. The global latitudinal diversity gradient for mammals (far left), and comparisons of (from left to right) estimated speciation rate, extinction rates, net diversification rate, and dispersal. Overall, tropical biomes show higher speciation rates, lower extinction rates, and higher dispersal. (Redrawn from Rolland et al. 2014.)

a significant time lag between typical tropical clade origination and subsequent appearance at higher latitudes some ~5 million years later (Jablonski et al., 2013). At higher latitude, these groups may have diversified further, but higher extinction rates at higher latitude and ongoing shifts in distribution limit total diversity.

## BIOGEOGRAPHIC PATTERNS AND 'RULES'

A number of other patterns in ecology and life history have been observed across latitudinal gradients, beyond the general trend in species richness (Table 11.1). Most of these 'rules' were coined as observations on a particular clade (almost all come from animals, and often only vertebrates), and in some cases there are as many papers disproving a 'rule' as demonstrating its proposed pattern. Some suggest morphological trends in variation for pigmentation (Gloger, 1833), limb proportions (Allen, 1877; Jordan, 1892) or total body size (Bergmann, 1847). These may not be extendable to generalities for all organisms, but in clades that show climatic or geographic gradients predicted by these rules, they provide a null hypothesis to test for deviations. Morphological changes are an early symptom of response to a changing environment. Such visible changes, in the form of morphological plasticity or acclimatisation, are directly relevant to these various rules. A latitudinal gradient in the present day can be compared to a temporal gradient of morphology in past climates (Millien et al., 2006). The geographic ranges of species are a heritable trait (Jablonski, 1987), either directly in terms of physical presence at birth or indirectly since the geographic range of ancestor and descendent lineages are autocorrelated.

A number of the ecological drivers that have been proposed in studying the LDG predict particular speciation patterns, such as niche specialisation and small range size promoting speciation by allopatry (Fine, 2015). Geography and geological history lead to high vicariance in Southeast Asia, so allopatric speciation via geographic separation would seem intuitive (Woodruff, 2010; Figure 11.3). This does not exclude other kinds of speciation. In individual studies, sympatric species complexes are apparently common in Southeast Asia (Stuart et al., 2006). Similarly,

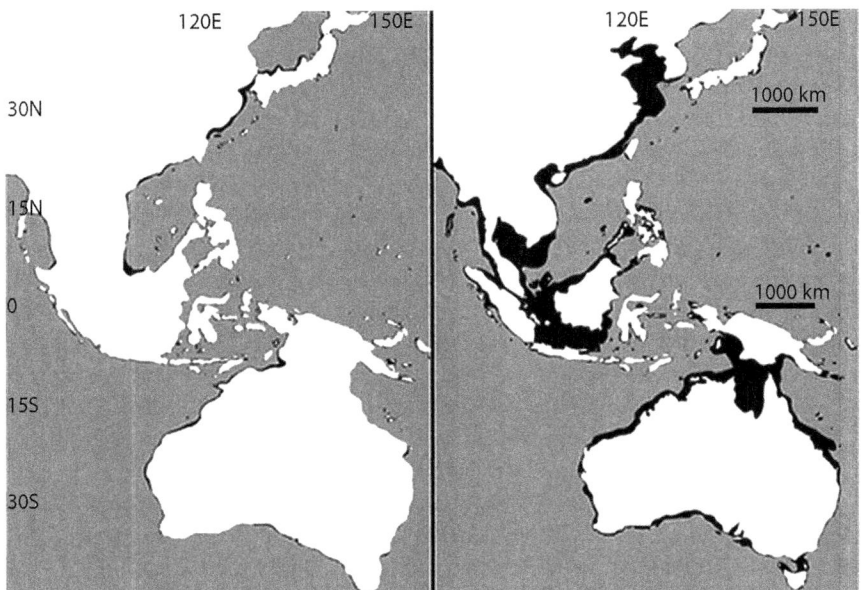

**FIGURE 11.3** Rising sea level separates land and creates more and more complex coastlines, increasing vicariance with new geographic barriers; (left) land area (white) and coastal shelf (black), at the Last Glacial Maximum, compared to the arrangement (right) today. (Redrawn with permission from Ludt and Rocha, 2015.)

other biogeographic 'rules' are informative as thought experiments and useful predictors, but they cannot be expected to act as exclusive or complete explanations.

Rapoport's rule states that tropical species tend to have narrower geographic ranges (latitudinally) than those at higher latitudes; moving from the tropics to the poles, species you encounter are more and more widespread in distribution (Rapoport, 1982). The increasing scale of environmental fluctuations at higher latitudes supports the evolution of lineages with broad tolerances that can thrive in varying environmental conditions (Stevens, 1989). Conversely, species with larger geographic ranges are also more robust to stochastic extinction, and so are buffered from increased extinction rates at higher latitudes. Species at higher latitudes occupy wider ranges and have broader morphological variation (Clarke, 1978). This is not universal; at extremely high latitude many polar species have narrow ranges. Many species thought to have 'bipolar' ranges in both Arctic and Antarctic latitudes have been separated by molecular investigations into separate northern and southern species (e.g., Hunt et al., 2010). In the fossil record of bivalves, which has been crucial to reconstructing LDG origination dynamics and the 'out of the tropics' paradigm, species with the broadest latitudinal ranges are tropical living species that cover a wide area with the same temperature profile (Jablonski et al., 2013).

Species ranges are related to the organisms' ability to disperse. Thorson's rule for marine organisms such as fish and others with dispersing planktonic larvae states that tropical species tend to produce relatively large numbers of planktotrophic larvae that are capable of feeding and living independently in the plankton to facilitate dispersal. By contrast, higher-latitude species more often have larger eggs and lecithotrophic larvae that disperse planktonically but for a potentially limited time since they cannot feed, or direct-developing or brooded larvae that do not disperse. The ecological principles of this idea were articulated by Thorson (1950) and other contemporaries (Laptikhovsky, 2006), but formalised and later credited to Thorson (Mileikovsky, 1971). Explanations for the trend mainly focus on food availability; limited resource availability in colder waters requires higher maternal investment to secure offspring survival. However, as noted above, primary productivity is not limiting in high latitude coastal marine systems, although it varies seasonally, and for some animals there may be relatively little difference in larval duration for feeding or lecithotrophic larval types in that context. The evolution of life history traits is influenced not only by latitude and nutrient flow. The overall trend of Thorson's rule may in part be a legacy effect of glaciation, as poleward range expansion for modern lineages largely occurred during glacial retreat when primary productivity was more limited (Poulin & Féral, 1996).

Planktonic larval duration is strongly predicted by temperature (O'Connor et al., 2007). Warmer temperatures increase metabolic rate and decrease the length of the larval phase (but many tropical feeding larvae have long dispersal potential). Short larval durations in warm tropical seas, and shifting to strategies like brooding at high latitude, would all favour autocyclic recruitment. Thus, both tropical and temperate species have strategies that would increase success but limit dispersal. Proximity between parents and offspring can have benefits, but competition for space (or other resources) can limit the growth of a population; this is sometimes called conspecific negative density dependence. Tropical trees show shorter distances between saplings and parent trees than temperate species; close proximity to healthy adult trees is credited to protecting seedlings from parasites and pathogens, which are more pervasive in the tropics, and this helps maintain higher density species richness (LaManna et al., 2017). This is an aspect of dispersion that could also potentially apply to tropical marine organisms. But clearly not all recruitment is autocyclic—some things disperse. Within established modern ranges, Thorson's rule suggests that species in tropical latitudes have more allocyclic recruitment. The tension between pressures toward allocyclic or autocyclic recruitment in the tropics reinforces the pattern of these areas as biogeographic lineage sources and sinks—both cradle and museum in the parlance of lineage origination—in contrast to high latitude sinks (Figure 11.4).

Temperate                          Low / Tropical                          Temperate
                                      Latitudes

**FIGURE 11.4**   Idealised schematic for dispersal of marine species in tropical and temperate latitudes. Arrows indicate dispersal tendencies, with both allocyclic and autocyclic recruitment in the tropics, and moderate dispersal in high (temperate) latitudes, and boxes are species ranges showing increasing breadth moving away from the tropics, according to some predictions of 'out of the tropics'.

## LIFE IS RESILIENT

Most species are rare; the frequency distribution of species abundance in any ecosystem is strongly left skewed with a few species at high abundance. In tropical systems, with higher species richness, we can expect a longer 'long tail' of less abundant species. Small populations, specialist niche requirements, and restricted geographic ranges all make a species more vulnerable to extinction. Human impacts have continued for millennia and shaped landscapes worldwide that appear 'wild' or 'pristine' (Willis et al., 2004), but those human activities also led to substantial prehistoric extinction of at least large animals (Martin & Klein, 1989).

Some early settlements from the 13th to 17th centuries in the Upper Xingu region of Brazil extended over 500 km$^2$ and indigenous agricultural practices enriched the soils in ways that impact forest communities today (Heckenberger et al., 2003). These pre-European developed areas are far larger than many people might expect, but that is less than half the area of modern Rio de Janeiro. Sustainable management of biodiversity is dependent on matters of scale.

Habitat destruction is clearly tied to human population growth; Wright and Muller-Landau (2006) showed that tropical forest cover is directly correlated to rural population density in Asia, Africa, and South America. There is a global trend toward declining rural populations and increasing urbanisation. Tropical forest clearance coincides with rural farming activity, whereas urban expansion spreads into areas that are already cleared. This meticulous but controversial study concluded that these patterns thus predict a decrease in tropical forest loss and concluded that projected deforestation would not extend to substantial future extinction (Wright & Muller-Landau, 2006). Aspects of these data are unassailably important to planning conservation measures and identifying future threats. Critics demonstrated that the conclusions were overly optimistic, mainly because forest cover is not of uniform quality (Laurance, 2007). 'Forests' can include old growth, or monoculture plantations of banana, oil palm, or other crops, and so that blanket category cannot account for habitat complexity and provisioning for diversity, including rare species.

Increasing awareness of the value of conservation has increased the area of tropical forests under national protection, but most of these hotspots with natural forests have lost up to 90% of their native tree cover (Laurance, 2007). Forest recovery is a long process; many tree species that are characteristic of old growth forest habitats will avoid early stage successional forest communities, and in reality, most replanted forest is re-felled for commercial activity (Gardner et al., 2007). Diverse assemblages of trees provide habitat for many other organisms; but tropical tree species themselves are poorly known, with many species undescribed (Dick & Cress, 2009).

Beyond the generality of the LDG, specific hotspots of terrestrial or marine biodiversity hold a disproportionate share of species diversity. Over 40% of vascular plant species and over 35% of vertebrates are distributed among 25 terrestrial biodiversity hotspots in the megadiverse countries (Myers et al., 2000; Chapter 12). In terms of functional, rather than taxonomic diversity, rare or less abundant species are no less important than those that produce more biomass (Violle et al., 2017). This is most easily illustrated with an example of mobile predators, which occur in low abundance but have strong top-down control on ecosystem dynamics. Habitat fragmentation by anthropogenic disturbance can create small, potentially isolated ecosystem patches. This favours generalist organisms of the sort that thrive in disturbed and urban landscapes, and that have high dispersal potential or mobility and may also become invasive species outside their native range. Among a natural species assemblage, patchiness or the unreliable availability of resources causes stress to specialists, but generalists as well (Hinsley et al., 2009). The general trend of tropical ecosystems is that they have relatively more specialists.

Disturbed habitats nonetheless also support biodiversity. There is mounting evidence that there is no place on Earth that is not disturbed by human presence. Increasing movement of humans increases transportation of our companion organisms including parasites (Stoll, 1947). Our activities over millennia have caused the extinction of many species, and speciation events in a few. Some have suggested that new species might just evolve to fill the gaps. This idea links to the false precept of a fixed number of species and niches, and a misunderstanding of time scales involved in speciation. Human-mediated speciation is a difficult proposition to quantify (Bull & Maron 2016). We have domesticated over 250 plant species and over 450 animal species including terrestrial, freshwater, and marine taxa. There are now around 500 aquatic species raised by farming, though not all of those would be considered domesticated, and the rate of genetic modification is accelerating. Artificial selection has shaped the morphology, physiology, and behaviour of domesticated organisms. Certain domestic animals, like the guinea pig *Cavia porcellus* (Linnaeus, 1758) are now clearly distinct species separate from any living wild population (Gentry et al., 2004). Most of the marine animals and seaweeds domesticated via aquaculture are effectively indistinguishable from their uncaged relations, but they have been domesticated far more recently (Duarte et al., 2007). Around eight of the world's crop plants are considered novel species created by humans and this could increase with new gene editing technology (Bourke, 2017). This is still in stark contrast to the level of modern extinction, comprising over 750 known species (Chapter 9). There is an additional uncalculated influence of habitat destruction and population loss, both on species extirpated before they were ever described, but also on the decreased diversity and hence opportunity for surviving species to move into new niches through speciation processes.

Because of the long history of human impacts on the present biota, there is arguably no modern control group to compare the 'natural' behaviour of ecosystems. The best approximation we have is the Pleistocene (from 2.58 million to 11,700 years ago), and the fossil record of that period preserves evidence of microevolutionary and ecological changes in response to changing climate (Bennett, 1997). Today, protecting landscapes uninhabited by humans is a highly effective strategy to preserve as many other species as possible; the recently suggested goal of setting aside 50% of the Earth for conservation would help not only wildlife but more importantly stem accelerating climate change (Wilson, 2016). It is equally important to make space for other species within the realms we also inhabit, for mutual benefit.

Acknowledging species as naturally dynamic entities is essential to predict, and assist, their future survival. Large-scale global patterns like the LDG that can be traced through the fossil record are the best data to track the expected distribution of biodiversity. Historically, and today, tropical latitudes are the source of most species diversity, and much of that diversity may be vulnerable by virtue of rarity and specialisation. Damage to the tropics will have a disproportionate impact on the long-term recovery and future evolution of global biodiversity.

## REFERENCES

Abrams MD, Nowacki GJ. 2008. Native Americans as active and passive promoters of mast and fruit trees in the eastern USA. *The Holocene.* 18: 1123–37.

Adams J. 2010. *Species Richness: Patterns in the Diversity of Life.* Springer.

Allen JA. 1877. The influence of physical conditions in the genesis of species. *Radical Review.* 1: 108–40.

Barnosky AD. 2001. Distinguishing the effects of the Red Queen and Court Jester on Miocene mammal evolution in the northern Rocky Mountains. *Journal of Vertebrate Paleontology.* 21: 172–85.

Belanger CL, Jablonski D, Roy K, Berke SK, Krug AZ, Valentine JW. 2012. Global environmental predictors of benthic marine biogeographic structure. *Proceedings of the National Academy of Sciences.* 110: 14046–51.

Benton MJ. 2009. The Red Queen and the Court Jester: Species diversity and the role of biotic and abiotic factors through time. *Science.* 323: 728–32.

Bergmann C. 1847. Über die Verhältnisse der Wärmeökonomie der Thiere zu ihrer Grösse. *Göttinger Studien*, 3, 595–708.

Bollache L, Dick JT, Farnsworth KD, Montgomery WI. 2008. Comparison of the functional responses of invasive and native amphipods. *Biology Letters.* 4: 166–9.

Bourke I. 2017. CRISPR: Can gene-editing help nature cope with climate change? *New Statesman.* 5 October 2017.

Briggs JC. 1974. *Marine Zoogeography.* McGraw-Hill.

Brown JH. 2014. Why are there so many species in the tropics? *Journal of Biogeography.* 41: 8–22.

Buckley YM, Catford J. 2016. Does the biogeographic origin of species matter? Ecological effects of native and non-native species and the use of origin to guide management. *Journal of Ecology.* 104: 4–17.

Bull JW, Maron M. 2016. How humans drive speciation as well as extinction. *Proceedings of the Royal Society B.* 283: 20160600.

Carlton JT, Chapman JW, Geller JB, Miller JA, Carlton DA, McCuller MI, Treneman NC, Steves BP, Ruiz GM. 2017. Tsunami-driven rafting: Transoceanic species dispersal and implications for marine biogeography. *Science.* 357: 1402–6.

Carpenter EJ, Smith KL. 1972. Plastics on the Sargasso Sea surface. *Science.* 175: 1240–1.

Cavender-Bares J, Ackerly DD, Hobbie SE, Townsend PA. 2016. Evolutionary legacy effects on ecosystems: Biogeographic origins, plant traits, and implications for management in the era of global change. *Annual Review of Ecology, Evolution, & Systematics.* 47: 433–62.

Child MF. 2009. The Thoreau ideal as a unifying thread in the conservation movement. *Conservation Biology.* 23: 241–3.

Clarke AH. 1978. Polymorphism in marine mollusks and biome development. *Smithsonian Contributions to Zoology.* 274: 1–14.

Clynes T. 2018. Laser scans reveal Maya 'megalopolis' below Guatemalan jungle. *National Geographic.* 1 February 2018.

Davidson JW, Lytle MH. 2004. *After the Fact: The Art of Historical Detection.* McGraw-Hill.

Davis N. 2018. Laser scanning reveals 'lost' ancient Mexican city 'had as many buildings as Manhattan'. *The Guardian.* 15 February 2018.

Dawkins R, Krebs JR. 1979. Arms races between and within species. *Philosophical Transactions of the Royal Society B.* 205: 489–511.

Denevan WM. 1992. The pristine myth: The landscape of the Americas in 1492. *Annals of the Association of American Geographers.* 82: 369–85.

Dick CW, Kress WJ. 2009. Dissecting tropical plant diversity with forest plots and a molecular toolkit. *BioScience*. 59: 745–55.

Dickson ML, Wheeler PA. 1993. Chlorophyll a concentrations in the North Pacific: Does a latitudinal gradient exist? *Limnology and Oceanography*. 38: 1813–8.

Dillehay TD, Rossen J, Andres TC, Williams DE. 2007. Preceramic adoption of peanut, squash, and cotton in northern Peru. *Science*. 316: 1890–3.

Duarte CM, Marbá N, Holmer M. 2007. Rapid domestication of marine species. *Science*. 316: 382.

Edie SM, Jablonski D, Valentine JW. 2018. Contrasting responses of functional diversity to major losses in taxonomic diversity. *Proceedings of the National Academy of Sciences*. 5: 201717636.

Edie SM, Smits PD, Jablonski D. 2017. Probabilistic models of species discovery and biodiversity comparisons. *Proceedings of the National Academy of Sciences*. 114: 3666–71.

Fine PV. 2015. Ecological and evolutionary drivers of geographic variation in species diversity. *Annual Review of Ecology, Evolution, & Systematics*. 46: 369–92.

Foster JB. 1964. Evolution of mammals on islands. *Nature*. 202: 234–5.

Fuller E. 2014. *The Passenger Pigeon*. Princeton University Press.

Gardner TA, Barlow J, Parry LW, Peres CA. 2007. Predicting the uncertain future of tropical forest species in a data vacuum. *Biotropica*. 39: 25–30.

Gentry A, Clutton-Brock J, Groves CP. 2004. The naming of wild animal species and their domestic derivatives. *Journal of Archaeological Science*. 31: 645–51.

Glazier DS. 2015. Is metabolic rate a universal 'pacemaker' for biological processes? *Biological Reviews*. 90: 377–407.

Gloger CL. 1833. *Das Abändern der Vögel durch Einfluss des Klimas*. Breslau, Breslau.

Harmon LJ, Harrison S. 2015. Species diversity is dynamic and unbounded at local and continental scales. *American Naturalist*. 185: 584–93.

Heckenberger MJ, Kuikuro A, Kuikuro UT, Russell JC, Schmidt M, Fausto C, Franchetto B. 2003. Amazonia 1492: Pristine forest or cultural parkland? *Science*. 301: 1710–4.

Hillebrand H. 2004. On the generality of the latitudinal diversity gradient. *American Naturalist*. 163: 192–211.

Hinsley SA, Hill RA, Bellamy P, Broughton RK, Harrison NM, Mackenzie JA, Speakman JR, Ferns PN. 2009. Do highly modified landscapes favour generalists at the expense of specialists? An example using woodland birds. *Landscape Research*. 34: 509–26.

Huang S, Roy K, Jablonski D. 2014. Do past climate states influence diversity dynamics and the present-day latitudinal diversity gradient? *Global Ecology & Biogeography*. 23: 530–40.

Hung CM, Shaner PJ, Zink RM, Liu WC, Chu TC, Huang WS, Li SH. 2014. Drastic population fluctuations explain the rapid extinction of the passenger pigeon. *Proceedings of the National Academy of Sciences*. 111: 10636–41.

Hunt B, Strugnell J, Bednarsek N, Linse K, Nelson RJ, Pakhomov E, Seibel B, Steinke D, Würzberg L. 2010. Poles apart: The 'bipolar' pteropod species Limacina helicina is genetically distinct between the Arctic and Antarctic oceans. *PLoS ONE*. 5: e9835.

Inderjit, Catford JA, Kalisz S, Simberloff D, Wardle DA. 2017. A framework for understanding human-driven vegetation change. *Oikos*. 126: 1687–98.

Jablonski D, Belanger CL, Berke SK, Huang S, Krug AZ, Roy K, Tomasovych A, Valentine JW. 2013. Out of the tropics, but how? Fossils, bridge species, and thermal ranges in the dynamics of the marine latitudinal diversity gradient. *Proceedings of the National Academy of Sciences*. 110: 10487–94.

Jablonski D, Huang S, Roy K, Valentine JW. 2017. Shaping the latitudinal diversity gradient: New perspectives from a synthesis of paleobiology and biogeography. *American Naturalist*. 189: 1–12.

Jablonski D, Roy K, Valentine JW. 2006. Out of the tropics: Evolutionary dynamics of the latitudinal diversity gradient. *Science*. 314: 102–6.

Jablonski D. 1987. Heritability at the species level: Analysis of geographic ranges of cretaceous mollusks. *Science*. 238: 360–4.

Jablonski N. 2012. *Living Color: The Biological and Social Meaning of Skin Color*. University of California Press.

Jones HD, Sluys R. 2016. A new terrestrial planarian species of the genus *Marionfyfea* (Platyhelminthes: Tricladida) found in Europe. *Journal of Natural History*. 50: 2673–90.

Jordan DS. 1892. Relations of temperature to vertebrae among fishes. *Proceedings of the United States National Museum*. 1891: 107–120.

Jordan DS. 1908. The law of geminate species. *American Naturalist*. 42: 73–80.

Kane WF De V. 1893. Is the frog a native of Ireland? *The Irish Naturalist*. 2: 95–8.

Kirby D. 2010. The northern Great Plains as viewed by the Lewis and Clark expedition. *Rangelands*. 32: 2–4.

Krug AZ, Jablonski D, Valentine JW. 2007. Contrarian clade confirms the ubiquity of spatial origination patterns in the production of latitudinal diversity gradients. *Proceedings of the National Academy of Sciences*. 104: 18129–34.

LaManna JA, Mangan SA, Alonso A, Bourg NA, Brockelman WY, Bunyavejchewin S, Chang LW, Chiang JM, Chuyong GB, Clay K, Condit R. 2017. Plant diversity increases with the strength of negative density dependence at the global scale. *Science*. 356: 1389–92.

Laptikhovsky V. 2006. Latitudinal and bathymetric trends in egg size variation: A new look at Thorson's and Rass's rules. *Marine Ecology*. 27: 7–14.

Laurance WF. 2007. Have we overstated the tropical biodiversity crisis? *Trends in Ecology & Evolution*. 22: 65–70.

Liow LH, Di Martino E, Voje KL, Rust S, Taylor PD. 2016. Interspecific interactions through 2 million years: Are competitive outcomes predictable? *Proceedings of the Royal Society B*. 283: 20160981.

Liow LH, Van Valen L, Stenseth NC. 2011. Red Queen: From populations to taxa and communities. *Trends in Ecology & Evolution*. 26: 349–58.

Ludt WB, Rocha LA. 2015. Shifting seas: The impacts of pleistocene sea-level fluctuations on the evolution of tropical marine taxa. *Journal of Biogeography*. 42: 25–38.

MacArthur R, Wilson EO. 1967. *The Theory of Island Biogeography* (reprinted 2001). Princeton University Press.

Mann CC. 2005. *1491: New Revelations of the Americas Before Columbus*. Alfred a Knopf Incorporated.

Martin PS, Klein RG (editors). 1989. *Quaternary Extinctions: A Prehistoric Revolution*. University of Arizona Press.

May RM. 1992. How many species inhabit the Earth? *Scientific American*. 267: 42–9.

Mileikovsky SA. 1971. Types of larval development in marine bottom invertebrates, their distribution and ecological significance: A re-evaluation. *Marine Biology*. 10: 193–213.

Millien V, Lyons SK, Olson L, Smith FA, Wilson AB, Yom-Tov Y. 2006. Ecotypic variation in the context of global climate change: Revisiting the rules. *Ecology Letters*. 9: 853–69.

Montgomery WI, Lundy MG, Reid N. 2012. 'Invasional meltdown': Evidence for unexpected consequences and cumulative impacts of multispecies invasions. *Biological Invasions*. 14: 1111–25.

Murray GG, Soares AE, Novak BJ, and 21 other authors. 2017. Natural selection shaped the rise and fall of passenger pigeon genomic diversity. *Science*. 358: 951–4.

Myers N, Mittermeier RA, Mittermeier CG, Da Fonseca GA, Kent J. 2000. Biodiversity hotspots for conservation priorities. *Nature*. 403: 853–8.

Neumann TW. 1985. Human-wildlife competition and the passenger pigeon: Population growth from system destabilization. *Human Ecology*. 13: 389–410.

O'Connor MI, Bruno JF, Gaines SD, Halpern BS, Lester SE, Kinlan BP, Weiss JM. 2007. Temperature control of larval dispersal and the implications for marine ecology, evolution, and conservation. *Proceedings of the National Academy of Sciences*. 104: 1266–71.

Pluess T, Cannon R, Jarošík V, Pergl J, Pyšek P, Bacher S. 2012. When are eradication campaigns successful? A test of common assumptions. *Biological Invasions*. 14: 1365–78.

Poulin R. 2014. Parasite biodiversity revisited: Frontiers and constraints. *International Journal for Parasitology*. 44: 581–9.

Poulin E, Féral JP. 1996. Why are there so many species of brooding Antarctic echinoids? *Evolution*. 50: 820–30.

Powell MG. 2009. The latitudinal diversity gradient of brachiopods over the past 530 million years. *Journal of Geology*. 117: 585–94.

Qian H, Ricklefs RE. 2006. The role of exotic species in homogenizing the North American flora. *Ecology Letters*. 9: 1293–8.

Rabosky DL, Hurlbert AH. 2015. Species richness at continental scales is dominated by ecological limits. *American Naturalist*. 185: 572–83.

Rapoport EH. 1982. *Aerography: Geographical Strategies of Species*. Pergamon.

Rejmanek M, Simberloff D. 2017. Origin matters. *Environmental Conservation*. 44: 97–9.

Rolland J, Condamine FL, Jiguet F, Morlon H. 2014. Faster speciation and reduced extinction in the tropics contribute to the mammalian latitudinal diversity gradient. *PLoS Biology*. 12: e1001775.

Rosenzweig ML. 1992. Species diversity gradients: We know more and less than we thought. *Journal of Mammalogy*. 73: 715–30.

Roy K, Jablonski D, Valentine JW, Rosenberg G. 1998. Marine latitudinal diversity gradients: Tests of causal hypotheses. *Proceedings of the National Academy of Sciences*. 195: 3699–702.

Simpson GG. 1944. *Tempo and Mode in Evolution*. Columbia University Press.

Stevens GC. 1989. The latitudinal gradient in geographical range: How so many species coexist in the tropics. *American Naturalist*. 133: 240–56.

Stoll NR. 1947. This wormy world. *Journal of Parasitology*. 33: 1–8.

Stuart BL, Inger RF, Voris HK. 2006. High level of cryptic species diversity revealed by sympatric lineages of Southeast Asian forest frogs. *Biology Letters*. 2: 470–4.

Stuckey ME. 2017. American elections and the rhetoric of political change: Hyperbole, anger, and hope in US politics. *Rhetoric & Public Affairs*. 20: 667–94.

Terborgh J. 1973. On the notion of favorableness in plant ecology. *American Naturalist*. 107: 481–501.

Thorson G. 1950. Reproductive and larval ecology of marine invertebrates. *Biological Reviews*. 25: 1–45.

Valentine JW, Jablonski D. 2015. A twofold role for global energy gradients in marine biodiversity trends. *Journal of Biogeography*. 42: 997–1005.

Van Cauwenberghe L, Vanreusel A, Mees J, Janssen CR. 2013. Microplastic pollution in deep-sea sediments. *Environmental Pollution*. 182: 495–9.

Van Valen L. 1973. A new evolutionary law. *Evolutionary Theory*. 1: 1–30.

Vermeij GJ. 1999. Inequality and the directionality of history. *American Naturalist*. 153: 243–53.

Violle C, Thuiller W, Mouquet N, Munoz F, Kraft NJ, Cadotte MW, Livingstone SW, Mouillot D. 2017. Functional rarity: The ecology of outliers. *Trends in Ecology & Evolution*. 32: 356–67.

Vitousek PM, Mooney HA, Lubchenco J, Melillo JM. 1997. Human domination of earth's ecosystems. *Science*. 277: 494–9.

Walker TD, Valentine JW. 1984. Equilibrium models of evolutionary species diversity and the number of empty niches. *American Naturalist*. 124: 887–99.

Wallace AR. 1855. On the law which has regulated the introduction of new species. *Annals and Magazine of Natural History.* 2nd Series. 16: 184–96.

Wiens JJ, Pyron RA, Moen DS. 2011. Phylogenetic origins of local-scale diversity patterns and the causes of Amazonian megadiversity. *Ecology Letters.* 14: 643–52.

Williams CM, Ragland GJ, Betini G, Buckley LB, Cheviron ZA, Donohue K, Hereford J, Humphries MM, Lisovski S, Marshall KE, Schmidt PS. 2017. Understanding evolutionary impacts of seasonality: An introduction to the symposium. *Integrative & Comparative Biology.* 57: 921–33.

Willis KJ, Gillson L, Brncic TM. 2004. How 'virgin' is virgin rainforest? *Science.* 304: 402–3.

Wilson EO. 2016. *Half-Earth: Our Planet's Fight for Life.* Liveright

Winsor L, Johns PM, Barker GM. 2004. Terrestrial planarians (Platyhelminthes: Tricladida: Terricola) predaceous on terrestrial gastropods. In: Barker GM (editor). *Natural Enemies of Terrestrial Molluscs,* pp. 227–278. CAB International.

Woodruff DS. 2010. Biogeography and conservation in Southeast Asia: How 2.7 million years of repeated environmental fluctuations affect today's patterns and the future of the remaining refugial-phase biodiversity. *Biodiversity & Conservation.* 19: 919–41.

Wright SD, Ross HA, Keeling DJ, McBride P, Gillman LN. 2011. Thermal energy and the rate of genetic evolution in marine fishes. *Evolutionary Ecology.* 25: 525–30.

Wright SJ, Muller-Landau HC. 2006. The future of tropical forest species. *Biotropica.* 38: 287–301.

# 12 Translating Biodiversity across Cultural Barriers

## MAKERS AND USERS OF SPECIES NAMES

Cross-cultural barriers, even those between scientists in different fields of study, can create scenarios where two well-meaning people both think they speak the same language and come away either having no idea what the other person meant, or worse, feeling confident they understand completely, and getting it utterly wrong. Business leaders have quite sincerely asked why we need so many species, 'Why don't you just pick the best ones and keep those?' Even among scientists nominally studying biodiversity there is foundational confusion over the idea of what species are (Chapter 2). Other scientific disciplines are often unaware of the scope of the taxonomic knowledge gap—the number of global undescribed species is staggering, but most of us only talk about species that are already named. A further pressing issue is the global distribution of biodiversity, as the species and the scientific infrastructure to describe them are disjointly distributed.

Although an increasing number of scientists are engaged in projects that lead to the publication of new species descriptions (Costello et al., 2013), taxonomy and systematics remain a minority discipline within life sciences (Chapter 7). The number of specialists that have the experience and technical skill to lead descriptive studies may be declining (Poulin, 2014). Meanwhile, artificial intelligence and computer vision are advancing on automated identification tools (Carranza-Rojas et al., 2017). Technology and interdisciplinary collaboration are poised to accelerate species discovery and overcome bottlenecks including mapping, data sharing, and access to reliable data about species (Wheeler et al., 2012). For the grand project to understand Earth's biodiversity there is ample scope for optimism.

The process of describing species is one of iteratively testing and refining hypotheses (Chapter 4, Chapter 7). This fluid process is not intuitive to our experience of species identification. Most users of species names are much more likely to get information about identification from field guides, dichotomous keys, and reference works like floras or checklists, rather than from primary literature in systematics and phylogenetics. We expect to be able to find diagnostic features, or step through a key, and come to a definitive, final answer about whether the organism in front of us is species A or species B. The background process of describing species is the process of weighing the balance of many different lines of evidence.

Species are complicated scientific problems, but the urgency to provide robust and rapid taxonomic assessments of biodiversity is increasing (Wilson, 2017). Taxonomy has been called the real 'oldest profession' (Hedgepeth, 1962). Practitioners tend to take a long-term view, connecting with data and results produced hundreds of years ago, by taxonomists from Linnaeus onward. Other fields of science, which depend on species implicitly or explicitly, need advice on much shorter timescales. Broadly dividing scientists into the makers and users of species names, each has separate perspectives that can lead to some common misconceptions.

## MISCONCEPTION: TAXONOMY IS DONE BY 'SOMEONE ELSE'

Taxonomy, in its sense as the categorical application of names to organisms, is practised by everyone. Everyone identifies species and needs names to call them (Chapter 3). This is one part of a larger project, of recognising differentness that may be indicative of separating evolutionary lineages that should be described as separate species.

Research activity leading to the description of species, species groups, and their inter-relationships (systematics and phylogenetics) is focussed in museums and botanic gardens, specialist scientific infrastructures that put rich primary data at your fingertips. This is exactly the same logic that explains why high-energy physics tends to happen at particle accelerators—large-scale, multi-billion dollar, publicly funded scientific infrastructure. In both fields, there are many scientists who do not work within those research institutes, but who work collaboratively on the data they produce, visit the infrastructure as research visitors, or access their public data products online. There is a strong correlation between the number of biodiversity collections per country and how many researchers are experts that have a taxon-focussed research programme or expertise (Paknia et al., 2015). The other broad similarity is that specialist institutes are usually very welcoming to collaboration and embrace innovative uses for their resources. The existence of specialist institutes that focus on taxonomic research does not imply those are the only places where such research can be done; on the contrary, though specialist resources are helpful, taxonomy can be done anywhere, and should be done in response to biodiversity needs.

## MISCONCEPTION: TAXONOMY IS EASY NOW, BECAUSE

### YOU CAN JUST SEQUENCE THINGS

Molecular tools do simplify part of the process of identifying new species. DNA provides another line of evidence for living species that can be combined with other data (morphology, geographic distribution, behaviour) to distinguish species. Rates of evolution are very different in various taxa, and even between relatively closely related species depending on interactions between species and environment, so it does not make sense to expect a simple percent-divergence test that would indicate the complex and subtle process of speciation (Sigwart and Garbett, 2018). Although mitochondrial markers are used as 'barcode' regions because of their supposedly neutral and clock-like evolution, there is a variable and unknown level of intra-specific variation among taxa (Galtier et al., 2009). Molecular tools frequently uncover diversity that is cryptic to traditional morphological approaches, and DNA barcoding can indeed produce rapid assessments to estimate species richness (Smith et al., 2009). However, information about the natural history may be the decisive evidence about whether an organism belongs to a new species.

The major impediment to molecular species identification for established species is the incomplete nature of reference databases. A sequence that has no strong match could be a new species, or a known species that was never previously sampled. Among fungi, the number of sequences from environmental samples dramatically outnumber those that are from voucher specimens that can be tied to a known species (Blackwell, 2011). If a new sequence matches another database entry for an unidentified sequence, you can know you are not alone, but it does not help pin down an identification. Previously published sequences could also have come from material that was misidentified, or contaminated. (This is anecdotally rampant among marine invertebrate taxa in GenBank but requires an advanced level of taxonomic knowledge to disentangle.) GenBank sequences are often not linked to a specimen at all so it is impossible to test whether an aberrant sequence is contamination, misidentification, or a new species.

The National Center for Biotechnology Information (NCBI) databases, GenBank and NCBI Genomes, now include annotations to indicate when sequences were obtained from name-bearing type material (Federhen, 2014). However, the steps to use these features are not obvious to a user depositing sequence data. Naming new species requires type specimens and comparative material. Type sequences may be derived from historical material or new species. New species descriptions of cryptic animal species have been named on the basis of diagnostic sequences (e.g., Jörger & Schrödl, 2013). But in the interpretation of the Zoological Code it is not the sequence itself, but the specimen that the DNA came from, that is the type (even if the type specimen is destroyed; Chapter 8). Connecting a GenBank record to the type specimen provides a strong, definitive connection between the taxon name and the sequence data.

It should be added that naturally this does not apply to palaeontological species, or indeed to any of the many species living or extinct that do not have specimens that preserve DNA.

## MISCONCEPTION: TAXONOMIC PAPERS NEVER GET CITED

Competitive journal metrics set a uniquely pernicious trap for species descriptions (Werner, 2006). The dilemma is that scientific journals emphasise citation, the number of times a contribution has been cited in other later papers. Yet, at the same time, the convention in most scientific journals is to omit credit for the key product of taxonomic papers: the species. Species authorities and dates are not usually included when a scientific name is reported outside of specialist systematics publications. The discovery of a fresh new species might be commented on and cited, but after a species becomes normalised by the development of a body of work around its biology, the species authority is dropped and the original description is not cited at all. Citing species authorities would redress the balance somewhat (Werner, 2006), integrating the history of taxonomic literature with digital identifiers is another complementary approach (Page, 2016).

Until recently it was routine to cite the relevant flora, or equivalent work, to explain the standard version of nomenclature used in a nontaxonomic paper ('accepted names according to …'). Anecdotally, this convention seems to have waned, which may be partly attributable to the rise of continuously updated online nomenclatural databases rather than periodically published standard references. The citation of standard systematic arrangements does not solve the citation gap for the individual descriptive papers, but it points to another related issue: species are hypotheses, subject to revision. Citing a reference for the systematic arrangement is an acknowledgement that the identification and classification is a working scientific project, in progress; citing nothing leans toward the trap of thinking that all named species and their classification are fully resolved.

Bibliometric analyses have not supported a significant difference between papers that focus on taxonomy or other aspects of well-studied groups (Steiner et al., 2015). So, it is not clear whether a lack of citations for descriptive papers is a phenomenon of taxonomy, or of the taxon. A new species of mammal is very exciting, because it is rare, and this is correlated with the fact that lots of people work on mammals. For example, a new orangutan was named in 2017 (Nater et al., 2017), in a high impact journal, and was immediately cited in several other papers about primates and conservation. This level of interest also happens with some taxonomic descriptive papers about important fossils. Papers on under-studied organisms are generally poorly cited, because almost by definition there is very little published work on those species. (As a corollary, it is possible that in some cases no one is working on them because no one can identify them.)

The motivation to be strategic about published outputs is usually to protect careers of the scientists involved, who are expected not just to publish, but publish material that is read and cited by others. There is a large community of academics who have strong interest in, and opinions about, taxonomy in general and its direction as a discipline, including reviews and commentaries in high impact journals (e.g., Garnett & Christidis, 2017; Wilson, 2017). The subject of taxonomy has a broad audience and garners cross-disciplinary interest. Yet there is an expectation that species descriptions should be put into context of some other biological aspect (as a contribution to ecology or phylogeny). Standard advice in some corners is that you should not publish 'just' a species description. This undermines the perceived legitimacy of descriptive work as a valuable career output. At the same time, aiming for broader context is undeniably a useful strategy to ensure that papers about otherwise obscure species are read and found in an increasingly suffocating and crowded world of online publication. The most constructive approach is, therefore, not to abandon taxonomy, but to hone multiple approaches in parallel.

## MISCONCEPTION: BIODIVERSITY IS WELL-KNOWN

The discovery gap is the disparity between total diversity and the fraction that has been studied by science. That frontier is broad by definition in under-studied taxa and relatively narrower in others, such as tetrapod vertebrates, or many other, small, well-studied clades across the eukaryotes. The concerns of working scientists are scaled relative to available knowledge about the problem at hand. At one end of that scale, research on model organisms can investigate finer and finer detail about biological processes and their mechanisms, within one species. From this perspective, it may be difficult to conceive of just how little is known about most other species, or even how many living species there are. At the other extreme, where taxonomy sits, it is swamped with new species and missing data. One project in an area of 450 km$^2$

on a single taxonomic family of tropical snails in the family Triphoridae, collected material for 259 species and found that around 70% of them were new to science (Albano et al., 2011).

Gaston and May (1992) documented the disparity in global taxonomic need and taxonomic effort in organisms and in geography. In their study, the dominant fraction of taxonomists was distributed among North America, Europe, or Australia. By contrast, most species, and most clades, are tropical (Chapter 11; Jablonski et al., 2013). The modern centres of human economic and scientific power are in countries at higher latitudes, and concentrated in the northern hemisphere, sometimes called the 'Global North'. The experience of working in a fauna that is relatively well described, such as in Europe, sets up a further latitudinal cultural gradient with the 'Global South' where floras and faunas are relatively poorly described.

## GEOGRAPHIC AND CULTURAL DIVIDES

Conservation International recognised 17 'megadiverse' countries; they do not contain all of the biodiversity hotspots nor the full pantheon of marine and terrestrial biomes (Marchese, 2015), but these 17 nations are home to the large majority of Earth's species, and all of them have at least some tropical territory (Mittermeier et al., 1998). There is a substantial misalignment between the geographic distribution of taxonomic capacity and taxa (Figure 12.1). Naturally, some areas are well studied, especially if they are adjacent to research institutes, for example a comprehensive study of the intertidal marine fauna in Plymouth, England, was published over 100 years ago (MBA, 1904). This does not mean the flora and fauna of the Global North are completely described. The discovery gap has many dimensions. Hortal and colleagues (2015) listed seven separate issues that hamper understanding large-scale biodiversity: the lack of knowledge about species regarding their taxonomic recognition, distribution, abundance, phylogeny, abiotic tolerances, species traits, and biotic interactions. The first aspect, the number of unrecognised or un-named species, is sometimes called the 'Linnean shortfall'. This is a separate problem from the second and other most critical issue, which is knowing the distributions of species, as some species are named without knowing much about their natural history; this is the 'Wallacean shortfall', so called for the pioneer of biogeography Alfred Russel Wallace.

New distribution records are still found, and new species are still discovered almost everywhere, but the flavour of these discoveries is different by region. New species described in Britain or Ireland, for example, are usually from poorly described taxa (e.g., fungi, Zamora et al., 2015) and/or invasive species first discovered there, but that had never originally been described in their native habitat before being introduced to Europe (e.g., Jones & Sluys, 2016). New discoveries in the tropics are often on the scale of the paper mentioned above with potentially hundreds of new species at a time (Albano et al., 2011). Local perspectives and experience shape our expectations, so the concern is that scientists who live and work in well-described ecosystems, may not be naturally empathetic to the overwhelming experience of other scientists who face a much larger local discovery gap.

Global species richness of fungi has the highest uncertainty among eukaryotic groups, a long-standing estimate of around 1.6 million global species was based

## Relative Biodiversity

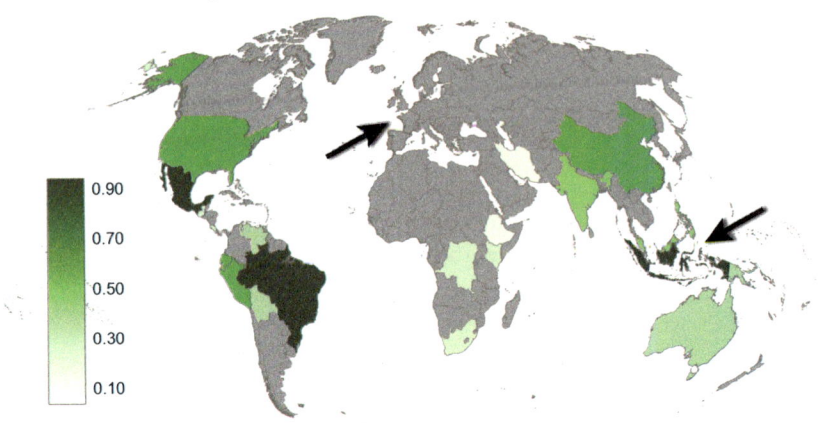

## Relative Wealth

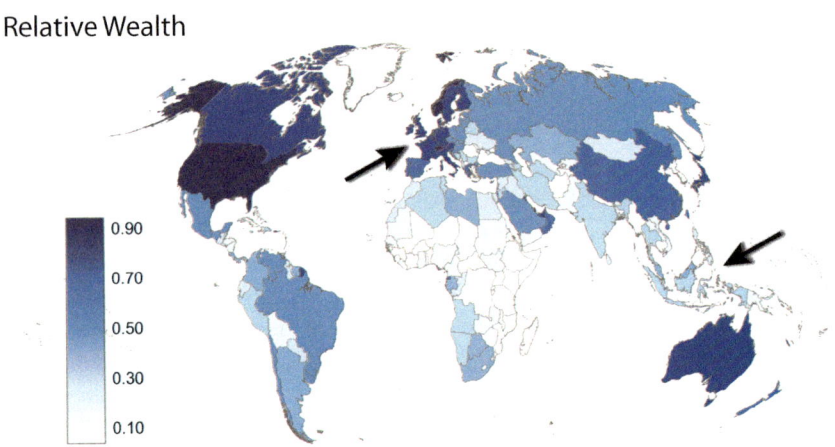

**FIGURE 12.1** A comparison of national biodiversity (estimated species richness of mammals, amphibians, reptiles, and birds) in megadiverse countries, and global national wealth. Each index was scaled to show relative values in a range between 0 and 1. (Redrawn from Lira-Noriega & Soberón, 2015.)

on extrapolating to a global number from a 6:1 ratio of fungi to vascular plants in temperate Britain (Hawksworth, 1991). This was criticised as unrealistic in tropical ecosystems, where such a ratio would imply that a seemingly unbelievable 95% of species are undescribed (Scheffers et al., 2012). Yet, molecular tools continue to find more new fungi even in temperate regions, and the global estimate has been extended to over 2.2 million anticipated fungal species (Hawksworth & Lücking, 2017).

Not being able to identify specimens in biodiversity assessments is a significant barrier to getting more interesting, far-reaching analyses. The issue of time and labour to process samples is familiar to ecologists working anywhere in the world, this is a common workflow bottleneck, an issue of limited resources in time and funding. But it is *different* when you are working in an environment where an unknown but potentially large proportion of species have never been described. At an international workshop, I once listened to a Malaysian colleague explain this problem, and a British colleague earnestly relating their experience of using a professional taxonomic service from a consultancy group. (Just pay someone to do the identifications for you, it is really worth the investment.) This was sincerely supportive but inappropriate advice; to anyone working in tropical Southeast Asia it might sound a little bit like 'Let them eat cake'. Environmental consultancies that provide identification services can generally only identify things with reference to established names and identities. There are no effective service providers that can provide comprehensive taxonomic identifications for marine invertebrates in Malaysia, because it is not yet possible. Many excellent companies exist for such work in Europe and other well-described faunas. The workflow bottleneck in the tropics is more fundamental, the number of species involved is much higher, and the uncertainty about their identification more frequent, and many species may be undescribed. Among a large sampling of modern terrestrial ecology publications, 90% of studies were performed in the wealthiest nations, and temperate biomes were dramatically over-represented (Martin et al., 2012). One problem that contributes to that disparity is the basic access to relevant data. Working in a megadiverse tropical ecosystem, scientists want to get on with doing serious experimental ecology, but the time and labour to identify the species collected is overwhelming, and even then there may be little confidence in the identifications.

There is more biodiversity, and less of it is described, in the tropics, areas of the world that are mostly populated by developing economies. Lira-Noriega and Soberón (2015) compiled a detailed quantitative analysis of factors for species richness, taxonomic infrastructure, economic wealth, and governance. The relationship is not a direct inverse correlation between metrics of species richness and economy, of course, but many countries with the greatest biodiviersty assets are less well equipped in terms of scientific expertise and government support for science (Figure 12.1).

The comparison between temperate and tropical ecosystems is obviously not a simple binary contrast between depauperate or lush diversity. Environmental sceptics in the USA have written disparaging comments comparing the rich and abundant shallow marine environment of Alaska to the apparently inferior diversity of damaged reef environments in Hawai'i. Indeed, the temperate North Pacific has a spectacular terrestrial and marine flora and fauna, some of the most beautiful and pristine places in the world, and its fair share of undiscovered species.

## QUANTIFYING SPECIES RICHNESS

Most species are rare, and most species are small. This creates a bottleneck to quantifying biodiversity, which gets worse in systems with increasing species richness. But large-bodied animals are often more sensitive to anthropogenic disturbance, and large-bodied species are also relatively easier to observe. In a natural system (not cultivated monocultures of oil palms or cows), species richness is a strong predictor of biomass (e.g., Duffy et al., 2016). Species richness or biomass are simple metrics that may not convey the complexity of biodiversity as it relates to ecosystem function and resilience. Diverse communities are overall more buffered from disturbance;

this resilience is clearly not a product of simply having a higher count or total weight species, but from their ecological interactions. Diversity, system complexity, is itself an asset that promotes biodiversity on multiple scales.

The higher density of species in tropical ecosystems results in a larger number of inter-species interactions. If there is a fixed number of 'functions' then a larger number of interactions may be equivalent, with an appearance that various species serve the same purpose to the ecosystem. (The idea that co-occurring species, from independent evolutionary histories, can be redundant to each other, is rather anthropocentric.) However, the distribution of functions among species in tropical fish assemblages follows a skew distribution, with some traits shared by many species and others by few species. The stochastic removal of species, for instance through fishing pressures driving species to local extirpation, removes not only rare species but also rare functional traits, which disproportionately disrupts total ecosystem functioning (Mouillot et al., 2014).

Comparative biodiversity can be quantified in many ways, by focussing on the richness and distribution of well-described and visible larger organisms like butterflies and birds (Balmford & Long, 1995; Koh & Wilcove, 2008), or taxonomic surrogacy (Naser, 2010), or functional diversity (Petchey & Gaston, 2006). These all aim to find achievable ways to distil data about species in a manageable way that will reveal predictive patterns. But even in a temperate ecosystem, different metrics to assess regional-scale biodiversity hotspots do not result in a complete consensus (Tolimieri et al., 2015). The specific question of prioritising conservation actions is markedly separate from the scientific study of biodiversity.

The fundamental metric of biodiversity is species richness; in the megadiverse tropics, an unknown number of species are undescribed. Undescribed species are likely to be comparatively rare (if only because, statistically, most species are rare), but species that occur in small areas or low abundance may have unique traits that increase ecosystem complexity. While other approaches are useful for urgent questions about conservation priorities, true understanding of tropical biodiversity depends on being able to identify species and their distributions.

## TAXONOMY IS SCIENTIFIC INFRASTRUCTURE

Assessing species-level biodiversity requires good taxonomic identification; robust species identifications depend on a knowledge base about the flora and fauna. That knowledge base includes the expertise of taxonomists and systematists, and others who use similar techniques in molecular biology and natural history. More pertinently, it includes taxonomic data products, in the form of reliable voucher collections, faunal inventories, identification keys, and field guides. In tropical megadiverse countries, these data resources may be unavailable, highly incomplete, and/or based on coarser assessments of a broader region. The larger region might contain two, or twenty, similar-looking species, but a general field guide does not indicate how to differentiate them.

The number of taxonomic descriptive papers being published is increasing in some megadiverse countries with strong economic growth (Paknia et al., 2015; Table 12.1). This has been highlighted as part of a larger, appropriate shift in the intellectual leadership around tropical biodiversity from a historical Euro-centric or colonial

## TABLE 12.1
## Where are Papers About New Species Being Published?

This table lists the geographic distribution from a set of published scientific articles reporting new species of animals, plants, and fungi, in the period 2001–2015; each paper ($n = 36,579$) is counted only once, based on the country of affiliation for the first author. Countries are listed in order of decreasing productivity, showing number of first-authored publications, frequency (% of global taxonomic papers), and the proportional output per million people in the country's population. These data are drawn from results in Web of Science for publications including 'sp. nov.' or 'n. sp.' or 'new species' in the title of the article, excluding topics, keywords, and sources related to microbiology. Region and population figures from the US Census Bureau (2013). Megadiverse countries determined by Conservation International (Mittermeier et al., 1997) are noted MC and in bold text, additional countries in the Group of Like-Minded Megadiverse Countries are noted LMMC. All these megadiverse countries are included in the map shown in Figure 12.1.

| Country | Papers 2001–2015 | Percent of All Papers | per million Population | Region | Megadiverse Nations |
|---|---|---|---|---|---|
| **United States** | **5589** | **16.8%** | **18.7** | **NORTHERN AMERICA** | MC |
| **China** | **4945** | **14.9%** | **3.8** | **ASIA** | MC/LMMC |
| **Brazil** | **3978** | **11.9%** | **21.2** | **LATIN AMERICA** | MC/LMMC |
| Japan | 1259 | 3.8% | 9.9 | ASIA | |
| **Mexico** | **1243** | **3.7%** | **11.6** | **LATIN AMERICA** | MC/LMMC |
| **Australia** | **1241** | **3.7%** | **61.2** | **OCEANIA** | MC |
| Germany | 1204 | 3.6% | 14.6 | WESTERN EUROPE | |
| Spain | 1104 | 3.3% | 27.3 | WESTERN EUROPE | |
| **India** | **1098** | **3.3%** | **1.0** | **ASIA** | MC/LMMC |
| Russia | 1098 | 3.3% | 7.7 | RUSSIA[a] | |
| Argentina | 911 | 2.7% | 22.8 | LATIN AMERICA | |
| Poland | 779 | 2.3% | 20.2 | EASTERN EUROPE | |
| United Kingdom | 774 | 2.3% | 12.8 | WESTERN EUROPE | |
| Turkey | 769 | 2.3% | 10.9 | NEAR EAST | |
| Italy | 742 | 2.2% | 12.8 | WESTERN EUROPE | |
| France | 713 | 2.1% | 11.7 | WESTERN EUROPE | |
| Czech Republic | 642 | 1.9% | 62.7 | EASTERN EUROPE | |
| Iran | 590 | 1.8% | 8.6 | ASIA | |
| **South Africa** | **549** | **1.6%** | **12.4** | **AFRICA** | MC/LMMC |
| Korea, South | 519 | 1.6% | 10.6 | ASIA | |
| Canada | 457 | 1.4% | 13.8 | NORTHERN AMERICA | |
| Taiwan | 438 | 1.3% | 19.0 | ASIA | |
| Belgium | 392 | 1.2% | 37.8 | WESTERN EUROPE | |
| Thailand | 374 | 1.1% | 5.8 | ASIA | |
| Netherlands | 356 | 1.1% | 21.6 | WESTERN EUROPE | |

*(Continued)*

**TABLE 12.1 (*Continued*)**
**Where are Papers About New Species Being Published?**

| Country | Papers 2001–2015 | Percent of All Papers | per million Population | Region | Megadiverse Nations |
|---|---|---|---|---|---|
| Colombia | 290 | 0.9% | 6.7 | LATIN AMERICA | MC/LMMC |
| Sweden | 286 | 0.9% | 31.7 | WESTERN EUROPE | |
| New Zealand | 284 | 0.9% | 69.7 | OCEANIA | |
| Singapore | 257 | 0.8% | 57.2 | ASIA | |
| ... | | | | | |
| Malaysia | 175 | 0.5% | 7.2 | ASIA | MC/LMMC |
| ... | | | | | |
| Venezuela | 99 | 0.3% | 3.8 | LATIN AMERICA | MC/LMMC |
| Indonesia | 85 | 0.3% | 0.3 | ASIA | MC/LMMC |
| ... | | | | | |
| Costa Rica | 82 | 0.2% | 20.1 | LATIN AMERICA | LMMC |
| ... | | | | | |
| Ecuador | 73 | 0.2% | 5.4 | LATIN AMERICA | MC/LMMC |
| ... | | | | | |
| Peru | 62 | 0.2% | 2.2 | LATIN AMERICA | MC/LMMC |
| Philippines | 58 | 0.2% | 0.6 | ASIA | MC/LMMC |
| ... | | | | | |
| Bolivia | 32 | 0.1% | 3.6 | LATIN AMERICA | LMMC |
| ... | | | | | |
| Kenya | 18 | 0.1% | 0.5 | AFRICA | LMMC |
| ... | | | | | |
| Madagascar | 8 | 0.0% | 0.4 | AFRICA | MC/LMMC |
| ... | | | | | |
| DR Congo | 3 | 0.0% | 0.0 | AFRICA | MC/LMMC |
| ... | | | | | |
| Guatemala | 3 | 0.0% | 0.2 | LATIN AMERICA | LMMC |
| ... | | | | | |
| Ethiopia | 1 | 0.0% | 0.0 | AFRICA | LMMC |
| ... | | | | | |
| Papua New Guinea | 0 | 0.0% | 0.0 | OCEANIA | MC |

[a] Broader region: Commonwealth of Independent States.

model to a stronger role for the scientists living in more species rich regions. There are regional shifts over the last several decades in the institutions where taxonomic activity is centred (Costello et al., 2013). Brazil and China may be exceptions rather than trend-setters; in developing economies there is little incentive to invest in basic research that does not have a clear direct connection to enabling economic growth.

Access to specimens, the raw data that validates species identification, is what makes the study of biodiversity repeatable (Lee et al., 2007). Where species diagnoses are well understood and clearly defined, the connection to physical evidence becomes less critical. This leads to another cultural divide of expectation: standard methods in many disciplines are developed in a context of well-described ecosystems where preserving and curating specimen vouchers may not be necessary to ensure robust identifications. This is potentially problematic, as less experienced researchers may not be judicious about identification or the resources used for species determinations. Taxonomic misidentifications are acknowledged to be common in marine surveys everywhere (Vecchione et al., 2000). Bortolus (2008) found that a majority, more than 60%, of a sample of papers in ecological journals contained no reference to any specific sources (people or literature) used for taxonomic identifications. Probably not all of those papers contain misidentifications. The more important point is that such practices are common and set a disciplinary norm (Martin et al., 2012). Trendsetting methods in any field will be taken up by others in different environmental contexts. The approaches used in the Global North set a standard that will be followed by the Global South. It is reasonable to do field identifications without voucher specimens in a habitat with relatively few and well-studied species, but that practice is not transferable to other regions with bigger discovery gaps. Taxonomic errors in a study can lead to a dramatic cascade of misinterpretation about biological phenomena as those data are built on in later work (Bortolus, 2008).

Historically, most specimen material collected from tropical megadiverse countries was transported to permanent museum collections in Europe or North America. Global museums are connected through international networks to share information (e.g., Weber, 2018). Taxonomic specialists interested in a taxon or a region are aware of these trends through the historical literature; some are obvious direct results of former colonial control (historical specimens from Indonesia are in collections in the Netherlands), some connections are perhaps less intuitive (substantial material from the Congo basin is in the American Museum of Natural History, New York; there is an important collection of Indian birds in the National Museum of Ireland). These major museums contain types, the definitive material that fixes the identification of a species (Chapters 4 and 8). Historical type specimens are part of these collections, but the majority of new species described from tropical megadiverse countries are still transferred to foreign museums (Paknia et al., 2015). The long history of museums has also accumulated material of many species that were not recognised at the time they were collected. Established museum collections may collectively hold up to 0.5 million undescribed species (Costello et al., 2013). Molecular sequence data extracted from museum specimens is revealing many specimens that were misidentified, or represent new species (Bickford et al., 2007). Some of these undescribed species are already extinct, so they cannot be recollected, and there is no other way to study them (Hawksworth & Cowie, 2013). That also does not account for fossils, but the same applies to palaeontological specimens and species, especially those collected from localities that may have been destroyed, or under threat from development (e.g., Lipps & Granier, 2009).

If wealthier countries allow their museum collections to fall into states of decreasing funding, or even close collections all together, this has a compounding effect in other countries. Taxonomic literature is founded in data, and those data are archived by museums in the form of physical specimens. Museums have been compared to specimen libraries, but that analogy is incomplete; comparison to libraries invokes an idea that each museum is a local branch that provides access to resources. When a library branch closes, it impacts the local community, but there are other copies of those books in other, bigger libraries, and anyway most things are online now. Similarly, museums are engaging in digitisation projects to improve access (Ang et al., 2013), but there are limits to the utility of image data. While some museum collections have some overlap in scope, their contents are not redundant. Specimens are the physical remains of organisms that each led a unique life.

Museum collections are the fulcrum for other scientific activities, and require ongoing maintenance, investment, and improvement (Suarez & Tsutsui, 2004). Like other well-built infrastructure, the foundational structures get taken for granted, neglect takes a long time to accumulate to a point of disaster, and maintenance costs are unattractive to funders because there may be no big impressive finale. Economists have shown that road maintenance has a higher return on investment but is more poorly funded than new construction (Gramlich, 1994). Museums can provide a centre for activity, including a base for international researchers, and interdisciplinary projects centred on the common resources in collections. But perhaps most importantly, they provide a public face to promote biodiversity to the general public, to teach the value of biodiversity as a natural resource (Chapter 8).

Rural or regional museums are essential to public education, connecting people with their local natural history, and these can be successful even without national funding (Qumsiyeh et al., 2017). Local people in many megadiverse countries are frequently unaware of how their national biodivesity compares to other (usually larger) countries even when their natural diversity may be iconic to worldwide audiences (Häussermann & Schrödl, 2017). This is frequently true in less diverse countries, too; everyone thinks that biodiversity lives 'somewhere else'. The presence of a museum provides taxonomic infrastructure for scientists, but it also creates a visible public presence to celebrate local identity, and to promote the study of species diversity.

## INTERNATIONAL LAW, BIODIVERSITY, AND BENEFIT SHARING

The most important international treaty for the global study of species, the United Nations Convention on Biological Diversity (CBD), has recognised the importance of museum collections to the study and preservation of biodiversity. Formal reports by the CBD secretariat stated: 'Parties and authorities responsible for museums and herbaria have been asked to strengthen reference collections in countries of origin, invest, on a long-term basis, in the development of appropriate infrastructure for their national collections ...' (CBD, 2006). Barriers to growing and developing permanent, reliable voucher collections include financial investment and technological resources for collections care, but the CBD and other national research programmes have funded substantial capacity building efforts. The more significant obstacles at present seem to be a lack of political motivation or understanding of the downstream benefits (Paknia et al., 2015).

The CBD was brought into force at the end of 1993, with 168 national signatories (Table 12.2). The original treaty was launched in 1992 at the 'Earth Summit', the United Nations Conference on Environment and Development in Rio de Janiero,

## TABLE 12.2
## Vocabulary about United Nations Treaties that Govern the Use and Protection of Biodiversity

Vocabulary from the UN Glossary of terms relating to treaty actions (United Nations, 2018).

| | |
|---|---|
| **CITES** (Convention on International Trade in Endangered Species of Wild Fauna and Flora) | Multilateral international treaty for the protection of threatened and endangered species.<br>Also known as: Washington Convention<br>*Entered into force:* 1 July 1975 |
| **CBD** (Convention on Biological Diversity) | Multilateral international treaty for conservation of biodiversity, and the sustainable and equitable use of biodiversity resources.<br>Also known as: Rio Convention<br>*Entered into force:* 29 December 1993 |
| **Cartagena Protocol** on Biosafety to the Convention on Biological Diversity | Supplementary agreement to the CBD. Provides protocols and rules for handling, use, and international transfer of genetically-modified living organisms (LMO).<br>*Entered into force:* 11 September 2003 |
| **Nagoya Protocol** on Access to Genetic Resources and the Fair and Equitable Sharing of Benefits Arising from their Utilization to the Convention on Biological Diversity | Supplementary agreement to the CBD. Provides a legal framework governing ownership and benefit-sharing for the utilisation of genetic resources from biodiversity (broadly interpreted).<br>*Entered into force:* 12 October 2014<br>n.b. the Nagoya protocol is not retroactive and does not apply to specimen material collected prior to 12 October 2014. |
| **COP** (Conference of the Parties) | An official meeting of all states party to a UN treaty; the CBD COP meets regularly every 2 years. |
| **LMO** (Living modified organism) | The UN legal designation of living organisms that have been genetically modified through the intervention of modern biotechnology. Organism is broadly interpreted to include microbes and viruses (UN, 2005: Art. 3). |
| **LMMC** (Group of Like-Minded Megadiverse Countries) | Cooperative agreement of 20 nations for the cooperative protection of biodiversity. The LMMC is a self-identified group, not all are parties to the CBD. |
| **Adoption** | UN treaties are 'adopted' when their text is finalised, that is, at the close of negotiations. States that participate in negotiations and agree to the final form of the treaty, are its signatories. After adoption the treaty is open for ratification. |
| **Entry into force** | A UN treaty will 'enter into force' shortly after a sufficient number of nations ratify the treaty. The time between *adoption* and *entry into force* may be several years. |
| **Signatories** | Countries sign a treaty at the time it is adopted. The initial signature qualifies, but does not require, a state to proceed to ratify the treaty. Being a signatory does not legally bind a state to the terms of the treaty, but does create an obligation to refrain from any action that would sabotage the treaty objectives. For example, the United States *signed* but did not *ratify* the CBD. |

*(Continued)*

**TABLE 12.2 (*Continued*)**
**Vocabulary about United Nations Treaties that Govern the Use and Protection of Biodiversity**

| | |
|---|---|
| **Ratify** | A state indicates its consent to be bound to a treaty. Ratification, approval, or acceptance makes a state legally bound to the terms of a treaty. |
| **Accession, Acceptance, Approval** | Accepting or approving the treaty is used to signify that the state consents to the terms but may indicate the relevant national law does not require the treaty to be ratified by the head of state. *Accession* specifically is the act of agreeing to treaty after it was adopted. These actions carry the same legal effect as ratification. |
| **Parties** | States that have ratified (or accepted/approved) a treaty. |
| **Protocol** | The protocols noted above are amendments to the CBD. Parties of the main treaty are not automatically parties of these accessory protocols; many countries that are parties to the CBD did not sign—or signed but did not ratify—the Nagoya Protocol. |

Brazil, and the CBD is often called the 'Rio Convention'. The signatories to the CBD increased to 196 parties (the European Union is counted as a single party) including the United States, which has signed but not ratified the treaty. As such, it impacts work on biodiversity around the world. The remit of CBD is quite separate, and far more wide-reaching that the more familiar Convention on International Trade in Endangered Species of Wild Flora and Fauna (CITES). CITES is a treaty on the protection of specifically endangered species of animals and plants. The CBD is not a replacement or extension of CITES, it is a separate framework. The CBD concerns the sustainable exploitation of biodiversity and defines 'biological diversity' very broadly. The remit of the CBD includes not only protected species, but all diversity, as well as intraspecific variability and 'ecological complexes' (UN, 1993: Art. 2). The CBD is a legally binding agreement that requires each party to prepare and implement a national biodiversity strategy and action plan.

The main tenets of the CBD are sustainable use and equitable benefit from biodiversity. It is a mechanism to curtail the tendency of so-called 'parachute science', where foreign researchers do not engage with local stakeholders. Imagine that a team of researchers from wealthier countries arrive in Anchuria*; they collect some organisms, leave, then publish their findings in *Nature* and never mention it to anyone in Anchuria. This might come across as rude. However, such behaviour is as much an issue of ignorance than prejudice; if the Anchurian scientists arrived in the US for their fieldwork and published nice papers about American flora or fauna without American co-authors, most people would not be concerned. The critical point here is that between the two countries, access to international travel and the infrastructure to complete the research is not balanced or equitable. The 'benefits' envisioned in the 'access and benefits sharing' mandate of the CBD includes commercial uses but also

---
* A fictional country, the original Banana Republic (Henry, 1904).

general research and development, access to participation in research, and access to results and downstream outputs.

One study of the prolific research outputs of international collaborations on natural product chemistry from Indonesian invertebrates examined authorship in over 100 papers: around half included Indonesian co-authors, but only 14% had an Indonesian first author, and 40% did not acknowledge any Indonesian collaborators at all (Januar, 2016). We might speculate about why this happens, in terms of the communication between collaborators who are separated by distance and potentially culture. First or senior authorship is indicative of leadership; this is an important metric of who is driving the research agenda. The problem implied by a lack of in-country first authors is that most of the science is being driven by outsiders. Rather than multiplying productivity through a multilateral effort among equal collaborators, it is stuck in a more colonial mode. The spirit of the CBD and modern international development is to empower a wider range of participants. Capacity building is not only about equipment and methods; the limits on scientific capacity depend on researchers having enough training and, critically, confidence in all elements of the publication pipeline. That pipeline includes everything from formulating relevant questions to developing a strategy and a plan to answer the question, and deputising tasks to other (sometimes foreign) colleagues, to writing, peer review, and maintaining good communication with faraway colleagues in spite of the pressures of teaching and other duties. The cure for parachute science is not to exclude foreigners or to patronise scientists in less developed countries—both of these are counter-productive—but to increase the leadership skills of the scientific workforce in less developed countries. New scientific leaders have to be supported by their mentors, colleagues, and the country they work in.

## INTERNATIONAL LAW AND THE MOVEMENT OF SPECIMENS

Participating countries to the CBD have representation within the convention's governing authority, called the Conference of the Parties (COP), which meets biannually. Delegates to the COP include environmental agencies or national biological records centres of member countries. In the course of implementing the CBD, the COP has agreed to two supplementary treaties: the Cartagena Protocol (UN, 2005) adopted by the majority of parties, and the Nagoya Protocol (UN, 2010). The Cartagena Protocol governs the international movements and trade of 'living modified organisms' (LMOs) or genetically-modified organisms, and it takes a precautionary approach to the potential impacts of LMOs on biodiversity. The definition of LMOs in the treaty text explicitly includes combining cells at the level of taxonomic family (a level chosen presumably because it was thought to be beyond the realm of natural hybridisation). This is an example of the importance of Linnean ranks, outside of the realm of systematics or phylogenetics: it governs the movement of LMOs.

The Nagoya Protocol was more controversial and was signed by a smaller subset of the CBD parties, to date only 104 countries of the 196 parties to the larger CBD have agreed to the Nagoya Protocol, which entered into force on 12 October 2014. The Nagoya Protocol recognised that genetic resources (sequence data and their potential commercial applications) fall under the sovereign control of the country of origin, in the same sense as other natural resources like mineral rights (Buck & Hamilton, 2011). The legal case for this is straightforward in principle. Moreover, it protects the interests of developing countries that disproportionately contain tropical species that are the main targets for discovery of bioactive chemical compounds. A group of 'Like Minded Mega-Diverse Countries' (LMMC) lobbied for these protections, in response to a perception that richer countries are systematically raiding tropical rainforests for commercially valuable drug precursors without sharing any profit with the country of origin. The Nagoya Protocol does not contain any force for material obtained before its implementation; specimens collected before October 2014 are not subject to the Protocol (Buck & Hamilton, 2011).

In the future, the protection of the Nagoya Protocol should lead to long-term benefits in international collaboration on taxonomy while protecting the rights of all nations (Wheeler et al., 2012). Keeping voucher specimens, and types, in the country of origin is essential to improve species identification and all the concomitant scientific benefits that come from robust knowledge of the local fauna. The application of this legal framework, however, is strained and there is immediate concern that it could stymie the use of genetic data for noncommercial scientific research as in discovery-driven studies of biodiversity. Greiber and colleagues (2012) wrote a clear guide to the treaty and its implications.

The drawback of the Nagoya Protocol is that it adds complexity and uncertainty to international collaborations and has unanticipated consequences for taxonomy (Watanabe, 2015). The nature of the treaty requires that international collaborations involving specimens, and genetic sequences, require formal 'prior informed consent' and 'mutually agreed terms' about the use and access to genetic resources (UN, 2010). The CBD is a government-to-government multilateral treaty, yet research projects are planned and executed by individuals (Buck & Hamilton, 2011). The authority to approve this 'consent' is often granted to research institutes, which should mean something as straightforward as having an administrator sign off on a grant application (but anecdotally, some administrators are understandably uncomfortable about approving things that sound like complex international treaty negotiations). What constitutes consent varies with local legislation and permitting requirements.

Tighter permit processes now demand comprehensive lists of species to be collected in advance, but fieldwork is usually discovery driven (Bouchet et al., 2016). Confusion over the Nagoya Protocol in biodiversity studies mostly stems from questions of intent: the law assumes some foreknowledge of what will be collected and why. Commercial timber harvest by a foreign agency is not covered by 'access and benefit sharing' under the CBD; that is a commercial activity governed by other mechanisms. However, if it was the original intention to use some of those samples to extract genetic material, that intent is then explicitly the purview of the Nagoya Protocol (Greiber et al., 2012). Museums are thus stuck in a grey area, since the research model of museums largely depends on scientists finding new and unanticipated ways to use old specimens. The initial agreement should lay out downstream ownership; for example it is not entirely clear what applies if specimens were collected with the honest intent to voucher specimens for morphological identifications and only later used to extract DNA. The great fear among many who work on biodiversity is that such confusion will encourage people to give up and destroy rare and precious specimens rather than face onerous paperwork and the intimidating threat of legal confusion.

Early collectors built valuable museum resources through collecting a wide spectrum of biodiversity, whereas restricted permits will narrow the scope of discovery. More than half of type specimens in each of four major herbarium collections were obtained by only 2% of the documented collectors (Bebber et al., 2012). Conservation actions leave out most biodiversity already, even among vertebrate animals (Sitas et al., 2009) and research efforts are disproportionately focussed on protected areas (Martin et al., 2012). While the aim is to retain specimen resources in the country of origin, many of the countries in question do not have appropriate facilities to house permanent collections. Scientists come up with workarounds to protect their research, such as sending specimen material on 'long term loan' to a museum in Europe or North America.

Species do not respect national boundaries. But, genetic sequences come from the tissue of individual organisms, collected in a specific place and time. The impact on museum collections of implementation of the CBD and the Nagoya Protocol will redistribute power and resources to institutions, some of which do not yet exist.

## THE FALLACY OF ECONOMICS

The idea of nature as a source of benefits that provide direct and indirect value is not new. Grove (1995) documented the precursor of the modern conservation movement in the mid-17th century. European colonial expansion rapidly and dramatically changed the landscape of other continents, and contemporary leaders knew that the expansion of capitalism and trade was pushing the limits of available natural resources. Colonial powers were aware of the need for conservation, and that deforestation and pollution could undermine agriculture and trade and even alter local climates (Redford & Adams, 2009).

The idea of ecosystem 'goods and services' has rapidly gained popularity as a tool to improve the balance in perceived value between economic development and environmental conservation. Access to clean air and water, and natural spaces, are important for human physical and mental health. The raw cost of some of these 'ecosystem services' can be calculated, for example pollination or water filtration. This approach resonates when it produces impressively large values in billions or trillions of dollars for activities that happen for 'free' if ecosystems are conserved. Economic valuation provides important context for the value of nature, but it will emphasise current interests over long-term value (Chee, 2004). Economics is inherently concerned with market value—the price—not the objective value of commodities, as demonstrated by the capricious history of trade in, for example, diamonds, tulips, and houses (Heal, 2000). Market forces fundamentally favour the unavailable and desirable, not the important.

Ecosystem services are valued through an anthropocentric lens, as there is limited context to evaluate a price point for other ecosystem functions that do not serve human needs. The real value of ecological processes is not the raw material value (such as the board feet of timber in a forest), or the indirect replacement cost (carbon sequestration), but how the removal of those same functions can create an unpredictable cascade of additional lost revenue (elevated temperatures in the area that was previously cooled by transpiration of the missing trees).

Some of the natural behaviour of ecosystems is contrary to human development. Wildfires devastate human and natural spaces indiscriminately (Koplitz et al., 2016). Yet, in dry, forested areas, controlled burning can be effective long-term mitigation against wildfire, mimicking the natural cycle of fires through human intervention (Bright et al., 1993). The cost/benefit analysis in strictly economic terms can be complicated.

Valuing ecosystem services places emphasis on the process over the constituent species in an ecosystem; natural coastal defences are valuable protection against flooding and erosion, whether it is wave-attenuating reefs of bivalves or corals, mangrove forests, or estuarine wetlands (Temmerman et al., 2013). But in the coarsest interpretation of these assets, it does not matter what sort of barrier it is—this could lead to entirely logical argumentation that it would be more efficient to select the best single mangrove species, or genetically engineer a super-mangrove that is cold tolerant, and plant that tree in monoculture everywhere. That would deliver the 'service' of wave attenuation, but it would be foolish to assume a service engineered to current requirements would have the stability or total value of a natural system.

That emphasis on service, dissociated from preserving native species assemblages, has historical parallels with early efforts at bio-control through introduced species that may be more efficient for specific functions than native counterparts (Redford & Adams, 2009). Kudzu (*Pueraria* spp.) was introduced in the US to prevent erosion of disturbed ground, and Tilapia (various species of cichlid fishes) have been introduced as food, or to control insects or aquatic plants; these organisms are considered to be highly destructive invasive species but they are in fact providing the ecosystem services as intended.

The economics of biodiversity are fundamentally different to other commodities: once a species is extinct, it is gone forever. Some spurious arguments have been made about reviving extinct species; but this returns to the idea of species as fixed created essences that could be brought back from the dead. Cloning a handful of individuals from reconstructed DNA, or selectively breeding for ancestral traits, is an experimental project that has no equivalency with real, complex, living population lineages (Shapiro, 2017).

Perhaps the overriding concern about the valuation of ecosystem services should be that it puts the ecosystem at a negotiating disadvantage in developing economies. The value of services depends on local economic conditions. Prices are cheap in less-developed countries, exactly the same forces that attract the interest of large foreign investors who can profit from investment in commercial development. The worst-case scenario is a pristine habitat in a megadiverse country with a price tag on it, and investors who are happy to buy.

## CULTURAL OWNERSHIP OF BIODIVERSITY

Paknia et al. (2015) proposed an international aid programme for biodiversity, with global contributions in the same way that we support humanitarian aid. This is the morally correct approach; people in crisis deserve our support regardless of their national identity. Biodiversity is in crisis, other species do not recognise national boundaries at all, and the only way to safeguard it is through international cooperation.

This is not the same as cries to save taxonomy or that taxonomists are facing extinction; this is about global biodiversity, in its natural habitat, which is overwhelmingly in countries that currently have to prioritise the basic needs of their human citizens. There is limited empirical evidence for the complaints about neglect of taxonomy as a discipline, which have been sounded continuously since before the 1970s (Isley, 1972). Perhaps that constant complaining is the only thing that has saved us. Work on other vertebrates has shown that the 'endangered' status seems to be effective at slowing the rate of population deterioration (Hoffmann et al., 2010), and it may be working for taxonomy as a discipline. Or perhaps now is an appropriate time to refocus on emphasising positive contributions and expand the population.

Working with species names, and scientific specimen collections, requires a vision that your contribution fits into a much longer term, larger narrative about global biodiversity. That narrative must include all nations. Partners in Western countries often have inadvertently patronising attitudes toward collaborators in less developed countries; a study on North American–South American knowledge exchange in health sciences stressed paternalism and deeply rooted assumptions about the superiority of North American approaches (Jentsch & Pilley, 2003). Implicit bias is bilateral; members of any minority group could easily conclude that their approaches, or themselves, are inferior. The 'Matthew effect' notes how scientific outputs of well-known names are more highly cited, disproportionate to quality or relevance; the corollary 'Mathilda effect' is that contributions of female scientists are less recognised relative to comparative contributions of male scientists (Rossiter, 1993). In the same vein, the 'matata effect' (Van der Stocken et al., 2016) downplays the contributions of authors from institutions in 'exotic' countries, or with non-Western names. Although academic research is no doubt a competitive field, all of us in positions of privilege have an ethical duty to make room at the table for others and to elevate voices that might not be heard. This is the long view; diversifying participation in science, like any diverse ecosystem, provides resilience.

Can we imagine a world without name-making? Isley (1972) followed a narrative thought experiment of a dystopian future without any systematic botanists or herbaria. Problems arise almost immediately without access to reliable identification of toxic plants and plant-derived anti-cancer agents, and cascade catastrophically. But Isley's short story did not extend to any impacts outside the United States. The frustration of not having good access to names is the current reality for megadiverse tropical countries. Improvements will only come from the political will of those nations' governments, and strong support from the wider community. Developing economies look to richer nations as role models for intellectual and physical infrastructural development. The onus is on all scientists—makers of names, and users of names—to communicate our findings and celebrate the positive contributions of taxonomy and museums to the grand project of understanding and protecting Earth's biodiversity.

# REFERENCES

Albano PG, Sabelli B, Bouchet P. 2011. The challenge of small and rare species in marine biodiversity surveys: Microgastropod diversity in a complex tropical coastal environment. *Biodiversity & Conservation.* 20: 3223–37.

Ang Y, Puniamoorthy J, Pont AC, Bartak M, Blanckenhorn WU, Eberhard WG, Puniamoorthy N, Silva VC, Munari L, Meier R. 2013. A plea for digital reference collections and other science-based digitization initiatives in taxonomy: Sepsidnet as exemplar. *Systematic Entomology.* 38: 637–44.

Balmford A, Long A. 1995. Across-country analyses of biodiversity congruence and current conservation effort in the tropics. *Conservation Biology.* 9: 1539–47.

Bebber DP, Carine MA, Davidse G, Harris DJ, Haston EM, Penn MG, Cafferty S, Wood JRI, Scotland RW. 2012. Big hitting collectors make massive and disproportionate contribution to the discovery of plant species. *Proceedings of the Royal Society B.* 279: 2269–2274.

Bickford D, Lohman DJ, Sodhi NS, Ng PK, Meier R, Winker K, Ingram KK, Das I. 2007. Cryptic species as a window on diversity and conservation. *Trends in Ecology & Evolution.* 22: 148–55.

Blackwell M. 2011. The Fungi: 1, 2, 3… 5. 1 million species? *American Journal of Botany.* 98: 426–38.

Bortolus A. 2008. Error cascades in the biological sciences: The unwanted consequences of using bad taxonomy in ecology. *AMBIO: A Journal of the Human Environment.* 37: 114–8.

Bouchet P, Bary S, Héros V, Marani G. 2016. How many species of molluscs are there in the world's oceans, and who is going to describe them? *Tropical Deep-Sea Benthos.* 29: 9–24.

Bright AD, Manfredo MJ, Fishbein M, Bath A. 1993. Application of the theory of reasoned action to the National Park Service's controlled burn policy. *Journal of Leisure Research.* 25: 263–80.

Buck M, Hamilton C. 2011. The Nagoya Protocol on Access to Genetic Resources and the Fair and Equitable Sharing of Benefits Arising from their Utilization to the Convention on Biological Diversity. *RECIEL.* 20: 47–61.

Carranza-Rojas J, Goeau H, Bonnet P, Mata-Montero E, Joly A. 2017. Going deeper in the automated identification of herbarium specimens. *BMC Evolutionary Biology.* 17: 181.

[CBD] Secretariat of the Convention on Biological Diversity. 2006. Guide to the global taxonomy initiative. CBD Technical Series. 30: 1–195.

Chee YE. 2004. An ecological perspective on the valuation of ecosystem services. *Biological Conservation.* 120: 549–65.

Costello MJ, May RM, Stork NE. 2013. Can we name earth's species before they go extinct? *Science.* 339: 413–16.

Duffy JE, Lefcheck JS, Stuart-Smith RD, Navarrete SA, Edgar GJ. 2016. Biodiversity enhances reef fish biomass and resistance to climate change. *Proceedings of the National Academy of Sciences.* 113: 6230–5.

Federhen S. 2014. Type material in the NCBI Taxonomy Database. *Nucleic Acids Research.* 43: D1086–98.

Galtier N, Nabholz B, Glémin S, Hurst GD. 2009. Mitochondrial DNA as a marker of molecular diversity: A reappraisal. *Molecular Ecology.* 18: 4541–50.

Garnett ST, Christidis L. 2017. Taxonomy anarchy hampers conservation. *Nature.* 546: 25.

Gaston KJ, May RM. 1992. Taxonomy of taxonomists. *Nature.* 356: 281–2.

Gramlich EM. 1994. Infrastructure investment: A review essay. *Journal of Economic Literature.* 32: 1176–96.

Greiber T, Moreno SP, Åhrén M, Carrasco JN, Kamau, EC, Medaglia JC, Oliva MJ, Perron-Welch F, Ali N, Williams C. 2012. An explanatory guide to the Nagoya Protocol on Access and Benefit-sharing. IUCN Environmental Policy and Law Paper No. 83

Grove RH. 1995. *Green Imperialism. Colonial Expansion, Tropicalisland Edens and the Origins of Environmentalism, 1600–1860.* Cambridge University Press.

Häussermann V, Schrödl M. 2017. *BiodiversiTOT—Die Globale Artenvielfalt Jetzt Entdecken, Erforschen und Erhalten.* Books on Demand.

Hawksworth DL. 1991. The fungal dimension of biodiversity: Magnitude, significance, and conservation. *Mycological Research.* 95: 641–55.

Hawksworth D, Cowie R. 2013. The discovery of historically extinct, but hitherto undescribed, species: An under-appreciated element in extinction-rate assessments. *Biodiversity & Conservation.* 22: 2429–32.

Hawksworth DL, Lücking R. 2017. Fungal diversity revisited: 2.2–3.8 million species. *Microbiology Spectrum.* 5: FUNK-0052-2016.

Heal G. 2000. Valuing ecosystem services. *Ecosystems.* 3: 24–30.

Hedgepeth J. 1962. *11th Annual University of the Pacific Research Lecture, 1961.* University of the Pacific.

Henry O. 1904. *Cabbages and Kings.* Project Gutenberg edition 2008.

Hoffmann M, Hilton-Taylor C, Angulo A, Böhm M, Brooks TM, Butchart SH, Carpenter KE, Chanson J, Collen B, Cox NA, Darwall WR. 2010. The impact of conservation on the status of the world's vertebrates. *Science.* 330: 1503–9.

Hortal J, de Bello F, Diniz-Filho JA, Lewinsohn TM, Lobo JM, Ladle RJ. 2015. Seven shortfalls that beset large-scale knowledge of biodiversity. *Annual Review of Ecology, Evolution, & Systematics.* 46: 523–49.

Isley D. 1972. The disappearance. *Taxon.* 21: 3–12.

Jablonski D, Belanger CL, Berke SK, Huang S, Krug AZ, Roy K, Tomasovych A, Valentine JW. 2013. Out of the tropics, but how? Fossils, bridge species, and thermal ranges in the dynamics of the marine latitudinal diversity gradient. *Proceedings of the National Academy of Sciences.* 110: 10487–94.

Januar HI. 2016. North-South biodiscovery research collaboration of Indonesian sponge and soft coral: A bibliographic analysis of publications over the last two decades. *Journal of Scientometric Research.* 5: 43–8.

Jentsch B, Pilley C. 2003. Research relationships between the South and the North: Cinderella and the ugly sisters? *Social Science & Medicine.* 57: 1957–67.

Jones HD, Sluys R. 2016. A new terrestrial planarian species of the genus Marionfyfea (Platyhelminthes: Tricladida) found in Europe. *Journal of Natural History.* 50: 2673–90.

Jörger KM, Schrödl M. 2013. How to describe a cryptic species? Practical challenges of molecular taxonomy. *Frontiers in Zoology.* 10: 59.

Koh LP, Wilcove DS. 2008. Is oil palm really destroying tropical biodiversity? *Conservation Letters.* 1: 60–4.

Koplitz SN, Mickley LJ, Marlier ME, Buonocore JJ, Kim PS, Liu T, Sulprizio MP, DeFries RS, Jacob DJ, Schwartz J, Pongsiri M. 2016. Public health impacts of the severe haze in Equatorial Asia in September–October 2015: Demonstration of a new framework for informing fire management strategies to reduce downwind smoke exposure. *Environmental Research Letters.* 11: 094023.

Lee WL, Bell BM, Sutton JF. 2007. Characterization of voucher specimens. In: Knell SJ (editor). *Museums in a Material World*, pp. 46–50. Routledge.

Lipps JH, Granier BRC (editors). 2009. *Paleoparks—The Protection and Conservation of Fossil Sites Worldwide.* Carnets de Géologie/Notebooks on Geology: Book 2009/03.

Lira-Noriega A, Soberón J. 2015. The relationship among biodiversity, governance, wealth, and scientific capacity at a country level: Disaggregation and prioritization. *Ambio.* 44: 391–400.

Marchese C. 2015. Biodiversity hotspots: A shortcut for a more complicated concept. *Global Ecology & Conservation*. 3: 297–309.

[MBA] Marine Biological Association. 1904. Plymouth marine invertebrate fauna. *Journal of the Marine Biological Association*. 7: 155–298.

Martin LJ, Blossey B, Ellis E. 2012. Mapping where ecologists work: Biases in the global distribution of terrestrial ecological observations. *Frontiers in Ecology and the Environment*. 10: 195–201.

Mittermeier RA, Gil PR, Mittermeier CG. 1997. *Megadiversity. Earth's Biologically Wealthiest Nations*. Conservation International, Washington, DC.

Mittermeier RA, Myers N, Thomsen JB, Da Fonseca GA, Olivieri S. 1998. Biodiversity hotspots and major tropical wilderness areas: Approaches to setting conservation priorities. *Conservation Biology*. 12: 516–20.

Mouillot D, Villéger S, Parravicini V, Kulbicki M, Arias-González JE, Bender M, Chabanet P, Floeter SR, Friedlander A, Vigliola L, Bellwood DR. 2014. Functional over-redundancy and high functional vulnerability in global fish faunas on tropical reefs. *Proceedings of the National Academy of Sciences*. 111: 13757–62.

Naser HA. 2010. Testing taxonomic resolution levels for detecting environmental impacts using macrobenthic assemblages in tropical waters. *Environmental Monitoring & Assessment*. 170: 435–44.

Nater A, Mattle-Greminger MP, Nurcahyo A, Nowak MG, de Manuel M, Desai T, Groves C, Pybus M, Sonay TB, Roos C, Lameira AR. 2017. Morphometric, behavioral, and genomic evidence for a new Orangutan species. *Current Biology*. 27: 3487–98.

Page RD. 2016. Surfacing the deep data of taxonomy. *ZooKeys*. 550: 247–60.

Paknia O, Sh HR, Koch A. 2015. Lack of well-maintained natural history collections and taxonomists in megadiverse developing countries hampers global biodiversity exploration. *Organisms Diversity & Evolution*. 15: 619–29.

Petchey OL, Gaston KJ. 2006. Functional diversity: Back to basics and looking forward. *Ecology Letters*. 9: 741–58.

Poulin R. 2014. Parasite biodiversity revisited: Frontiers and constraints. *International Journal for Parasitology*. 44: 581–9.

Qumsiyeh M, Handal E, Chang J, Abualia K, Najajreh M, Abusarhan M. 2017. Role of museums and botanical gardens in ecosystem services in developing countries: Case study and outlook. *International Journal of Environmental Studies*. 74: 340–50.

Redford KH, Adams WM. 2009. Payment for ecosystem services and the challenge of saving nature. *Conservation Biology*. 23: 785–7.

Rossiter MW. 1993. The Matthew/Matilda effect in science. *Social Studies of Science*. 23: 325–41.

Scheffers BR, Joppa LN, Pimm SL, Laurance WF. 2012. What we know and don't know about Earth's missing biodiversity. *Trends in Ecology & Evolution*. 27: 501–10.

Shapiro B. 2017. Pathways to de-extinction: How close can we get to resurrection of an extinct species? *Functional Ecology*. 31: 996–1002.

Sigwart JD, Garbett A. 2018. Biodiversity Assessment, DNA Barcoding, and the Minority Majority. *Integrative and Comparative Biology*. doi:10.1093/icb/icy076

Sitas N, Baillie JEM, Isaac JB. 2009. What are we saving? Developing a standardized approach for conservation action. *Animal Conservation*. 12: 231–37.

Smith M, Fernandez-Triana JO, Roughley R, Hebert PD. 2009. DNA barcode accumulation curves for understudied taxa and areas. *Molecular Ecology Resources*. 9: 208–16.

Steiner FM, Pautasso M, Zettel H, Moder K, Arthofer W, Schlick-Steiner BC. 2015. A falsification of the citation impediment in the taxonomic literature. *Systematic Biology*. 64: 860–8.

Suarez AV, Tsutsui ND. 2004. The value of museum collections for research and society. *AIBS Bulletin*. 54: 66–74.

Temmerman S, Meire P, Bouma TJ, Herman PM, Ysebaert T, De Vriend HJ. 2013. Ecosystem-based coastal defence in the face of global change. *Nature*. 504: 79–83.

Tolimieri N, Shelton AO, Feist BE, Simon V. 2015. Can we increase our confidence about the locations of biodiversity 'hotspots' by using multiple diversity indices? *Ecosphere*. 6: 1–3.

[UN] United Nations. 1993. Convention on Biological Diversity (with annexes). Concluded at Rio de Janeiro on 5 June 1992. *UN Treaty Series*. 1760(30619): 79–307.

[UN] United Nations. 2005. Cartagena Protocol on Biosafety to the Convention on Biological Diversity. Montreal, 29 January 2005. *UN Treaty Series*. 2226(30619): 208–372.

[UN] United Nations. 2010. Decision X/1. Access to genetic resources and the fair and equitable sharing of benefits arising from their utilization. *Decision adopted by the Conference of the Parties to the Convention on Biological Diversity at its tenth meeting*. UNEP/CBD/COP/DEC/X/1.

Van der Stocken T, Hugé J, Deboelpaep E, Vanhove MP, de Bisthoven LJ, Koedam N. 2016. Academic capacity building: Holding up a mirror. *Scientometrics*. 106: 1277–80.

Vecchione M, Mickevich MF, Fauchald K, Collete BB, Williams, AB, Munroe TA, Young RE. 2000. Importance of assessing taxonomic adequacy in determining fishing effects on marine biodiversity. *ICES Journal of Marine Science*. 57: 677–81.

Watanabe ME. 2015. The Nagoya Protocol on Access and Benefit Sharing. *BioScience*. 65: 543–50.

Weber C. 2018. National and international collection networks. In: Beck LA (editor). *Zoological Collections of Germany*, pp. 29–36. Springer.

Werner YL. 2006. The case of impact factor versus taxonomy: A proposal. *Journal of Natural History*. 40: 1285–6.

Wheeler QD, Knapp S, Stevenson DW, Stevenson J, Blum SD, Boom BM, Borisy GG, Buizer JL, De Carvalho MR, Cibrian A, Donoghue MJ. 2012. Mapping the biosphere: Exploring species to understand the origin, organization and sustainability of biodiversity. *Systematics & Biodiversity*. 10: 1–20.

Wilson EO. 2017. Biodiversity research requires more boots on the ground. *Nature Ecology & Evolution*. 1: 01590.

Zamora JC, Calonge FD, Martín MP. 2015. Integrative taxonomy reveals an unexpected diversity in *Geastrum* section *Geastrum* (Geastrales, Basidiomycota). *Persoonia*. 34: 130–65.

# Species and Systematics

**Phylogenetic Systematics**: Haeckel To Hennig, *by Olivier Rieppel*

**Evolution By Natural Selection**: Confidence, Evidence and the Gap, *by Michaelis Michael*

**The Evolution of Phylogenetic Systematics**, *edited by Andrew Hamilton*

**Molecular Panbiogeography on the Tropics**, *by Michael Heads*

**Beyond Cladistics**, *edited by David M. Williams and Sandra Knapp*

**Comparative Biogeography**: Discovering and Classifying Bio-Geographical Patterns of a Dynamic Earth, *by Lynee R. Parenti and Malte C. Ebach*

**Species**: A History of the Idea, *by John S. Wilkins*

**What Species Mean**: Understanding the Units of Biodiversity, *by Julia D. Sigwart*

For more information about this series, please visit:
https://www.crcpress.com/Species-and-Systematics/book-series/CRCSPEANDSYS

# Index

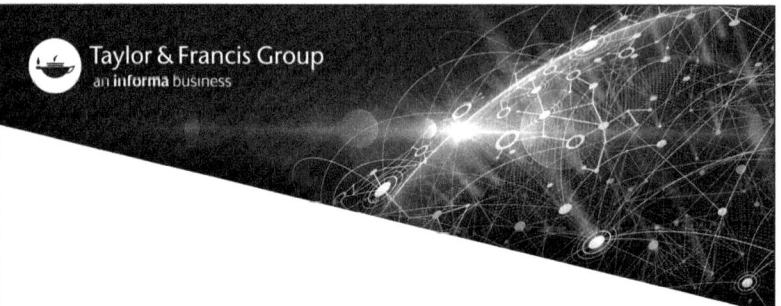